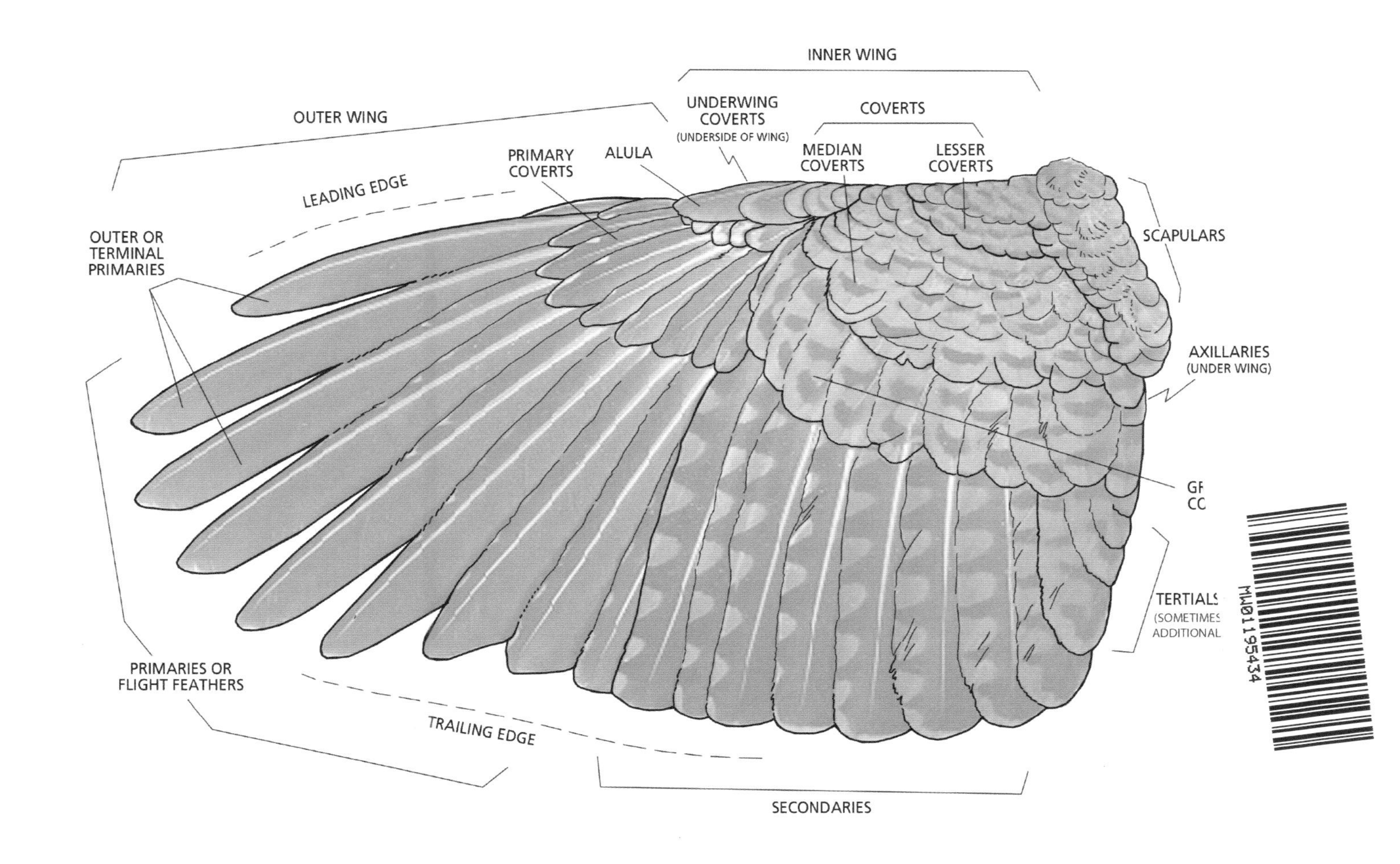
INNER WING
OUTER WING
UNDERWING COVERTS
(UNDERSIDE OF WING)
COVERTS
MEDIAN COVERTS
LESSER COVERTS
PRIMARY COVERTS
ALULA
LEADING EDGE
SCAPULARS
OUTER OR TERMINAL PRIMARIES
AXILLARIES
(UNDER WING)
PRIMARIES OR FLIGHT FEATHERS
TRAILING EDGE
SECONDARIES

Peterson Reference Guide to

# WOODPECKERS of North America

Peterson Reference Guide to

# WOODPECKERS OF North America

STEPHEN A. SHUNK

HOUGHTON MIFFLIN HARCOURT
BOSTON NEW YORK
2016

**Sponsored by
the Roger Tory Peterson Institute
and the National Wildlife Federation**

To Larry -
Thanks for letting me show you
around our Woodpecker Wonderland!

www.hmhco.com



Library of Congress Cataloging-in-Publication Data is available.
ISBN 978-0-618-73995-0

Endpaper illustrations by Barbara B. Gleason / BGleason Design & Illustration

Book design by Eugenie S. Delaney

Printed in China

SCP 10 9 8 7 6 5 4 3 2

The legacy of America's greatest naturalist and creator of the field guide series, Roger Tory Peterson, is kept alive through the dedicated work of the Roger Tory Peterson Institute of Natural History (RTPI). Established in 1985, RTPI is located in Peterson's hometown of Jamestown, New York, near the Chautauqua Institution in the southwestern part of the state.

Today RTPI is a national center for nature education that maintains, shares, and interprets Peterson's extraordinary archive of writings, art, and photography. The institute, housed in a landmark building by world-class architect Robert A. M. Stern, continues to transmit Peterson's zest for teaching about the natural world through leadership programs in teacher development as well as outstanding exhibits of contemporary nature art, natural history, and the Peterson Collection.

Your participation as a steward of the Peterson Collection and supporter of the Peterson legacy is needed. Please consider joining RTPI at an introductory rate of 50 percent of the regular membership fee for the first year. Simply call RTPI's membership department at (800) 758-6841 ext. 226, or e-mail membership@rtpi.org to take advantage of this special membership offered to purchasers of this book. For more information, please visit the Peterson Institute in person or virtually at www.rtpi.org.

*It is generally agreeable to be in the company of individuals*
*who are naturally animated and pleasant.*
*For this reason, nothing can be more gratifying than*
*the society of Woodpeckers in the forest.*

—JOHN JAMES AUDUBON, 1836

Dedicated to my late father, Gordon A. Shunk, who taught me that life sometimes doesn't go the way we want it to, but we persevere through our challenges and celebrate our victories. This one's for you, Dad.

Peterson Reference Guide to

# WOODPECKERS of North America

# CONTENTS


Preface xi

**INTRODUCTION 1**

Anatomy and Adaptation 2

Behavior 14

Ecology and Conservation 28

How to Use This Book 38

**SPECIES ACCOUNTS 43**

Genus *Melanerpes*

- Lewis's Woodpecker 44
- Red-headed Woodpecker 52
- Acorn Woodpecker 60
- Gila Woodpecker 71
- Golden-fronted Woodpecker 78
- Red-bellied Woodpecker 86

Genus *Sphyrapicus*

- Williamson's Sapsucker 95
- Yellow-bellied Sapsucker 102
- Red-naped Sapsucker 112
- Red-breasted Sapsucker 123

Genus *Picoides*

- Ladder-backed Woodpecker 133
- Nuttall's Woodpecker 140
- Downy Woodpecker 149
- Hairy Woodpecker 159
- Arizona Woodpecker 168
- Red-cockaded Woodpecker 177
- White-headed Woodpecker 188
- American Three-toed Woodpecker 199
- Black-backed Woodpecker 206

Genus *Colaptes*

- Northern Flicker 214
- Gilded Flicker 226

Genus *Dryocopus*

- Pileated Woodpecker 234

Genus *Campephilus*

- Ivory-billed Woodpecker 244

Acknowledgments 255

APPENDIX 1. Measurements of North American Woodpeckers 256

APPENDIX 2. Nest Site Data for North American Woodpeckers 257

APPENDIX 3. Parenting Data for North American Woodpeckers 258

APPENDIX 4. The Woodpecker "Family Tree" 260

APPENDIX 5. Woodpecker Conflicts with Humans 261

Glossary 262

Bibliography 264

Index 294


**OPPOSITE:** As a proficient hunter of aerial insects, the Red-headed Woodpecker spends more time in the air than most North American woodpeckers. *James Vellozzi, June 2009; Sarasota Co., FL.*

# PREFACE

> This heavy bird, with straight, chisel bill, and sharp-pointed tail-feathers; with his short legs and wide, flapping wings, his unmusical but not disagreeable voice, and his heavy, undulating, business-like flight, is distinctly bourgeois, the type of bird devoted to business and enjoying it. No other bird has so much work to do all the year round, and none performs his task with more energy and sense. . . . Above all other birds he is the friend of man and deserves to have the freedom of the hills.
>
> —Fannie Hardy Eckstorm, 1900

**None has expressed better** than early naturalist and historian Fannie Hardy Eckstorm the senses of industry, joy, and inspiration brought to the forest by woodpeckers. Another early naturalist (and pioneer of bird photography), Herbert Job, placed woodpeckers in their very own order, the "International Order of the Knights of the Chisel."

Woodpeckers play a singular role in the world's forests and woodlands. In North America, for example, more than 40 species of birds depend on woodpecker carpentry for their nest and roost cavities. Hundreds of insect species branded as curses on the nation's timber fall prey to the "knights." And what would birding be without woodpeckers? Attractive, curious, and charismatic, woodpeckers open a window on nature accessible to birders and nonbirders alike. Their presence affords four seasons of observation, and even in their absence we can study their handiwork in our wooded backyards. In one way or another, woodpeckers present challenging puzzles and enter millions of human lives across the continent.

Comprising the subfamily Picinae, at least 200 woodpecker species inhabit the globe, including most wooded areas, many islands, some deserts and grasslands, and a few small patches of alpine tundra. Classified among 24 or more genera, they exhibit their greatest diversity in Southeast Asia and the Neotropics. Despite their absence from Australasia, Antarctica, Madagascar, and the great deserts and remote ocean lands, woodpeckers hold fast to their place in the world's avifauna, standing near the evolutionary gap between the non-passerines and the passerines, with unique forms of communication and a host of astonishing physiological adaptations.

Since the early twentieth century, we have studied woodpeckers to assess the health of our continent's forests. The stories woodpeckers tell depict the complicated relationships between humans and habitats, and by listening to them more carefully, we may one day restore our woodlands and forests to their formerly magnificent ecological grandeur.

Enjoy this book, and use these stories of our magnificent woodpeckers to help foster a culture of conservation.

Lewis's Woodpecker is generally considered a weak excavator, but removing the rotten, fungus-plagued heartwood from a mature juniper does not require much strength. *Brent McGregor; June 2010, Deschutes Co., OR.*

**OPPOSITE:** Northern Flickers bear the largest broods of all North American woodpeckers. This large cavity entrance allows all five of these nestlings to beg for food simultaneously. Excavating such an entrance may be an adaptation that leads to increased fledgling rates for this widespread species. *Scott Carpenter; June 2012, Deschutes Co., OR.*

# INTRODUCTION

## ANATOMY AND ADAPTATION

**For centuries,** the anatomy of woodpeckers has captivated naturalists and scientists alike (see Figure 1). From seventeenth-century comparative anatomy to modern-day research to develop safer football and motorcycle helmets, woodpeckers have been the subject of a broad range of studies examining their anatomical adaptations for their specialized lifestyles.

### THE TONGUE AND THE HYOID APPARATUS

Woodpeckers possess a tongue and supporting hyoid apparatus that are perfectly adapted for each species' primary feeding behaviors. Variably extensible and subtly diverse, the fleshy tongue is operated by a system of bones and muscles that allows the bird to effectively gather its favored forage.

The Northern Flicker, for example, can extend its tongue 2 inches (5.1 cm) or more beyond the bill tip, allowing it to probe into the deep tunnels of underground ant dwellings. The sapsuckers, by contrast, possess a tongue only long enough to penetrate the deepest sap wells, a mere ½ to ¾ inch (1–2 cm) into the tree. Regardless of their lengths, all woodpecker tongues have the same basic anatomical structure and operate in the same manner.

#### *The Extensible Tongue*

The fleshy part of the woodpecker tongue shows three distinct sections: the base, center, and tip. The base is covered with a smooth sheath of skin that "accordions" when the tongue is retracted. The center appears as rough, rubbery skin to the naked eye, but a microscope reveals this roughness to be tiny backward-pointing granules, decreasing in size from the base of the tongue toward the tip. The character of these granules—especially the extent, number, and size—varies widely among species (see Figure 2).

The greatest variability in woodpecker tongues is right at the tip, which is stiff and hornlike with variable barbs, most of which angle back toward the base. These barbs may resemble sharp spines, and they number from just a few in the flickers to dozens in the Red-headed Woodpecker. In the sapsuckers, they resemble fine hairs, some directed outward from the surface and some angled backward. These barbs are particularly well developed on the tongues of the most specialized excavators—the Black-backed and American Three-toed Woodpeckers—allowing individuals to slide their tongue tip be-

The Northern Flicker possesses the most extensible tongue of all North American woodpeckers, and the species' conspicuous presence affords observation of its extended tongue in a variety of circumstances. *Paul Bannick; Dec. 2008, Seattle, WA.*

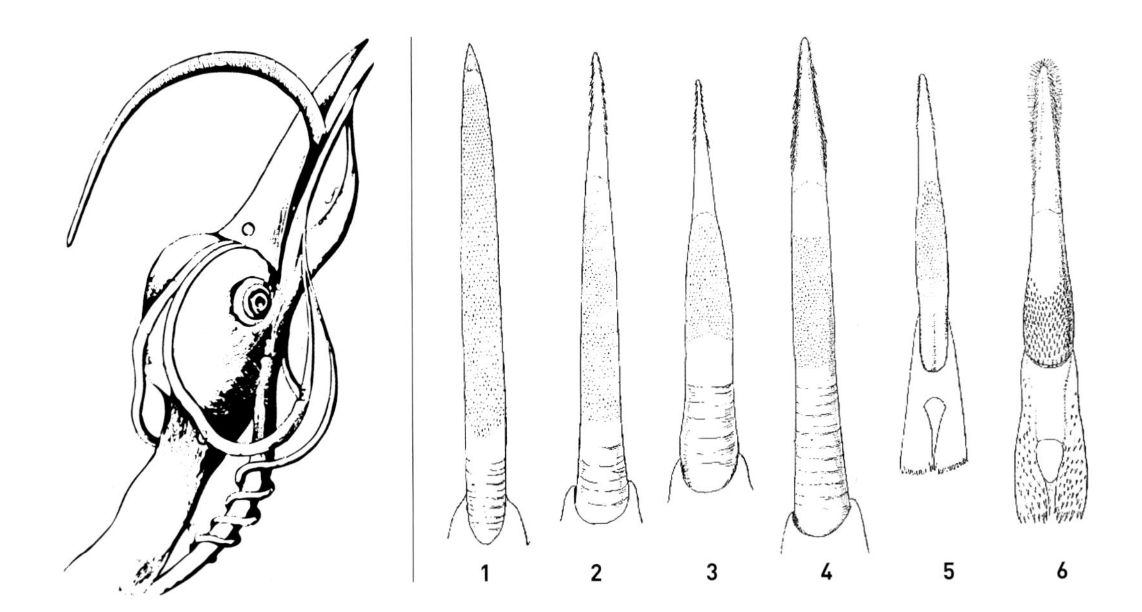


**FIGURE 1.** Borelli's Tongue. The first known drawing of the woodpecker tongue came from Giovanni Alfonso Borelli in his 1680 text *De Motu Animalium*, or *On the Motion of Animals*. Here Borelli depicts one of the extensor muscles wrapping around the trachea, which is not the case in all woodpeckers. *Borelli 1680.*

**FIGURE 2.** Diversity in Woodpecker Tongues. The tongue of each woodpecker species is specially adapted for its primary food-gathering behaviors. Those depicted here were studied and drawn in 1895 by Frederic Lucas for the U.S. Department of Agriculture; all are shown at 2.25 times their natural size. (1) Northern Flicker, (2) Hairy Woodpecker, (3) Black-backed Woodpecker, (4) Red-headed Woodpecker, (5) Ladder-backed Woodpecker, (6) Red-naped Sapsucker. *Lucas 1895.*

tween a grub and its gallery wall and then to hook the grub upon retraction. In adult woodpeckers, the barbs wear with use and constantly regrow. Fledgling woodpeckers have nearly smooth tongue tips.

To assist in prey retraction, woodpeckers possess a larger submaxillary salivary gland than most birds. This gland is largest in the flickers, reaching from the base of the skull into the lower mandible and coating the entire tongue with sticky mucus on its way out of the bill. As the flicker extends its tongue into an ant burrow, this mucus effectively glues ants to the flicker's tongue along most of its length. In species with more heavily barbed tongues, the mucus may be concentrated along the bumpy center of the tongue, enhancing the stickiness of this section but allowing the tongue tip the flexibility to seek and secure prey without getting "stuck."

### *The Hyoid Apparatus*

The hyoid apparatus comprises two sets of narrow bones, the medial and the paired hyoids. If the whole apparatus were laid flat, the medial bones and fleshy tongue would form the stem of a narrow, elongated Y. The paired hyoids, known as the hyoid horns, would form the flared part of the Y (see Figure 5). The forwardmost of the medial bones, the ceratohyals, are fused at the cartilaginous base of the fleshy tongue. The very flexible terminal bones of the hyoid horns are called the epibranchials. In most woodpeckers, the epibranchials wrap completely around the rear of the skull, coming together (but not fusing) atop the skull and typically veering toward the right orbital, or eye socket (see Figure 3). In those species with especially long hyoids, such as the flickers, the epibranchials extend through the right nasal opening, under the right nostril, or nare, and along the upper mandible nearly to its tip. Sapsuckers possess the shortest hyoids, proportionally equal in length only to those of some passerines. Fledgling woodpeckers have short hyoid horns, which grow rapidly once a bird begins feeding independently.

In species with bulging frontal bones (see Figure 8), the hyoid horns are obstructed from reaching the right nostril, so they curve backward after rounding the front of the right orbit.

The woodpecker's hyoid apparatus requires specialized muscles to render it functional (see Figure 5). Woodpeckers lack the specialized tongue retractor muscle, the stylohyoideus, found in other birds. Instead, they have two pairs of muscles: the protracting geniohyoideus and the retracting ceratotrachealis.

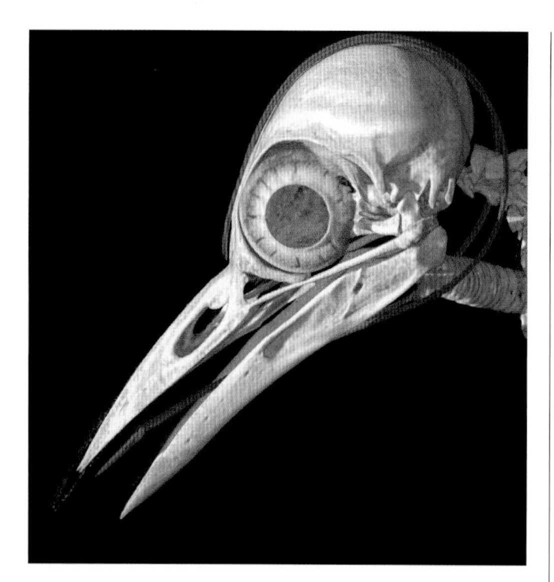

**FIGURE 3.** Retracted Hyoid Position. This high-resolution X-ray computed tomography (CT) scan highlights the in situ position of the hyoid bones of a Golden-fronted Woodpecker. *Digital reconstruction courtesy of DigiMorph.org.*

The paired geniohyoideus muscles completely encase the hyoid horns, which themselves comprise the ceratobranchial and the epibranchial bones. In the flickers, these muscles and the encased bones may dip downward along the trachea a short distance from the base of the skull before curving up and over the cranium.

In most birds, the ceratotrachealis originates at the larynx, but this would not fulfill the function of retrieving a woodpecker's long tongue. In some woodpecker species, these retractors may spiral down around the trachea several times. (See Figure 1.) In others, such as the Pileated Woodpecker, they may simply run parallel along the trachea, attaching about seven rings down from the larynx.

Another muscle uniquely adapted in woodpeckers, the esophagomandibularis (not pictured in Figure 5), connects the front of the esophagus to the upper mandible, pulling the larynx and trachea forward and enhancing the tongue's extensibility.

Two additional pairs of muscles, the ceratohyoideus and ceratoglossus, originate near the base of the tongue and, when used singly, allow a woodpecker to manipulate its tongue tip, much the way the strings of a marionette independently move each of the puppet's arms. These muscles allow a woodpecker's tongue to follow the curving gallery of a beetle larva in whatever direction necessary to retrieve the prey and to manipulate just the tongue tip inside the gallery. When used together, the paired ceratohyoideus allow a woodpecker to depress its tongue.

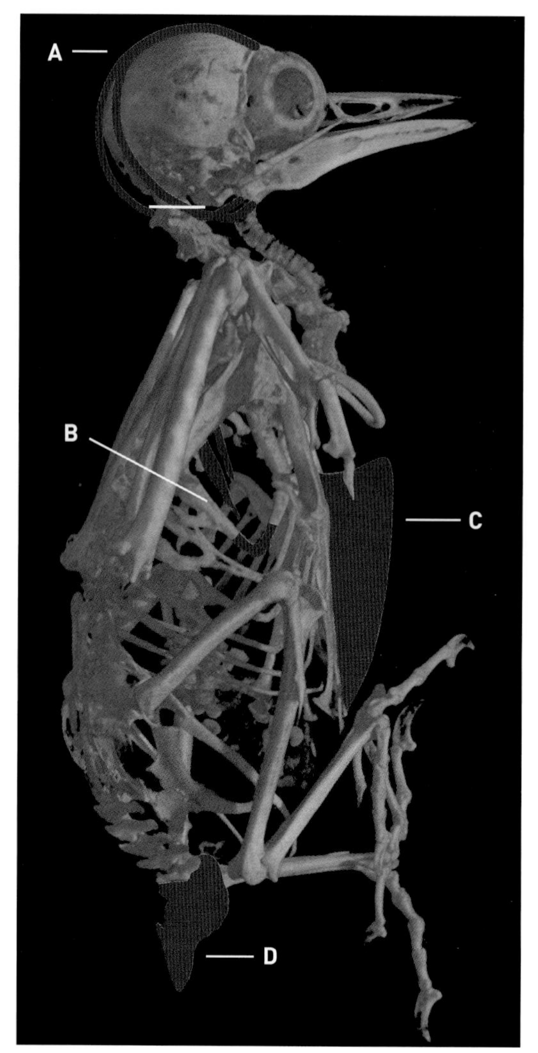


**FIGURE 4.** The Woodpecker Skeletal Structure. This CT scan shows the skeleton of a White-headed Woodpecker. Highlighted sections show (A) distal hyoid bones; (B) first and second floating ribs, and third vertebral rib connection with first sternal rib; (C) shallow keel; and (D) large pygostyle. See the text for adaptive significance of each. *Digital reconstruction courtesy of DigiMorph.org; highlights and labels by Christine Elder.*

## THE CRANIUM

A Pileated Woodpecker may slam its head on a tree up to 20 times a second, with as many as 12,000 individual strikes in a single day, at a deceleration of up to 1,200 g. A cough subjects a person to 3.5 g; for us to generate a similar 1,200 g, we would have to slam our head into a brick wall at 16 miles (26 km) per hour.

The cranial anatomy of woodpeckers has served as a model for safer sports helmets, and it has helped

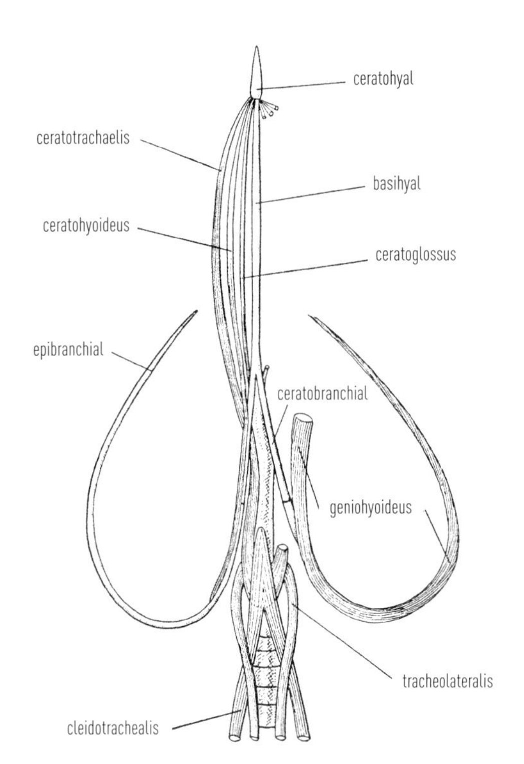


**FIGURE 5.** The Hyoid Apparatus. In 1930, William Burt drew this schematic of the hyoid apparatus from a Pileated Woodpecker. Bones are depicted as solid white and muscles are shown with striations. The fleshy part of the tongue (examples pictured in Figure 2, p. 3) attaches at the ceratohyal, with its basal "accordion" sheath extending distally to partly encase the basihyal (see The Extensible Tongue, p. 2). See the text for descriptions of most bones and muscles depicted here. The tracheolateralis and cleidotrachealis run along the trachea, between the syrinx and larynx, and their functions are not well understood vis-à-vis the hyoid apparatus. *Burt 1930, Burton 1984.*

neurologists better understand Shaken Baby Syndrome. Ophthalmologists have analyzed the woodpecker skull while researching treatments for retinal detachment and hemorrhages. The woodpecker head and neck structure may be among the best examples of resilient animal anatomy on the planet. This protection system involves several complex adaptations working in concert.

### *The Spongy Skull*

Thick, spongelike bone comprises the woodpecker skull, and it is especially thick and spongy at its rear base (the occiput) and at the frontal bone. As a woodpecker hammers a tree, a complex interaction between the tongue and protracting muscles of the maxilla pulls the brain casing backward and the pal-

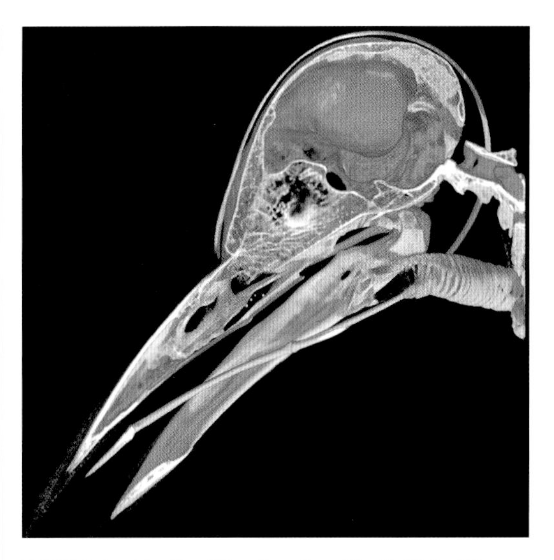

**FIGURE 6.** Inside the Woodpecker Skull. This CT scan shows the cushioning frontal and occipital bones in the skull of a Golden-fronted Woodpecker. *Digital reconstruction courtesy of DigiMorph.org.*

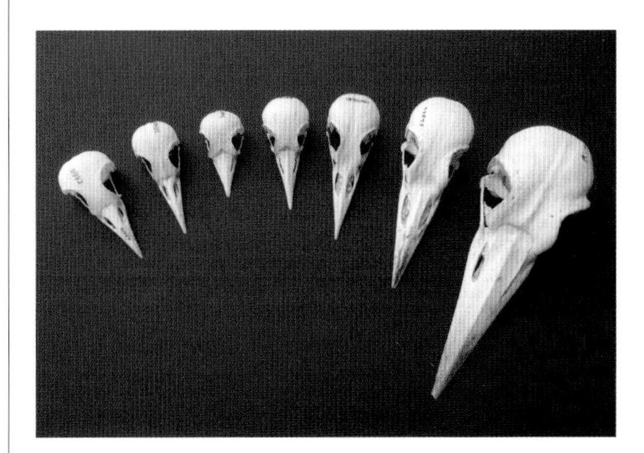

**FIGURE 7.** Skulls representing all six genera of North American woodpeckers; left to right: Lewis's Woodpecker, Red-bellied Woodpecker, Yellow-bellied Sapsucker, Hairy Woodpecker, Northern Flicker, Pileated Woodpecker, and the Ivory-billed Woodpecker's closest cousin, the Imperial Woodpecker of Mexico. Note the frontal overhang in the sapsucker and Hairy Woodpecker. *Kimball Garrett, courtesy of Natural History Museum of Los Angeles County.*

ate forward at specialized naso-frontal hinges with the upper jaw, seating the brain against the spongy occiput. Aiding in the distribution of impact forces is the near absence of cerebrospinal fluid in a very thin subarachnoid space surrounding the brain (traumatic brain injuries and subarachnoid hemorrhages occur most often as a result of head trauma in organisms with higher levels of cerebrospinal fluid, including humans).

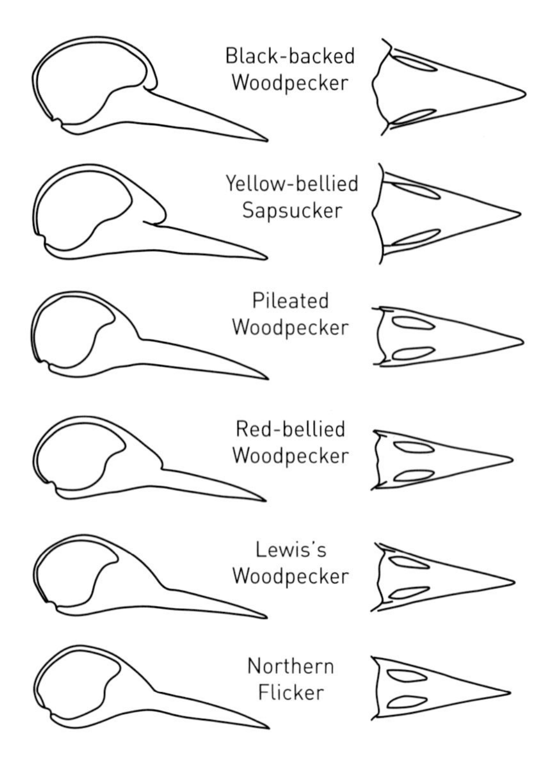


**FIGURE 8.** Skull and Bill Specialization. Sagittal cross sections of the skulls and dorsal views of the maxillae (upper half of the bill structure without the rhamphotheca, or sheath) of the five extant genera of North American woodpeckers; shown at equal length for comparison. Note the adaptations, from top to bottom, that show decreasing specialization for excavating. *Redrawn by Christine Elder after Burt 1930.*

In the woodpeckers that spend the most time hammering on trees, especially the Black-backed and American Three-toed Woodpeckers and the sapsuckers, the outer frontal bones of the skull (just forward of each orbit) extend farther forward than in other species. This gives the impression of an indented central forehead when viewed from above, but when viewed from the side these bones appear to bulge slightly over the upper mandible (see Figures 7 and 8). This "frontal overhang" prevents potentially damaging upward movement of the maxilla during blows.

### *Bill and Jaw Structure*

Woodpecker species that specialize in excavation have a straighter, broader, and flatter bill than species with more generalized feeding habits (see Figure 8). This bill shape spreads the distribution of impact forces, reducing stress on the naso-frontal hinges and minimizing shearing forces that might otherwise result in brain injury. Perfect perpendicular strokes against the tree surface create a direct line of force that sends the shock waves to the base of the skull, below the brain. In addition, the maxilla is longer and more compressible than the mandible, preventing impact on the latter. Species with even a slight curve to the bill, especially the flickers, cannot withstand the forces of heavy pounding without potentially damaging the entire bill and jaw structure. These birds tend to excavate nest cavities in soft or rotten wood, often reusing and enlarging existing cavities, and they feed by harvesting, flycatching, or gleaning, and even probing into ant burrows.

### *The Eye*

Several adaptations in the skull protect a woodpecker's eyes during impact, and the eye sockets themselves may help protect the brain.

Woodpeckers possess a thicker nictitating membrane, or nictitans, than most birds. High-speed photography reveals that a woodpecker closes these "third eyelids" immediately before its bill contacts a tree. This protects the eyes from airborne wood chips and may literally help prevent the eyes from popping out of the head.

Behind the nictitans, the space around the eyes is tightly filled with a complex polymer cushioning. Inside the eyes, two vascular structures, the pecten and choroid, increase intraocular pressure. Other birds possess these two structures, but they are uniquely adapted in woodpeckers. The pecten extends forward from near the optic nerve and fills with blood before impact, expanding nearly to the back of the lens. The choroid, a lining in the back of the eye, also swells with blood and an unknown mucopolysaccharide (a complex carbohydrate polymer). The blood- and fluid-engorged pecten and choroid greatly increase pressure in the eye, holding the lens and retina in place, and likely preventing retinal hemorrhage or detachment.

Also aiding in the protection of the eyes, the woodpecker orbital septum is thicker than that of most birds. This bony plate between the orbital cavities provides a strong substrate for the muscles that rigidly attach to the eyes, preventing potentially damaging movement during impact.

### *The Ear*

Woodpeckers possess special modifications in the ears for withstanding both forceful blows to the head and the noise produced by those blows. The middle ear of all birds comprises only one small bone, the columella, which transmits vibrations from the eardrum to the inner ear. In most birds, a single basal plate in the inner ear supports the columella. In woodpeckers, the columella is strengthened by two plates. The vestibular window is also much smaller, and it is covered by a thicker mem-

A male "Yellow-shafted" Northern Flicker perched on a branch and showing the zygodactyl toe configuration. Toes 2 and 3 are visible on each foot; toes 3 and 4 are in an opposing grip behind the branch. *Doug Backlund; Sept. 2007, Hughes Co., SD.*

brane in woodpeckers than in other birds of comparable size. Combined, these two features likely minimize the intensity of vibrations sent by the columella, thereby "turning down the volume" of a woodpecker's hammering sounds.

### MUSCULOSKELETAL FEATURES

From the neck down, subtle anatomical characters reflect the woodpeckers' penchant for heavy construction (see Figure 4). A woodpecker's rib cage comprises two cervical, or upper, ribs (also known as floating ribs, attached to the 13th and 14th vertebrae) and six thoracic ribs (lower ribs attached to both the sternum and vertebrae). Attached to the second floating rib is a set of thick neck muscles that serve multiple functions. They pull the head forward with sufficient energy to chisel a tree surface; they transmit the opposing forces downward to protect the brain; and they stabilize the base of the neck by inhibiting lateral and backward motion.

The specialized excavators have specially adapted versions of these anatomical features. These species have a particularly small first floating rib, which leaves extra room for the powerful neck muscles. Their second floating rib is especially broad, providing a stronger surface to which the neck muscles can attach. And the point of contact between the first sternal and third vertebral ribs is much broader in these species than it is in the generalist feeders. In addition, these excavating specialists have longer and more widely spaced thoracic vertebrae; these allow extra room for flexion compression, decreasing the stress to the cervical spine and transferring the impact forces downward and away from the brain casing.

Woodpeckers also have a shallower keel on the sternum than most birds, allowing them to lean their body close to the tree trunk. Minimal keel depth provides a relatively small surface area on which the flight muscles can attach, requiring that these muscles in woodpeckers attach at the scapula and ribs.

Finally, the lowest avian caudal vertebrae are fused into the pygostyle, the attachment site for all the muscles of the lower vertebral region and those controlling the woodpeckers' stiff tail feathers. Woodpeckers possess the largest and flattest pygostyle of all birds. This large pygostyle offers increased support for the tail, which props against a tree during blow delivery and helps balance the forces of impact throughout the skeletal structure.

### LEGS AND FEET

Hammering against a tree trunk involves three points of contact, and each maintains its own level of specialization. The head and tail (and connecting spinal structure) bear the greatest burden of distributing the impact forces, whereas the legs and feet provide a fulcrum of support for the body. Woodpecker feet, in particular, are highly adapted for life in the vertical plane.

Most birds possess four obvious toes, arranged

A male Downy Woodpecker clinging to a vertical surface with toe 4 swung nearly perpendicular to the bird's posture and toe 1 (the hallux) swung outward. *Kristine Falco; Oct. 2009, Deschutes Co., OR.*

with three in front and one in back: the anisodactyl foot. Toe number 1 points backward and is called the hallux. Toe 2 is the innermost forward toe, toe 3 the center forward toe, and toe 4 the outer forward toe. The anisodactyl foot is perhaps best represented in the order Passeriformes, sometimes called the passerines or "perching" birds. With their opposing toes, passerines can easily grip a perching surface.

Most species in the "woodpecker" order, the Piciformes (see Appendix 4, p 260), possess a zygodactyl toe configuration, with toes 2 and 3 pointing forward and toes 1 and 4 pointing backward. The cuckoos (Cuculiformes) and most parrots (Psittaciformes) also possess zygodactyl feet. The opposing toes of these birds allow them to strongly grip a perching surface. Parrots, for example, are frequently observed hanging upside down from a branch to reach a nearby tree fruit; this requires a decisive grip on the branch above. When a woodpecker perches on a branch or a flat surface (or is held in the hand by a bird bander), its toes align zygodactylly. One important fact, however, complicates our traditional understanding of the woodpecker foot: the zygodactyl toe arrangement represents a perching, not a climbing, adaptation, and most of the "true" woodpeckers (subfamily Picinae) are tree climbers.

When arborists climb trees, they may wear a climbing gaff, armed with a sharp downward-pointing spike, on each leg. Rock climbers typically cling to vertical surfaces with the fingers gripping the rock face between the 0-degree angle (fingers forward) and 90 degrees outward to either side. Woodpeckers easily climb trees without any safety ropes or hardware, and their toes have adapted accordingly.

Given access to the proper genetic toolkit, an organism may evolve morphological adaptations that support certain favorable behaviors, such that "form follows function." When most woodpeckers climb a tree, toe 4 swings forward, often up to or greater than 90 degrees from the rear-pointing zygodactyl position. Aligned as such, toe 4 provides an additional stabilizing force to support the woodpecker's vertical posture on the tree. In this case, "zygodactyl" does not properly describe the feet of true woodpeckers that spend most of their lives in the vertical plane. In 1959, Walter J. Bock employed the term "ectropodactyl" to describe the toe configuration of the scansorial, or climbing, woodpeckers, with their "outer movable" fourth toe.

Some woodpeckers have achieved extreme levels of foot adaptation. When clinging to a vertical surface, the hallux does not provide any support for climbing. It is typically the shortest toe and is often laid flat against the tree when the bird is clinging vertically to a tree surface. Some woodpecker species have even lost the hallux completely. In North America, this group includes the Black-backed and American Three-toed Woodpeckers, our continent's two most specialized excavators. On the other end of the extreme specialization spectrum, the Ivory-billed Woodpecker (like other *Campephilus* species; see photo, p. 9) has a very long hallux, which it swings outward with the fourth toe when climbing, so that all four toes point forward into a modified pamprodactyl toe configuration (the best-known birds with pamprodactyl feet are the swifts). Unlike most woodpeckers, which climb with their feet directly below their belly, *Campephilus* species spread their legs outward, which brings their bodies closer to the trunk. The entire undersurface of their foot makes contact with the tree, requiring special pads to minimize abrasive damage. Their "tree-hugging" posture, flattened feet, and forward-pointing toes combine to oppose the gravitational force working against these large woodpeckers. (See photo, facing page.)

Indeed, some woodpeckers spend more time on the ground or perching on a branch than climbing the vertical surfaces of trees. Among these, the flickers, and to a lesser extent the *Melanerpes* species, retain the definitive zygodactyl feet, having not adapted to the scansorial lifestyle. What about the Brown Creeper or the New World woodcreeper family? These birds also spend their lives in the vertical plane, so why do they not possess ectropo-

By swinging all of its toes forward into a modified pamprodactyl position, this Pale-billed Woodpecker (*Campephilus guatemalensis*)—likely the closest extant relative to the Ivory-billed Woodpecker—is able to hang nearly upside down while excavating a cavernous trough in this nearly horizontal branch. *Stephen Shunk; Mar. 2013, El Tuito, Jalisco, Mex.*

dactyl feet? These species are derived from the ancestral passerine lineage and so evolved with anisodactyl feet. Their adaptations for climbing include strongly curved claws on the three forward toes and long, stiff tail feathers to support them on the tree surface. The tree-climbing woodpeckers combine their specialized tails (see The Tail Prop, p. 12) with ectropodactyl toe configuration and strongly curved claws to be expert climbers.

## WOODPECKER FEATHERS

Feathers distinguish birds from all other animals on Earth, and the colors and styles of birds' feathers are as diverse as the birds themselves. From the powder downs of herons and parrots to the rictal bristles of flycatchers and nightjars, specialized feathers often help define the order, family, genus, and even species of birds.

Woodpeckers possess their own special feathers, some of which are shared among the Picidae family and others of which belong only to the "true" woodpeckers of the Picinae subfamily. In addition to their distinctive feather structures and growth patterns, woodpeckers also experience distinct molt cycles. Here is a little about what we know.

### *General Woodpecker Plumage*

Woodpeckers possess 10 primary and 10 to 12 secondary remiges (which include three tertials) and 12 rectrices (the remiges and rectrices together comprise the flight feathers of the wing and tail, respectively). The 10th, or outermost, primary is just barely longer than the primary coverts and serves little known function.

The two outer rectrices of all woodpeckers may be very small, leaving only 10 visible. The two central rectrices are among the woodpeckers' most specialized adaptations. They are typically the longest of the tail feathers, each with a very stiff rachis, or shaft, and they taper near the end of the arrow-shaped tip. (See The Tail Prop, p. 12, for a discussion of the tail's adaptive significance.)

Perhaps the most puzzling feature of woodpecker plumage is not the feathers the birds possess but those they lack. From hatchlings to adults, woodpeckers lack the insulating layer of down feathers present in most birds.

We can only speculate on the reason for this. At some point in time, ancestral woodpeckers began to exploit cavities as nest and roost sites, which may or may not have preceded their ability to excavate their own cavities. Life inside a tree cavity is well sheltered from the elements, and it is possible that woodpeckers evolved without the need for thermally protective down feathers.

### *Hatching Through Fledging*

Most woodpeckers are altricial, meaning they hatch completely naked, although a few fine white points may be seen where the rectrices will emerge. Without down feathers, a nestling must begin growing its contour (body) feathers as quickly as possible in order to regulate its own temperature. This leads to the prejuvenile molt, followed by a very unusual preformative primary molt that begins prior to fledging. See p. 11, under Other Feather Features of Juveniles, for further discussion on the preformative molt.

The difficulty of surveying woodpecker nest cavities has prevented researchers from acquiring detailed information about the timing of the first prebasic molt in many species. Among those species studied, eruption of the first feathers occurs as early as 2 days after hatching (Golden-fronted Woodpecker) and as late as 10 days (Acorn Woodpecker), with the average being about 5 days. In general, between the first and second full weeks after hatching, all feather tracts are visible, and the prejuvenile molt is completed just before or after fledging.

### *Reduced Inner Primaries*

As part of the prejuvenile molt, at least 12 woodpecker species, and possibly all woodpeckers, grow one or two stunted inner primaries, which are replaced before fledging. Primaries molt from the innermost feathers outward, and the first primaries are much shorter than the others. The extended time that a nestling woodpecker remains in the

A female Golden-fronted Woodpecker in flight, showing full ventral plumage, including the following: (A) the tips of two reduced outer remiges (primary flight feathers of the outer wing); (B) both alulae (proximal to the wing coverts); and (C) a pair of greatly reduced outer rectrices (flight feathers of the tail). *Alan Murphy; Jan. 2014, Hidalgo Co., TX.*

cavity allows it to grow and replace these greatly reduced primaries during a period when full-size feathers would be of no advantage. Smaller feathers save energy during this critical growth phase, and they make it easier for the nestling to fold its wings in the confines of a crowded cavity. Another advantage of these small primaries, observed in flickers, is that a nestling is able to extend its head up through the gap in its siblings' wings, improving its chance of receiving the next food delivery.

Lewis's Woodpecker and the four sapsuckers do not exhibit reduced inner primaries, and in fact they may not even exhibit the postjuvenal molt. Instead, they may undergo a protracted preformative molt that can continue into the winter (see below).

### *Other Feather Features of Juveniles*

As with other birds, the rectrices of woodpeckers are the last feathers to emerge, and they remain sheathed at their bases until after fledging. All of the

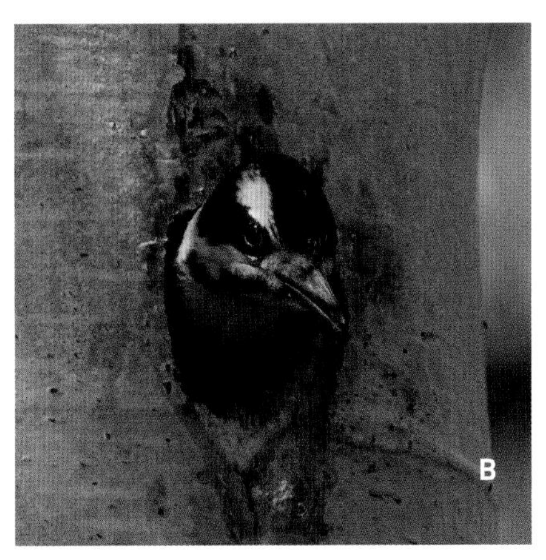

These three Hairy Woodpecker siblings show the variation in head plumage that can be present before the preformative molt: (A) a female with no apparent red in its crown, (B) a female with some red in its crown, and (C) a male with a typical red crown. *Kristine Falco; July 2009, Deschutes Co., OR.*

juvenal rectrices are strongly pointed (see photo, p. 12), and the 10th primaries are longer and broader in juveniles than they are in adults.

Juveniles of many *Picoides* species, often of both sexes, exhibit some red feathers in the crown (or yellow in the case of Black-backed and American Three-toed Woodpeckers). Juvenile females typically show much less extensive and duller red (or yellow) than males—or none at all—but this is highly variable among individual birds.

The first fall, or preformative, molt generally occurs from shortly before fledging through about October. The Acorn and Lewis's Woodpeckers and the Yellow-bellied Sapsucker may extend their first complete molts later into the fall, and in the case of the sapsucker, well into its first spring. In fact, the extension of the preformative molt in the three members of the *varius* superspecies (the Yellow-bellied, Red-breasted, and Red-naped Sapsuckers) correlates with each species' migratory behaviors. That is, the species that migrates the farthest, the Yellow-bellied, extends this molt the longest, whereas the generally sedentary Red-breasted Sapsucker completes this molt very early in its first fall. Thus, a wintering sapsucker that has yet to acquire its full basic plumage can be typically identified as "part" Red-naped or Yellow-bellied. Furthermore, an apparent Red-naped Sapsucker that has not acquired its full basic plumage by early spring could possibly be a hybrid Red-naped × Yellow-bellied. (See pp. 107–109 for further discussion on sapsucker hybridization.)

The juvenile Yellow-bellied Sapsucker replaces all of its primaries before the end of its first summer, well ahead of its contour feathers and before its rather extensive migration. Lewis's Woodpecker and some Acorn Woodpeckers typically do not shed their inner primaries until their second fall, and the outer primaries look almost identical in juveniles and adults (in most species, the outer primaries are more tapered and longer in juveniles than in adults). These two species also keep their juvenile rectrices through their second summer. Juvenile Lewis's, Acorn, and Red-headed Woodpeckers are all prone to a suspended molt of the primaries during migration or fall dispersal.

### *Adults*

Subsequent molts (birds in their second summer and beyond) in most species occur annually from late summer to early fall, with Lewis's Woodpecker consistently maintaining the most drawn-out molt cycle of North American woodpeckers. Woodpeckers generally arrive on their breeding territories in fresh plumage, but through the nesting period, after hundreds of trips in and out of the nest cavity, their feathers become extremely worn and sometimes discolored; in this tattered plumage, birds often do not fit their profiles in most field guides. Adult sapsuckers of the *varius* superspecies observed breeding anywhere near the interspecific contact zones may not be identifiable to species.

### *Feathers and Aging*

Aging woodpeckers in the field (as opposed to in the hand) can be problematic. However, careful observation of a more or less stationary bird or a crisp series of photos may produce results. The sequences of feather replacement through the molt cycle are fairly consistent among all species, but the timing varies. Most woodpeckers replace all their primaries and rectrices in their preformative molt, but a given individual may not replace all its juvenal primaries for 2 years. The secondaries may last up to another year, with primary coverts occasionally hanging on until the fourth or later fall.

The variability in molt timing for the secondaries and primary coverts can be perplexing. For a detailed discussion of these molt patterns, see Howell (2011), Pyle (1997), and Pyle and Howell (1995).

## FEATHER FUNCTIONALITY

### *The Tail Prop*

Woodpeckers possess some of the stiffest tail feathers among birds. The rachis, or shaft, of the two central rectrices in particular extends beyond the outermost vanes. The last few vanes become symmetrically shorter near the end, tapering the feather to a point. The central rectrices of the Pileated and Ivory-billed Woodpeckers even curve inward, offering extra resistence to support these large woodpeckers while climbing or propped against a tree trunk.

The dynamic posture and hitching motion of woodpeckers have been well studied. In general, a longer tail helps push the bird's center of gravity closer to the tree, minimizing the outward force against the body and placing it directly over the supporting base of the tail. In addition to the modifications of the central rectrices described above, nearly all woodpeckers have a mostly dark tail. The melanin that gives feathers their dark coloring also increases their strength and improves their longevity by minimizing wear and breakage.

The practical significance of the central rectrices can be seen in the unique molt sequence of the woodpecker tail. The feathers immediately outside the central pair fall first, progressing outward, with the central feathers dropped only after all the others have been replaced. In other birds, the paired first and second rectrices are dropped first, but the woodpeckers need rectrices 2–5 to support them on a vertical surface before they drop the crucial central pair.

The fully spread tail of a fledgling Red-breasted Sapsucker, showing the strongly tapered outer rectrices (pairs 3–5, counting from the central pair) that are typical of juvenile woodpeckers. *Stephen Shunk, courtesy of Institute for Bird Populations; July 2012, Yosemite National Park, CA.*

### *Bristled Feathers*

In 1866, Elliott Coues coined the genus name *Asyndesmus* solely for the Lewis's Woodpecker, calling the genus the "bristle-bellied woodpeckers." Coues described the breast feathers of Lewis's Woodpecker as "enlarged in caliber, bristly, of silicious hardness, loosened and disconnected, being devoid of barbicels and hooklets." The species has since been classified in the genus *Melanerpes,* but Lewis's Woodpecker does indeed possess bristled feathers on its underparts and partly on its gray collar. The adaptive significance of these bristled belly feathers is unclear.

Most woodpeckers possess a row of feathers on their inner abdominal region that are similar to those of Lewis's Woodpecker and intermediate in structure between those of down and typical contour feathers. These are lost during incubation and

An adult female White-headed Woodpecker removing fecal material from the nest cavity. Note the long nasal tufts covering the nares at the base of the maxilla. *Kristine Falco; June 2010, Deschutes Co., OR.*

An adult Lewis's Woodpecker, showing its bristlelike breast and belly feathers. *Paul Bannick; June 2005, Yakima Co., WA.*

replaced before the typical fall molt cycle of the adult. These unusual feathers inspired the naming of the Hairy Woodpecker.

Similar to the belly bristles are the specialized feathers known as "nasal tufts," which cover the nares, or nostrils, and prevent flying wood chips from obstructing the breathing passage. All woodpeckers possess nasal tufts.

### *Feather Colors*

The black-and-white pattern of most North American woodpeckers is an example of disruptive coloration. Because most of these birds live in the woods, they continually course through and perch in very contrasty filtered light. This bane of the nature photographer works in favor of the woodpeckers. Their plumage is the perfect camouflage against the disrupted light under the canopy.

In contrast to the melanin that gives feathers their black tone, carotenoid pigmentation gives woodpeckers their red or yellow coloration. In seven of our nine *Picoides* species, the male differs from the female in having some red on the head. This ranges from the tiny and inconspicuous red "cockade" of the male Red-cockaded Woodpecker to the mostly red crown and nape of the Ladder-backed Woodpecker. Sapsuckers also show varying amounts of red in their plumages, with Williamson's displaying a small amount of red in the throat and the "Northern" Red-breasted Sapsucker (of both sexes) showing more red in the head and breast than even a Red-headed Woodpecker. Male Black-backed and American Three-toed Woodpeckers show a lemon yellow disc on the crown and no red in their plumage whatsoever.

If the pied coloration in woodpeckers helps camouflage them in the forest, what is the purpose of the carotenoid pigments? Early in the breeding season, males may flare their red feathers when defending territory or attracting a mate. Females that also exhibit some red may play a larger role in territory defense than those that do not possess any red plumage. The flickers display varying levels of carotenoid pigmentation throughout their bodies, with the Gilded and "Yellow-shafted" Northern Flickers expressing less color (and hence yellow rather than red) than the "Red-shafted" Northern Flicker. Juvenile Northern Flickers may show varying amounts of red in the crown, the function of which is unknown. (See photo p. 14.)

Woodpeckers are often subject to environmental discoloration. When a white or lightly colored belly rubs constantly against a pitch-covered tree, the staining is unavoidable and conspicuous. The charred surfaces of burned trees also tint the feathers of woodpecker bellies. When woodpeckers nest in charred trees, their facial and body feathers can even make the birds appear melanistic. Other color-changing agents may include pollen, soils, ultraviolet radiation, and anthropogenic pollutants. In some cases, a dark woodpecker may actually be "melanistic," or darkened by extra melanin. On the opposite

A fledgling male Northern Flicker (window collision fatality), with extensive red in its crown, nape, and malar stripe. The combination of the red nape and malar results from mixed "Yellow-shafted" x "Red-shafted" parentage (see Northern Flicker, Subspecies, p. 219). *Stephen Shunk; June 2009, Deschutes Co., OR.*

This adult male intergrade Northern Flicker (note red in nape and see p. 214) has been dramatically darkened in the breast and belly by some environmental substance, possibly from dust bathing or from pitch near a cavity entrance. *Tom Taylor; June 2015, Lake Co., OR.*

end of the spectrum, woodpeckers are not exempt from the aberrations known as leucism and albinism—the partial or complete loss of melanin, respectively, in some or all of their plumage. In some cases, a bird may be amelanistic while still expressing its carotenoid pigmentation.

## BEHAVIOR

From cavity excavation to their signature flight style, woodpeckers exhibit a host of fascinating behaviors. This section discusses the most notable behaviors shared by most or all North American woodpeckers; those limited to one or a few species appear in the appropriate species accounts.

### BREEDING BEHAVIOR

#### *Courtship and Pair Bonding*

The courtship dance of a woodpecker pair resembles two spring-loaded puppets stuck to a tree, bobbing up and down and shuffling around the trunk, chortling and cooing to each other with crests raised and wings aflutter. This animated ritual brings new life to a chilly spring forest. Each species owns a variation on the theme, but the general dance is surprisingly similar throughout the Picinae subfamily.

At the height of the season, the dance takes to

An adult female Ladder-backed Woodpecker with whitish pollen obscuring its black facial stripes—a result of collecting food for nestlings in saguaro cactus blossoms. *Kristine Falco; May 2006, Pima Co., AZ.*

A pair of Nuttall's Woodpeckers at the onset of their brief copulation ritual. *Eleanor Briccetti; Apr. 2009, Santa Clara Co., CA.*

the air, and the pair may engage in chasing bouts through the woods or butterfly-like displays that often culminate in copulation. (See photo p. 104.) Sedentary species often maintain their bond throughout the year, drifting apart for occasional feeding bouts in winter but rekindling the pair bond with drumming and aerial displays well before the nesting season.

Woodpecker copulation postures often require adroit acrobatics. Possibly because of the stiffness of the tail feathers, most woodpeckers do not practice the mounting position of most birds. Instead, the male may flutter onto the female's back, where she waits in a flattened horizontal posture, often perpendicular to a large limb. The male proceeds to slide backward, typically falling off to one side, with his outer wing or foot braced beside the female on the limb. The male then tucks his vent underneath the female's while she tilts sideways slightly, consummating the act, and the two fluff off in different directions while calling back and forth and often drumming upon their first postcoital alightment.

### *The Nesting Calendar*

Most North American woodpeckers do not migrate, instead remaining in their respective permanent ranges. Breeding season begins early for these resident, or sedentary, species, especially in the desert Southwest. Ladder-backed Woodpeckers may begin courtship as early as January, followed closely by Gila and Arizona Woodpeckers and Gilded Flickers. Even the Williamson's Sapsucker has resident populations in Arizona that begin renewing pair bonds and staking out territories by late February and early March.

As with many birds, the length of the woodpecker nesting calendar is generally shorter at higher latitudes and elevations. For example, Yellow-bellied Sapsuckers in the central Yukon Territory compress their entire breeding season into June and July. This contrasts with populations breeding in Michigan, where excavation ensues by mid-April and fledglings may remain dependent through mid-August.

A pair of woodpeckers typically spends 1 to 2 weeks excavating a nest cavity, depending on the substrate. Hollowing out a live saguaro cactus doesn't take as much work as hollowing out a cottonwood snag. The saguaro nest, however, has to be started much earlier so that the inner lining of the cavity has time to dry before egg laying begins.

Woodpecker incubation periods are short compared with those of other altricial species, averaging a little less than 2 weeks. Conversely, the nestling period is longer than that of most altricial birds, even though it varies substantially from an average of 21 days in the Downy Woodpecker to 31 days in Lewis's and Acorn Woodpeckers.

*Nest Site*

Woodpeckers certainly do not have a corner on cavity nesting, but they do rule the roost on the practice of excavating, which involves careful selection of just the right tree. In general, woodpeckers do not choose live, healthy trees if alternatives are available; most species favor dead or dying trees. Many species select trees that are alive but inflicted with heart rot or sapwood fungus, facilitating more efficient excavation. Others may choose a large, decadent snapped-off limb in a tree that is otherwise perfectly healthy. Woodpeckers often excavate on the underside of a leaning tree or limb.

The best known of the live-tree excavators is the Red-cockaded Woodpecker. Many species will occasionally excavate in live trees, but the Red-cockaded does so more than any other. This allows the bird to maintain active, long-term resin flow around the cavity entrance as a barrier to predation from its primary enemy, the tree-climbing rat snake. Excavating in live trees involves an extra challenge, however, and the Red-cockaded Woodpecker may require months or even years to get through the living sapwood before reaching the rotten interior of mature southeastern pine trees.

Whether a woodpecker selects a hardwood or coniferous tree for its cavity varies widely by region. The preferred species of tree varies among the different woodpeckers and their locally preferred habitats.

Whatever the substrate, a woodpecker must chisel its way to the inside of the tree and dispose of the displaced material. Observing a woodpecker in the ritual "tossing of the chips" is nothing short of entertaining. In a quiet forest, the attentive birder might hear an incessant hammering and look around for the source of the sound. Then the sound will stop, and a woodpecker will stick its head out of a cavity entrance with a bill full of wood chips, tossing it pell-mell to the forest floor. (See photo, opposite page.) Some species will fly away from the cavity to dispose of the chips, possibly to avoid alerting predators to the location of the cavity tree.

Many species will excavate a new cavity in the same vicinity or even the same tree as in prior years, but most cavities see only one season of use for a given pair of woodpeckers. Where few snags exist, however, woodpeckers may reuse an old nest site. Also, if a pair fails in their first nesting attempt, because of weather or predation, for example, they may not have time to reexcavate and will resort to using an existing cavity. The *Melanerpes* and *Colaptes* woodpeckers are considered weak excavators, lacking the specialized anatomy of the others, and they rely mostly on existing cavities for their nest sites.

In the last step of nest preparation, the adults leave a layer of wood chips on the cavity floor, the equivalent of a songbird's nesting material, upon which they lay their eggs.

An adult male Acorn Woodpecker exiting a nest cavity. Note the entrance aspect on the underside of a leaning snag. *Kristine Falco; May 2006, Pima Co., AZ.*

As a survival strategy, hole nesting has distinct benefits, limiting nest predators to those able to climb or fly into the entrance. (See Predators, p. 26, for predation threats outside the cavity.) Tree cavities also shelter the nest from the elements, allowing some woodpeckers to begin nesting early in the season.

*Egg Laying*

All woodpeckers can presumably breed in their first full summer and annually thereafter, though not all individuals do so. Once nest excavation is complete and before the female begins laying, the male takes over the nest cavity as his roost site. If a male has already excavated a new roost cavity before selecting a mate, and if his mate approves of the cavity, it becomes the nest site. Throughout the nesting period, the female typically roosts at night in her own cavity. She lays 1 egg per day in the male's cavity until the clutch is complete, laying 4 to 5 eggs per clutch, on average.

*Incubation and Brooding*

Beginning with the first egg laid, one parent constantly attends the nest, but incubation generally begins with the laying of the penultimate or final egg. Both sexes share incubation duties in the

An adult Downy Woodpecker tossing wood chips from a cavity entrance after excavating the interior of the cavity. *Marie Read; May 2013, Tompkins Co., NY.*

daylight hours, and the male typically covers the entire night shift. During the daytime, the setting adult generally waits until its mate returns for the exchange to occur. Later in the incubation period or in warm weather, adults are more likely to leave the nest before the typical nest exchange. When the weather is cool or wet, adults spend more time on the eggs each day. Woodpecker eggs are generally similar in color, texture, and shape for all species; they are often described as creamy in color, slightly glossy in texture, and subelliptical.

Entering and exiting the cavity is not easy. The entrance tunnel is narrow and generally angles downward, and the adult enters the nest headfirst. This requires walking down the entrance tunnel upside-down, then turning around inside the chamber, baring the brood patch, and descending onto the eggs. Experiments with cavity cameras and one-way glass windows show that the setting adult rests its bill either on the cavity wall or tucked under its back feathers, often with its eyes closed. During extremely warm weather, the attending adult may roost inside the cavity but off the eggs.

Once the young hatch, one adult initially remains in the nest at all times, with the two parents taking turns brooding almost constantly for the first few days. This schedule becomes more sporadic as the young begin growing their first feathers and as the feeding demands increase. The male continues to night-roost in the nest cavity with the young birds until the last few days before fledging.

### *Nestling Development*

As with most altricial birds, woodpecker nestlings are nidicolous, meaning they remain in the nest for some days or weeks after hatching. Weighing only a few grams, newly hatched young are completely dependent on their parents for food and warmth. Most woodpeckers hatch asynchronously, usually over 12 to 24 hours. The chicks that hatch first are fed immediately and quickly gain in strength and size, giving them an advantage over their later-hatching siblings. Some of these late-hatching young never catch up in size and may die in the nest.

The hatchlings' eyes are fused shut, making the birds essentially blind, although they are photonic—able to detect changes in light—so they know when the parents return with food. The pinkish mandibles are out of proportion; the mandible is narrower at the base and shorter than the maxilla, and it has a glossy white egg tooth that wears off as the bird develops (although Northern Flickers are known to retain their egg-tooth throughout the nestling period). At the base of the maxilla, a bulging fleshy gape flange protrudes from each side. The legs and feet are pale pinkish with tiny white claws.

The bird rests on its lower leg bones, which lie flat on prominent callous pads at the upper joint, allowing the nestling to prop itself upright for feeding after about day 4.

A recent hatchling can hold its head up for only a few seconds, but by day 2 it can extend its neck completely upward. During the first 10 days, the young face each other with their necks intertwined, unwinding them with each feeding visit. At first, the adult must peck at the sensitive gape flanges to entice the young to feed, but this does not last long. Soon, the vocalizations of an approaching adult, the vibration of the parent landing on the outside of the nest tree, and the blocking of light as the adult enters the cavity all induce the begging reflex in the nestlings.

The chick that stretches its head highest gets the first morsel, subsequently allowing the others to raise their heads. After receiving its food delivery, the first chick lowers into the bottom spot while the others are fed, allowing them to continually alternate feeding positions. During hot weather, older chicks will sprawl out inside the cavity to regulate their body temperatures.

At the start of the third week, the young begin climbing the cavity walls, and individual dominance is asserted among the siblings. The dominant chick is consistently the most vocal and the one most often present near the nest entrance. Flickers and Pileated Woodpeckers generally excavate large cavity entrances, allowing multiple nestlings to peer out and beg simultaneously (see photos on p. x and in species accounts). By this time, the young have already reached their maximum weight, and in the last few days before fledging, the parents reduce the frequency of their feeding visits substantially. The nestlings then begin losing weight, and their hunger becomes intolerable. In order to be fed, the young soon have only one option: to leave the nest.

### *Feeding*

The sounds of an active woodpecker nest are unmistakable and often betray its presence from some distance. The nestlings may simply hiss persistently at first, but in the last 2 weeks before fledging, the sound from the nest is incessant. Calls alternate between a dull chirping when the adults are away on a long food-collecting bout and a loud squawking when the food arrives. Even at night, the young may maintain a quiet purring sound.

The adults enter the cavity completely to feed until the young are about 7 to 10 days old. During the second and third weeks, feeding continues with half the adult's body hanging out of the nest entrance. In the final week or few days before fledging, the active and persistent nestlings wait at the entrance and the adults feed them from outside the cavity.

Like most adult birds charged with the task of feeding developing nestlings, woodpecker parents select food opportunistically. Ants constitute the overwhelming majority of nestling foods, possibly because of the limited effort required to obtain large quantities. The strong excavators frequently bring beetle larvae, and the flycatching species, especially Lewis's Woodpecker, feed their young almost entirely on aerial insects, with the occasional caterpillar, grasshopper, or small fruit. (See photo, p. 20.)

### *Sanitation*

The adult woodpeckers typically leave broken eggshells and unhatched eggs at the bottom of the cavity. Occasionally they drop the shells at the base of the nest tree or carry them a short distance away.

Initially, an adult prods the nestlings near the

Nestling development from hatching through fledging in the Northern Flicker: (A) day 1 after hatching; (B) day 4; (C) day 8; (D) day 11; (E) day 15; (F) day 18; (G) day 21; (H) day 26, fledging date. *Tina Mitchell; May–June 2012, Lakewood, CO.*

This just-fledged juvenile sapsucker, the offspring of a mixed pair of Red-breasted and Red-naped Sapsuckers, clings tightly to the first branch it found upon leaving the nest cavity. *Stephen Shunk; July 2006, Deschutes Co., OR.*

anus, or vent, to stimulate defecation. For the first few days, the fecal material is often partially undigested and may be eaten by the adult. Also during this period, the adults continue to whittle chips from the cavity wall, which aids in the absorption of fecal material (adults can often be seen exiting the cavity with what looks like a goopy clump of wood chips; see photo on p. 12). The continued production of chips may also help increase the size of the cavity as the young grow. The male handles most of the sanitation duties, carrying fecal sacs out and away from the nest to be dropped, often in one or two preferred locations.

As the young grow, it becomes more difficult for the adults to collect the excrement, and no more wood chips are added. About the last week before fledging, nest sanitation ceases, and the young may become quite soiled. The fecal material may start decaying at the bottom of the nest, inviting a host of invertebrates, which is sometimes evidenced by an increasing number of insects flying around the cavity entrance.

### *Fledging*

After about 4 weeks, the parents effectively starve the young out of the nest by withholding food and then persistently calling from near the nest entrance with food in their bills. The hungry young begin hanging farther and farther out of the cavity until they eventually fall downward and take their first flight, landing on a nearby trunk or low branch. All the surviving young typically fledge in this manner over a few hours or into the following day.

Once they have fledged, juvenile woodpeckers usually do not reenter the cavity. They may loaf silently on a tree trunk for long periods or follow their parents around, while continually begging for food. In some double-brooded species, however, young from the first brood may reenter the cavity, easily outcompeting the smaller nestlings for the next food visit. For their first few nights after fledging, the young typically roost outside available cavities, after which they begin searching for their own evening roost sites. (See Predators, p. 26).

Until they become accustomed to their new world, the juveniles may clumsily fall from the side of a tree, swooping up to the next one nearby, or they may get their tail tip stuck in the bark while trying to hop downward toward their parent. They may also take some time learning to use their long tongue to its best advantage.

In the first week, the fledglings will often glean insects from inside bark crevices. They continue to beg food from their parents for up to a few weeks, but the parents become increasingly aloof. In species that raise multiple broods in a season, the first brood of fledglings may be abandoned after about 2 to 3 weeks to fend for themselves. In single-brooded species, the parents may start chasing off fledglings after 1 to 2 months, becoming increasingly aggressive and even pecking at the young to drive them to independence and prevent inter-family competition for food sources.

## FEEDING BEHAVIORS

Woodpeckers employ a wide variety of feeding habits, from gleaning to excavating to collecting nut mast. Each species has its own specialized style, many of them complementary, allowing them to partition out diverse forest and woodland resources. In a single stand of conifers, the White-headed Woodpecker probes the deepest furrows of the ponderosa pine bark; the Black-backed Woodpecker bores into the cambium of a recently burned larch; the Pileated Woodpecker gathers carpenter ants from a rotting log; and Lewis's Woodpecker sallies from snag to snag collecting aerial insects.

Woodpeckers possess relatively large forebrains—smaller on average than those of only the corvids, or crow family, and the psittacids, or parrot family—and some woodpecker feeding habits have been correlated with large brain size. Many woodpecker species use anvil sites to break up food that is too large to feed their young or to eat themselves. When they collect oversized food, such as nuts or large insects, they may jam the food item into a bark crevice or a snapped-off treetop and then use their bill to break the chunks into bite-sized pieces. A few species even use anvil sites to smash insects into boluses, making the food easier for nestlings to swallow.

The species accounts explore in detail the feeding practices of each woodpecker. What follows here is a broad-brush treatment of a woodpecker's foraging pursuits.

### *Subsurface Excavation*

The woodpecker's best-known feeding habit involves whittling through a tree's bark or cambium in search of insect prey. Woodpeckers feeding on wood-boring and bark beetle larvae leave behind conspicuous excavations in the bark or the inner wood of a tree. The cavernous excavations of a Pileated Woodpecker are excellent signs of carpenter ant activity.

A woodpecker's sense of hearing may also assist in its search for insect larvae. Persistent tapping over the bark surface may yield the telltale hollowness of a larval gallery or the weakness of a bark plate that would encourage scaling it for the prize underneath. Tapping may also stimulate insects to increase their movements underneath the surface. Large grubs gnawing deep inside the timber can even be heard by the attentive birder as a mantra-like crunching tone in an otherwise quiet forest.

### *Drilling of Sap Wells*

Many woodpecker species are known to consume tree sap, either from wells of their own construction or from natural wounds in a tree, but sapsuckers are notorious for the patterns of shallow holes, or wells, they excavate in tree trunks. The characteristics of the wells drilled depend on the season and the productivity of individual wells.

Early in the season, before the leaves appear on deciduous trees, sapsuckers tap the xylem tissue of a tree, which carries food upward to the branches to fuel leaf production. Xylem wells are typically small round holes, often drilled in rings around the trunk, that penetrate into the sapwood.

Once the trees leaf out, the leaves produce food

In a light summer rain, an adult Lewis's Woodpecker prepares to deliver a billful of ripe currants to its nestlings. *Stephen Shunk; June 2013, Bend, OR.*

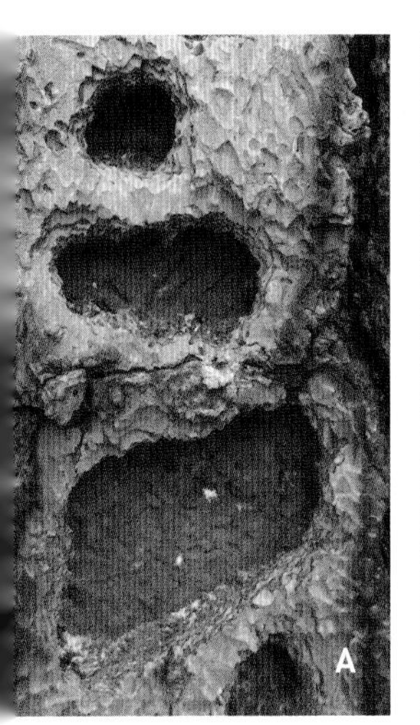

Woodpecker feeding signs, depicting varied feeding styles: (A) excavations only through the outer bark in search of bark beetle larvae; (B) subsurface excavations beyond the cambium in search of wood-boring beetle larvae; (C) the cavernous excavation of a Pileated Woodpecker in search of wood-boring carpenter ants; (D) the bark scaling of (likely) a Pileated Woodpecker in search of large bark beetle larvae—note the horizontal marks created by glancing blows used to pry bark from the tree surface. *Stephen Shunk, A, B, and C—Deschutes Co., OR; D—Arkansas Co., AR.*

Fresh xylem wells in a Florida palm tree, with very old, scarred wells above; in this region, these were certainly excavated by a Yellow-bellied Sapsucker. *Stephen Shunk; Feb. 2009, Polk Co., FL.*

A Red-breasted Sapsucker feeding at freshly excavated phloem wells in mountain mahogany. *Marie Read; June 2013, Mono Co., CA.*

and send it back down the trunk to fuel the growth of the tree. The sap carrying these nutrients flows in the outer phloem tissue. A sapsucker's phloem wells are shallower than xylem wells, and they are more geometrically shaped, often in puzzlelike vertical patterns that may cover large sections of the trunk. The sapsucker uses its bill like a jigsaw, etching the outline of a small oval or rectangle and then plucking out the bark patch and tossing it aside (frequently eating at least part of the woody material, or bast).

After drilling a ring or section of wells, a sap-

A Lewis's Woodpecker flycatching, aiming for its prey. *Marie Read; June 2014, Mono Co., CA.*

sucker often moves to a nearby tree to repeat the process. Individuals often maintain numerous feeding stations in a circuit, returning to each site after the sap has had a chance to exude from the previously drilled wells—similar to hummingbirds that wait for a flower to produce more nectar before revisiting the same blossom.

### *Surface Feeding and Flycatching*

Most woodpeckers glean and probe for insects and other arthropods on a wide variety of surfaces. Some have become specialists at reaching deep into the furrows of tree bark. The White-headed Woodpecker is particularly adept at twisting its head nearly 180 degrees to reach into the tight bark crevices of large pine trees in the western states, and its white head may even be an adaptation for reflecting light into deep bark furrows.

The "zebra-backed" *Picoides* woodpeckers are also adept gleaners, flakers, and chiselers, prying away bark pieces to extract the prey items underneath. Red-cockaded Woodpeckers can often be located in southeastern pine forests by watching for showers of bark flakes falling to the ground from high in the canopy. Flickers are master ground gleaners, snapping up ants and even probing into underground tunnels with their extra-long tongue.

Many woodpeckers flycatch opportunistically, but the solid-colored *Melanerpes* woodpeckers (Lewis's, Acorn, and Red-headed) are equipped with special flight adaptations for proficient aerial pursuit of flying insects. Lewis's Woodpecker in particular spends the majority of its time flycatching, with an individual frequently catching multiple insects on a single foray.

### *Food Collection and Storage*

The *Melanerpes* species have the most generalized feeding behaviors of the woodpeckers, and outside the breeding season many of them collect and store plant foods. The Acorn Woodpecker earned its name from its habit of gathering and storing the fruits of oak trees that it eats in the fall and winter. After drilling special storage holes in thick tree bark (or a nearby telephone pole or building), individuals will pluck fresh acorns from a productive tree and then tap each acorn into its own tight little storage cavity. This species' social structure centers around these virtual pantries of stored acorns, known as granaries. Other woodpecker species also collect and store fruits or nutmeats of many varieties, occasionally storing them in tree cavities or bark crevices.

### *Artificial Feeding Stations*

Woodpeckers will visit a wide variety of backyard bird feeders, preferring suet, shelled peanuts, and hulled sunflower seeds. Some species will take sunflower seeds in the shell, despite the energy required to extract the nutmeat (juveniles likely find this method easier than excavating into tree bark with their barely developed bills). Some woodpeckers will even attend a hummingbird feeder, using their extensible tongue to draw up the high-energy sugar water. (See photos, opposite page.)

Suet is also a favorite, but it should be provided primarily in cool weather—in warm weather, the melted fat can damage feathers, potentially leading to feather loss on the face. It is also important to provide a water source at any feeding station; most woodpeckers will drink from artificial water sources, including bird baths, stock tanks, and recirculating water features. (See photo, opposite page.)

## COMMUNICATION

Woodpeckers lack the anatomical capacity to sing. Instead, they have evolved other effective means of communicating. The signature element of a woodpecker's nonvocal repertoire is the drumroll.

### *Drumming*

Other than a single cockatoo species in Australasia, woodpeckers and the closely related piculets are the only birds in the world that employ a mechanical or instrumental form of communication. Many birds, such as snipe, grouse, manakins, nighthawks, and others, use various forms of nonvocal communication, but a woodpecker generates a "drumming" sound by using its bill to strike an external surface.

A Golden-fronted Woodpecker drinking on the bank of a small pond. Though the behavior is not reported very often, woodpeckers will certainly drink when water is available. *Cissy Beasley; Mar. 2014, Hidalgo Co., TX.*

Gila Woodpeckers are well-known visitors at hummingbird feeders in the desert Southwest. *Stephen Shunk; Aug. 2011, Cochise Co., AZ.*

A juvenile female Red-bellied Woodpecker at a suet feeder. *Gene McGarry; June 2012, Woodstock, NY.*

The purpose of a given drumming bout typically falls into one of five categories: establishing or defending a territory; attracting a mate; maintaining contact with a mate; promoting or strengthening the pair bond; or signaling readiness for or success at copulation. The woodpecker drumroll may have evolved as the birds heard the sounds of their own feeding and excavating behaviors. Drumming may even represent the precursor and evolutionary counterpart to singing. As with many aspects of bird behavior, the true origins of drumming remain a mystery. We do know about its context, however, and its use in species recognition.

In most North American woodpecker species, both sexes drum, males usually more often than females, and the drums are typically not sex-specific. Most woodpeckers begin drumming by 6 to 8 months of age, or around the start of their first breeding season, and some species drum throughout the year. Drumming usually occurs at the top of a dead, and often snapped-off, tree, though sometimes on dead branches or even large flakes of bark or the remaining bark on dead trees. Drumming is frequently heard by humans, especially when it is performed on a resonant artificial surface, such as the metal casing of an electrical transformer (see Appendix 5). The only real requirement of a drumming surface is its ability to resonate sound.

At a much slower cadence than the typical drumroll, and in shorter bouts, woodpeckers occasionally practice deliberate tapping rituals. Sometimes called "drum-taps," these are usually performed equally often by both sexes of most species, and more frequently by *Melanerpes* species than by species in other genera. Tapping usually accompanies courtship or pair-bonding behaviors, almost always occurs at the cavity entrance (often right on the edge), and can be given from either inside or outside the cavity.

In addition to their trademark drumming, woodpeckers make a variety of other nonvocal sounds—especially feeding sounds—most of which betray a woodpecker's presence but few of which are identifiable to species.

### *Vocalizations*

Woodpeckers are generally chatty, and their vocalizations are fairly distinctive, often easily separable from those of other birds. Each species has its own characteristic repertoire of calls. Some species can easily be identified by voice, but the calls of some congeners (species in the same genus) can be difficult to separate. Calls that are unique to a species are discussed in the individual species accounts. Below is a simple overview of woodpecker vocal sounds.

Most woodpecker vocalizations fit into three basic categories: (1) general call notes, typically given as position announcements; (2) rattles, or collections of notes, often given as more emphatic declarations during courtship or territoriality; (3) and contact calls given in close proximity to other individuals, which may include mates, offspring, or challengers. *Melanerpes* and *Sphyrapicus* woodpeckers also give

A "Northern" Red-breasted Sapsucker drumming on the metal cap of a wooden lamppost. Note the worn paint where this individual has been drumming frequently. *Stephen Shunk; May 2012, Marion Co., OR.*

a squeal or scream call, and sapsuckers also give a descending whine. Rattle calls may occasionally play the same role as drumming; species that drum often may rattle less so, and vice versa. Drumming and rattle calls—and the sapsuckers' territorial scream calls—are also important long-distance signals.

The best way to learn woodpecker sounds is the same as that for learning any bird sounds: spend more time in the field listening to the birds. Several references describe and compare woodpecker vocalizations and drumming in detail. For further study, see Dodenhoff et al. (2001), Stark et al. (1998), Winkler and Short (1978), and Winkler et al. (1995).

## SOCIAL INTERACTIONS

Woodpeckers are dynamic and conspicuous members of the forest and woodland community. Here are a few of the ways they interact with each other and other wildlife.

### *Living with Other Woodpeckers*

Woodpeckers can be either territorial toward or tolerant of each other, depending on the season and situation, and interactions may be inter- or intraspecific.

Individuals may defend a feeding or nesting territory, and some species also guard favored local food sources. Territorial displays closely resemble courtship behaviors, the two being indistinguishable in most species. (See Courtship and Pair Bonding, pp. 14–15.) In addition to typical displays, aggressive encounters may include chasing and, rarely, physical contact.

Woodpeckers may also forage in close proximity to others, especially their mates, and multiple species may feed together among a particularly productive mast crop or a stand of insect-infested trees. Occasionally, small groups of multiple sapsucker species migrate together; Lewis's and Red-headed Woodpeckers and Northern Flickers are regularly observed migrating in loose aggregations. Different woodpecker species have even been known to nest in the same tree, especially when snags or other nest substrates are at a premium.

True intrapecific sociality occurs among Acorn and Red-cockaded Woodpeckers, both of which live in communal family groups. These two species may be the best-studied woodpeckers in the world, due in part to our fascination with their dynamic social lifestyles.

### *Interactions with Non-Woodpeckers and Other Animals*

Especially outside the breeding season, woodpeckers will occasionally join mixed flocks of songbirds or other woodpeckers on feeding forays. More often, other non-woodpecker species join the foraging woodpecker to take advantage of nut or fruit fragments left behind or to glean remnant arthropods exposed by the woodpecker's probing and flaking of tree bark.

Male and female Acorn Woodpeckers displaying to each other at a granary site, with a second male—possibly an offspring from a prior season—looking on. This is a typically dramatic greeting between individuals of this species. *Marie Read; Feb. 2006, Orange Co., CA.*

Some woodpeckers will occasionally enter a cavity currently in use by another species, such as a wren or bluebird, and toss out the eggs or young in order to occupy the cavity as their own nest site. These nest usurpers may even eat the eggs or nestlings of another species. As weak excavators, the *Melanerpes* species are best known to employ this strategy in their quest for a pre-excavated nest or roost site. However, the number of birds that benefit from woodpeckers far outweighs the number lost to their sometimes aggressive behavior; more than 40 bird species in North America have been

A White-winged Dove and an adult male Ladder-backed Woodpecker perched together, waiting their turns to approach a feeding station. *E. J. Peiker; May 2012, Amado, AZ.*

documented nesting in former woodpecker cavities (see Cavity Excavation below, under Ecology and Conservation).

### *Predators*

Perhaps one of the greatest benefits of cavity nesting is the protection it affords from predators. Only a small number of predacious birds are able to enter a woodpecker cavity, but small mammals and some reptiles are well adapted for this. In the Southeast, rat snakes climb trees and enter cavities with some regularity. In the western states, chipmunks and small ground squirrels readily enter cavities and opportunistically depredate eggs, young, and rarely roosting adults. One of the few predators that can tear into a nest cavity is the black bear, which has been known to depredate active nests. (See photo at right.)

Outside the cavity, accipiters represent the most formidable threat. Adult woodpeckers may occasionally elude a Cooper's or Sharp-shinned Hawk, but the limited flight skills of recently fledged juveniles make them far more vulnerable. For their first few nights out of the nest, fledglings roost outside a

Very few predators are strong enough to tear into a woodpecker nest cavity, except for bears. This Black-backed Woodpecker nest was depredated by a black bear. *Kristen Hein Strohm, courtesy of Institute for Bird Populations; June 2012, Lassen Co., CA.*

protective cavity, making them further susceptible to nocturnal owl attacks.

## LOCOMOTION

Woodpeckers can move up, down, or around the surface of a tree, and they can skillfully fly through a densely forested canopy. The combination of each species' locomotion behaviors depends on the season and situation.

### *Climbing*

Their hooked claws, movable outer toes, and stiff tails make most woodpeckers ballet dancers in their scansorial lifestyles. The hitching or shuffling motion employed to ascend a tree involves a series of rapid sequential movements, allowing a woodpecker to quickly climb a tree of any height in a matter of seconds.

The upward hopping motion involves a crouch, a jump, and a landing. The bird first crouches against the trunk surface, pointing its bill in the direction to be traveled and lifting its tail off the trunk. The legs then apply a strong downward force, propelling the bird upward into a floating jump. In the jump phase, the woodpecker surrenders all contact with the tree surface. The feet and tail then quickly return to the tree to prepare for the next cycle. Woodpeckers move down a vertical surface in one of two primary ways: by releasing the tail, lifting the feet, and recovering from a brief controlled freefall; or by aligning the body parallel to the ground and shuffling sideways downward.

### *Flight*

Akin to its trademark climbing posture is the woodpecker's signature intermittent flight style, known as undulation or flap-bounding. The typical flight often begins with a crouch, spring, and release from a vertical surface. The body then turns to point downward into a controlled freefall before a full extension of the wings. After a short downward glide, the bird flaps its wings, climbing to the apex of the flight path before tucking its wings tightly for the next downward swoop. The landing usually ends with a steeper dive and an equally steep ascent up to the new surface, as in the return approach to a nest site (see photo, p. 28).

Woodpecker wing sounds are fairly distinctive and can easily be heard when a woodpecker flies closely overhead. The characteristic undulating flight style produces bursts of rapid wing flutters separated by long pauses. In close proximity, a *whoosh* can sometimes be heard during the glide phase of a woodpecker's flight.

Lewis's and Pileated Woodpeckers, and occasionally the flickers, fly with a slower, almost continuous flapping and a more direct flight path. Individual birds of all species may fly short distances from one tree to another or in an extended series of undulations between perches. In order to cover their large territories, Pileated and Ivory-billed Woodpeckers may fly above the canopy with a very direct flight style.

Lewis's, Red-headed, and Acorn Woodpeckers, and to a lesser extent the sapsuckers, are quite adept at flycatching for aerial insects. This behavior requires an infinite variety of flight styles, depending on the insects being hunted. The pursuit generally includes a powerful but labored ascent from a perch, the capture, and a glide-bounding return to the same or another perch.

## MISCELLANEOUS BEHAVIORS

A few sundry behaviors balance out a day in the life of a woodpecker. Individuals often perform "direct" head scratching, that is, scratching without lowering the wing, as opposed to passing the foot over the wing, as in passerines. Woodpeckers also stretch their wings differently than most birds, usually dropping one wing at a time at an extreme angle without moving a foot, and often simultaneously spreading the tail feathers on the same side as

This adult female "Yellow-shafted" Northern Flicker is using her wings and tail to brake as she approaches a cavity. *Paul Bannick; May 2007, Richmond Co., NC.*

the outstretched wing (see photo on p. 227). In the case of the bilateral, or two-winged, stretch, both wings may be lifted perpendicular to the body or dropped so the primaries extend below the tail.

The preening ritual can go on for extended periods at any time of year and may involve rubbing the sides of the head on a branch or trunk surface. Perching can occur in any number of locations: on a power line (especially among the solid-colored *Melanerpes*), parallel to a horizontal branch, tucked into the crotch of a tree, or simply by gripping a vertical surface and resting against the propped tail. Most species have been observed sunning, occasionally even in a state of pseudotorpor. Water- and dust-bathing probably occur more frequently than reports would indicate. (See photo p. 242.) Anting—the practice of allowing ants to crawl among the bird's feathers—occurs in a few species, although its exact purpose in woodpeckers is not clear.

I'll conclude this section with one final behavioral note. Many gaps exist in our documentation of woodpecker behaviors. Whether observed during a formal study or at your home feeding station, please share any interesting woodpecker behaviors at www.woodpeckerwonderland.com. We will all benefit from our huge collective body of observational experience.

## ECOLOGY AND CONSERVATION

Woodpeckers occur across North America just about everywhere that trees grow. Conversely, and with few exceptions (see photo on p. 137), if there are no trees, there are no woodpeckers. In the last century, our knowledge of forest and woodland ecology has grown exponentially. Today, our understanding of how these diverse ecological systems operate places us in a position of responsibility. As we learn to assess the value of intact, healthy ecosystems, we become responsible for their conservation. And woodpecker conservation is paramount to healthy forest and woodland habitats.

### WOODPECKERS AS KEYSTONE ORGANISMS

A healthy ecosystem comprises a web of organisms, each of which plays some role in that system's ecological balance. Some organisms play larger roles than others, not necessarily according to their total biomass but by the services they provide. We call these keystone organisms. Removing a keystone from the center of a stone arch will cause the arch to collapse; removing keystone organisms can cause serious damage to ecosystems.

Woodpeckers may represent one of the most important keystone organisms in a healthy forest or woodland, serving four essential ecological roles: cavity excavation, insect control, decay facilitation, and food provision.

#### *Cavity Excavation*

By far the most important ecological role that woodpeckers play is that of housing developer. By excavating their own nest and roost cavities, woodpeckers provide nest sites and shelter for more than 40 species of North American birds, not including the 23 woodpecker species themselves. In addition to many bird species, numerous small mammals and a few reptiles also depend on woodpecker cavities for shelter, as do innumerable invertebrates.

We often speak of the primary cavity excavators, such as the Black-backed and Hairy Woodpeckers, when touting the benefits of woodpeckers to other

An adult Prothonotary Warbler—with Lucy's Warbler, one of only two cavity-nesting warbler species—exiting a nest cavity. The small size of the entrance indicates that this is likely a former Downy Woodpecker cavity. *Jim Braswell, Show-Me Nature Photography; May 2011, Cass Co., MO.*

species, but in some habitats even a weak excavator can support an entire community of cavity-nesting birds. Our two flicker species are the least well adapted among woodpeckers for digging holes in trees, yet the Gilded Flicker performs a critical service to the Sonoran Desert ecosystem by excavating cavities in saguaro cacti. Likewise, by targeting mature trees with advanced heart rot, the Northern Flicker provides nest sites for a wide variety of cavity-nesting species across the continent.

### *Insect Control*

In 1901, entomologist A. D. Hopkins estimated that the average beetle-infested tree could harbor 100 larvae per square foot (9.29 sq. m) of bark, and that each tree averaged 60 square feet (5.57 sq. m) of infested bark, making it possible for a single tree to host 6,000 beetle larvae. Hopkins credited woodpeckers with the ability to consume half or more of these insects. At least through the 1970s, the hickory borer was a formidable pest of hickory trees in most of the Atlantic Coast states and southern provinces; woodpeckers consumed an estimated one-third of hickory borer larvae throughout the region. Since

An Elf Owl peering out of a former Gilded Flicker cavity in a saguaro cactus. *Paul Bannick; Apr. 2014, Pima Co., AZ.*

An adult emerald ash borer above its trademark D-shaped exit hole. *Phil Nixon; July 2004, Wayne Co., MI.*

A nestling Hairy Woodpecker peering out of its nest cavity in a mature aspen tree. Note the large fugal conch above the cavity entrance, indicating the presence of a heart-rot fungus. *Stephen Shunk; June 2009, Jefferson Co., OR.*

2002, the exotic emerald ash borer (*Agrilus planipennis*) has wrought havoc on ash trees across the Great Lakes region, killing more than 50 million trees in nine states and two provinces. Woodpecker populations have climbed in the region, especially those of the Red-bellied Woodpecker, and studies show that woodpeckers preferentially forage on afflicted ash trees over non-ash and healthy ash trees in infected regions, with up to 85 percent of beetle larvae in a given tree consumed by woodpeckers.

American Three-toed, Black-backed, Hairy, and Downy Woodpeckers in particular are well-documented "first responders" to bark- and wood-boring beetle infestations across the continent. Generally resident throughout their respective regions, these species will wander outside their normal habitats or home ranges to exploit beetle-plagued timber stands. By eating the insects that harm trees, the woodpeckers prolong the lives of the trees, consequently enhancing their own long-term foraging and nesting opportunities.

### *Decay Facilitation*

Woodpeckers prolong the lives of individual trees by eating injurious beetle larvae. In contrast, by flaking bark or excavating for grubs or cavities, they also expose the tree to a wide variety of decay organisms, and in many cases the woodpeckers themselves transport these organisms from tree to tree.

As early as 1913, Downy Woodpeckers were well-known transporters of chestnut blight (*Cryphonectria parasitica*) fungal spores. In one Pennsylvania study, a single Downy Woodpecker was found to be carrying more than 750,000 fungal spores from its contact with the tree bark. In 2001, at least half of the Hairy, White-headed, and Black-backed Woodpeckers tested in the Ochoco and Lassen National Forests (in central Oregon and northern California, respectively) were documented carrying spores of filamentous fungi responsible for sapwood decay in ponderosa pines.

Woodpeckers frequently excavate cavities in trees suffering from some form of fungal decay. This decreases the amount of energy required to create a nest or roost cavity. By creating openings in the trees through which decay vectors can enter, and by transporting some of those vectors themselves, woodpeckers ensure the long-term presence of trees that are suitable for cavity excavation. Passing fungal spores also benefits the greater forest ecosystem by catalyzing natural decay processes in the biomass. And fungi are a critical link in the forest life cycle, facilitating decay and ensuring that vital nutrients are returned to the soil.

### *Food Provision*

Rufous Hummingbirds often arrive on their nesting grounds well before the first flowers begin to bloom. They accomplish this by following sapsuckers on their northward migration routes. Some migrant hummers use sap wells as their primary source of energy; up to five individual hummingbirds have been observed following a single sapsucker in migration. Other woodpecker species will also exploit the resource provided by the sapsuckers, and a few, including the Acorn Woodpecker, perform their own fair share of sap mining. Other birds observed feeding at sap wells include kinglets, spar-

An Anna's Hummingbird feeding at an active phloem well, likely excavated by a Red-breasted Sapsucker. *Laura Osteen; Sept. 2008, Inyo Co., CA.*

rows, finches, and warblers; even insects and small rodents take advantage of the nutritive food source.

## WOODPECKERS AND FIRE

### *First Responders*

Wildfires attract wood-boring insects by the bushel, and the more intense the fire, the more insects that respond. Firefighters frequently report the arrival of beetles en masse—especially wood-boring beetles (Cerambycidae and Buprestidae) and bark beetles (subfamily Scolytinae)—even while the trees are still smoking. These adult beetles lay their eggs in the bark of the charred trees. The eggs hatch, often within just a few days, and the larvae bore either into the cambium of the tree (wood borers) or behind the bark (bark beetles). In turn, excavating and bark-scaling woodpeckers—primarily those in the genus *Picoides*—quickly move into these high-severity burns, and several species can be found in

A red-spotted purple (*Limenitis arthemis astyanax*)—a brushfooted butterfly of the Nymphalidae family—feeding at phloem wells excavated by a Yellow-bellied Sapsucker. *Marie Read; June 2006, Presque Isle Co., MI.*

A nestling male American Three-toed Woodpecker. This species and the Black-backed Woodpecker are among the first woodpeckers to exploit the abundant food resource in a freshly burned forest. *Kristine Falco; July 2008, Deschutes Co., OR.*

startling concentrations among recently burned forests.

In one Alaska burn site from 1983, the white-spotted sawyer (*Monochamus scutellatus*), a common and proficient wood-boring Cerambycid beetle, arrived immediately after the fire and began laying high densities of eggs on charred portions of spruce trees within the burn perimeter. Full-grown larvae of this species can grow to 2 inches (50 mm) in length, providing substantial food value to woodpeckers. In the burn, Black-backed Woodpeckers foraged primarily on moderately to heavily burned trees and only in areas of high sawyer concentrations. Two to 3 years after the fire, the adult beetles had emerged from the trees, subsequently dispersing and not reproducing again in the area. Black-backed Woodpeckers declined locally during this period, and once the last adult sawyers emerged, the woodpeckers had left the region.

In North America, Black-backed and American Three-toed Woodpeckers are among the bird species most closely associated with burned and beetle-infested forests, their local populations rising and falling with fire episodes and insect outbreaks. Up to 95 percent of the diet of the Black-backed Woodpecker consists of wood-boring beetle larvae, and the American Three-toed Woodpecker is equally dependent on larval bark beetles.

***Successional Responders***

Within the first year after a high-severity fire, ground dwelling ants recolonize the charred forest floor, and Northern Flickers respond quickly to the abundance of their favorite prey item. As weak excavators, flickers also take advantage of the nest cavities excavated earlier by their *Picoides* relatives. In successive years after a fire, the newly created open canopy invites abundant aerial insect populations, which attract the fly-catching Lewis's Woodpecker. Local Lewis's Woodpecker populations can increase dramatically in burned forests, and they will continue to nest there as long as suitable snags remain standing. However, after several years of

Pileated Woodpeckers typically prefer carpenter ants among downed logs and rotten timber, but they will opportunistically exploit the abundance of beetle larvae in a freshly burned forest. Note the much smaller feeding holes created by other woodpecker species in the closer tree, on the right. *Stephen Shunk; June 2009, Deschutes Co., OR.*

(A) This adult velvet long-horned beetle (*Cosmosalia chrysocoma*)—here on blue dicks (*Dichelostemma capitatum*)—is a common member of the Cerambycidae family across northern and western North America. (B) This adult *Chrysobothris* beetle—here on a burned yellow pine (*Pinus jeffreyi* or *P. ponderosa*)—belongs to this widespread North American genus of the Buprestidae family. The larvae of these two wood-boring beetle families provide an abundant food source for woodpeckers in burned forests. *Stephen Shunk; A—Lassen Co., CA; B—Deschutes Co., OR.*

winter winds, the rotting snags begin to fall, often prompting this nomadic species to move on. Conservation efforts pioneered on the eastern slope of central Oregon's Cascade Mountains may help sustain local Lewis's Woodpecker populations in older burns and other marginal habitats with the installation of artificial next structures.

***Forests Maintained by Fire***

Pine forests in the western and southeastern states evolved with low-severity fires that cleared the understory of competing vegetation. Co-evolving among these forests were the White-headed and Red-cockaded Woodpeckers. Both species favor mature pine forests that are maintained by frequent ground-clearing understory fires. In the southeastern states, these understory fires limit the encroachment of hardwoods, the presence of which invites other woodpecker species, exposing the Red-cockaded to unfavorable competition. Further competition results from dense understory conditions that favor forest rodents such as the yellow pine chipmunk (in the West) and the southern flying squirrel (in the East). These small mammals are well known to usurp nest cavities from woodpeckers,

Birders in central Oregon installed nest boxes to support a local population of Lewis's Woodpecker. *Kristine Falco; June 2006, Deschutes Co., OR.*

The White-headed (top) and Red-cockaded (below) Woodpeckers (adult males are pictured here) evolved among fire-maintained pine forests on opposite sides of the continent. *A, Paul Bannick; May 2005, Yakima Co., WA; B, Tom Grey; Mar. 2012, Osceola Co., FL.*

Our largest woodpecker, the Pileated (an adult male is shown here), favors carpenter ants among rotten snags and downed wood. *Gene McGarry; Nov. 2012, Woodstock, NY.*

occasionally even depredating woodpecker eggs and nestlings in the process.

### *The Matrix of Burn Severity*

Forests come in many forms—from bottomland hardwood to subalpine mixed conifer—and each evolved with its own natural fire regime. Regardless of the frequency with which fire occurs in a given forest or woodland, most fires burn in a diverse matrix of fire intensities, leaving live stands interspersed with those of varied burn severity. This diversity of burned forest types supports the retention and recruitment of live trees, and it leaves sustainable levels of woody debris on the forest floor. Live trees support the foraging needs of local sapsucker populations; downed wood and rotting snags allow carpenter ants to flourish, feeding the Pileated Woodpecker. All woodpecker species benefit from intensely burned stands, either from the beetle larvae on which they depend for food or from the newly created snags on which they depend for nesting substrates.

Our long legacy of silviculture and more than a century of aggressive fire suppression have permanently changed the way fire behaves in our forests. These forests and their inhabitants co-evolved with fire as a natural disturbance event, and much current research explores the relationships between woodpeckers and fire. We have learned that fires are paramount to healthy forest ecology and that woodpeckers fulfill keystone roles in these burned forests. But the health of the burned forest ecosystem requires that the trees remain standing, and postfire salvage logging prevents woodpeckers from performing their keystone roles in healthy forests.

## HABITAT CONSERVATION

Since the earliest settlement by Europeans, population growth in North America has required trees for wood products and open land for agriculture, precipitating the clearing of forests. Since then, swaths of forest in the eastern United States have begun to recover, but millions of acres of western coniferous forests have been irreparably converted to timber plantations. Today, presettlement forest conditions across the continent are few and far between.

Not surprisingly, the decline of some woodpecker populations has been correlated with the loss of forest habitats. Perhaps the best examples are the Red-cockaded and Ivory-billed Woodpeckers. Tragically, these two species remain critically endangered, and the Ivory-billed may no longer remain at all. These species' declines are discussed further in the respective species accounts. Below is a basic summary of the consequences of our legacy, along with some general prescriptions for healthy forests.

Most birders know the North American boreal forest for its importance to migratory songbirds, but it also supports more than 80 percent of the continent's Black-backed and American Three-toed Woodpeckers and more than 50 percent of the Yellow-bellied Sapsuckers. *Stephen Shunk; July 2002, Lesser Slave Lake, AB.*

This small isolated stand of trees near southern Florida's Fisheating Creek includes several standing snags, each of which contains multiple woodpecker cavities. *Stephen Shunk; Apr. 2013, Highlands Co., FL.*

### *Habitat Loss*

As with most of our native birds, the greatest threat to woodpeckers is habitat loss. By destroying its native habitat we have almost pushed the Red-cockaded Woodpecker beyond the brink, and we may well have extinguished the Ivory-billed. Other species, such as the Red-headed, White-headed, and Lewis's Woodpeckers, face extirpation from parts of their ranges, withdrawing from historic limits as forests fall to agricultural conversion and urban development.

### *Maintenance of Snags and Cavities*

All North American woodpeckers nest in cavities, and the availability of trees suitable for excavating is a critical element of healthy woodpecker habitats. In order for these habitats to sustain long-term populations of woodpeckers, a succession of dead or dying trees of diverse species at various stages of decay must remain.

Short-sighted vegetation management practices have reduced the availability of snags, countermanding woodpecker habitat conservation efforts. Developing monoculture forests in evenly aged stands and clearing forests for development severely limit snag availability. Snags are removed at their bases in urban parks and neighborhoods in the name of aesthetics and public safety, when they could often be topped at safe heights, leaving useful snags for woodpeckers (see photo, p. 38). Indiscriminate salvage logging of charred timber for its

Two trees remain in the center of this salvaged swath of burned forest, left standing as designated wildlife trees (note the orange W painted on the charred bark of the left tree). Such forest management practices completely ignore the ecological importance of the burned forest ecosystem, and especially its keystone woodpecker species. *Stephen Shunk, courtesy of Institute for Bird Populations; June 2012, Modoc Co., CA.*

purported economic potential eliminates any snag succession that could benefit entire snag-dwelling wildlife communities.

### *Cavity Competition*

The European Starling was introduced to eastern North America around 1890, providing one of the best examples anywhere of unintended consequences. It took just over 50 years for starlings to reach the West, arriving in Arizona by 1946 and in British Columbia by 1947. An opportunistic and aggressive cavity-nesting species, the starling began exploiting woodpecker carpentry anywhere woodpeckers occurred. Studies have correlated the declines of local Red-bellied and Gila Woodpecker populations with the expansion of starlings, the woodpeckers frequently being outcompeted by starlings for available nest sites. The Northern Flicker is experiencing a rangewide population decline, much of it attributed to the more aggressive starlings overtaking available nest sites. Lewis's and occasionally Red-headed are the only woodpeckers known to consistently dominate starlings when battling for previously excavated cavities. Starlings are now naturalized across North America, and most control efforts waste time and money.

The lack of snags on the landscape has also increased competition with small mammals. In healthy forests, woodpecker cavities are abundant enough to support all cavity dwellers, but fewer cavities lead to nest occupation and usurpation by squirrels and chupmunks.

### *Prescribing and Managing for Healthy Habitats*

Habitat restoration and management can support long-term woodpecker populations and serve as a proactive approach to conservation. However, it is impossible to set blanket prescriptions for woodpecker habitats across the continent. Different species thrive in different habitats in different parts of their respective ranges. Management of these habitats must take into consideration the localized needs of each species.

For example, Pileated and Red-headed Woodpeckers are species of concern in the Holly Springs National Forest of Mississippi, where one study showed the Pileated as almost absent in clear-cut areas and the Red-headed as common in cutover stands with remnant snags. Another study in suburban Seattle found that Red-breasted Sapsuckers occurred with higher frequency where more snags remained, that Northern Flickers and Downy Woodpeckers thrived in highly interspersed forest and developed areas, and that Hairy and Pileated Woodpeckers preferred areas dominated by forest.

Several recent studies have elucidated the importance of intact severely burned forests, especially in the western states. Territory sizes of Black-backed Woodpeckers are inversely correlated with the density of fire-killed trees on the landscape, with more

Forest rodents, such as the Douglas Squirrel, compete with local woodpeckers for available cavities. *Stephen Shunk; June 2013, Deschutes Co., OR.*

woodpeckers nesting in smaller areas where salvage logging has been curtailed.

Urban planning efforts can provide woodpecker habitat by retaining large contiguous patches of trees. In general, parks with the largest stands host the greatest diversity of woodpeckers and other cavity nesters. Snag retention and recruitment are also critical to healthy woodpecker habitat. In areas without snags, snag-creation efforts—including the baiting of trees with insect pheromones, injection of heart-rot fungi, girdling, and topping—have been employed to jump-start the decay process. Even suburban neighborhoods can be managed to retain existing snags that are not a hazard to homes and homeowners. As discussed earlier, snags that are topped rather than cut at their bases generally pose little safety risk but provide excellent benefits to local wildlife. Long-term public education will be required to shift the image of snags from being eyesores to icons of healthy woodlands.

### *Current Protections and Directions*

Species protection mechanisms vary widely among the states and provinces and between the United States and Canada. Some broad protections cover all species to limited extents, and formal listings for imperiled species elevate them in the management and conservation hierarchy.

The two most general protections are the United States Migratory Bird Treaty Act of 1918 and the Canadian Migratory Birds Convention Act of 1994. The U.S. Endangered Species Act of 1973 prioritized funding for the Red-cockaded Woodpecker, and the purported rediscovery of the Ivory-billed Woodpecker funneled significant habitat dollars into the southeastern bottomlands. The White-headed Woodpecker and Williamson's Sapsucker both rank as endangered in Canada, garnering special conservation plans and management priorities.

In addition to and often in concert with governmental listings, the American Bird Conservancy, National Audubon Society, and Partners in Flight all maintain special conservation designations. Spe-

Several snags remain standing in this southern Florida patch of pine flatwoods, providing optimal habitat for the Red-headed Woodpecker. *Stephen Shunk, courtesy of Florida Panther National Wildlife Refuge; Apr. 2013, Collier Co., FL.*

Snags can be retained in public places without jeopardizing public safety. Topping dead trees, rather than cutting them at their bases, leaves valuable snag habitat and maintains the natural character of the landscape. Note the Northern Flicker cavity near the top of the snag in the foreground. *Stephen Shunk; Nov. 2012, Bend, OR.*

cies covered under these programs include those with very restricted ranges, such as the Nuttall's and Arizona Woodpeckers, which would be extremely vulnerable to catastrophic environmental change within their respective ranges.

In 2000, wildlife management in the United States took a new turn when the U.S. Fish and Wildlife Service implemented the Wildlife Conservation and Restoration Program. This novel approach provided grants to U.S. states and territories to develop Comprehensive Wildlife Conservation Strategies. Historically, government funding has overwhelmingly favored endangered species management, but this new effort aims to "keep common species common," recognizing the importance of keystone organisms such as woodpeckers. The plans were all completed by about 2006, and some state wildlife management agencies have begun to take their plans from the paper to the field—with on-the-ground habitat restoration and conservation. However, federal funding covered only the development of these strategies, and it remains to be seen which states will be the most proactive when allocating resources to habitat projects.

### *Time for a Change*

North American land managers—both public and private—face an opportunity. Current wildlife management strategies allocate many millions of dollars to rescue, protect, and restore a handful of endangered species and their habitats. These are certainly worthy causes, as habitat conservation for these few species contributes to the body of scientific knowledge, provides many jobs, and benefits entire ecosystems. In addition to funding for at-risk species, agencies spend massive fire suppression budgets to protect forest-urban interfaces (and timber value). Meanwhile, land managers struggle to fund conservation and restoration over millions of collective acres falling outside of these priority habitats. These lands still need attention, however, and long-term habitat conservation will require sweeping changes in government and corporate spending priorities. For cavity-dwelling communities across the continent, only by monitoring and conserving all woodpecker species and their habitats can we help sustain the keystone roles of these woodland carpenters for perpetuity.

## HOW TO USE THIS BOOK

This volume covers the 23 species of woodpeckers (including the possibly extinct Ivory-billed) that nest in North America; this area is defined as Canada and the mainland United States, including Alaska. Beyond these borders, I offer some limited

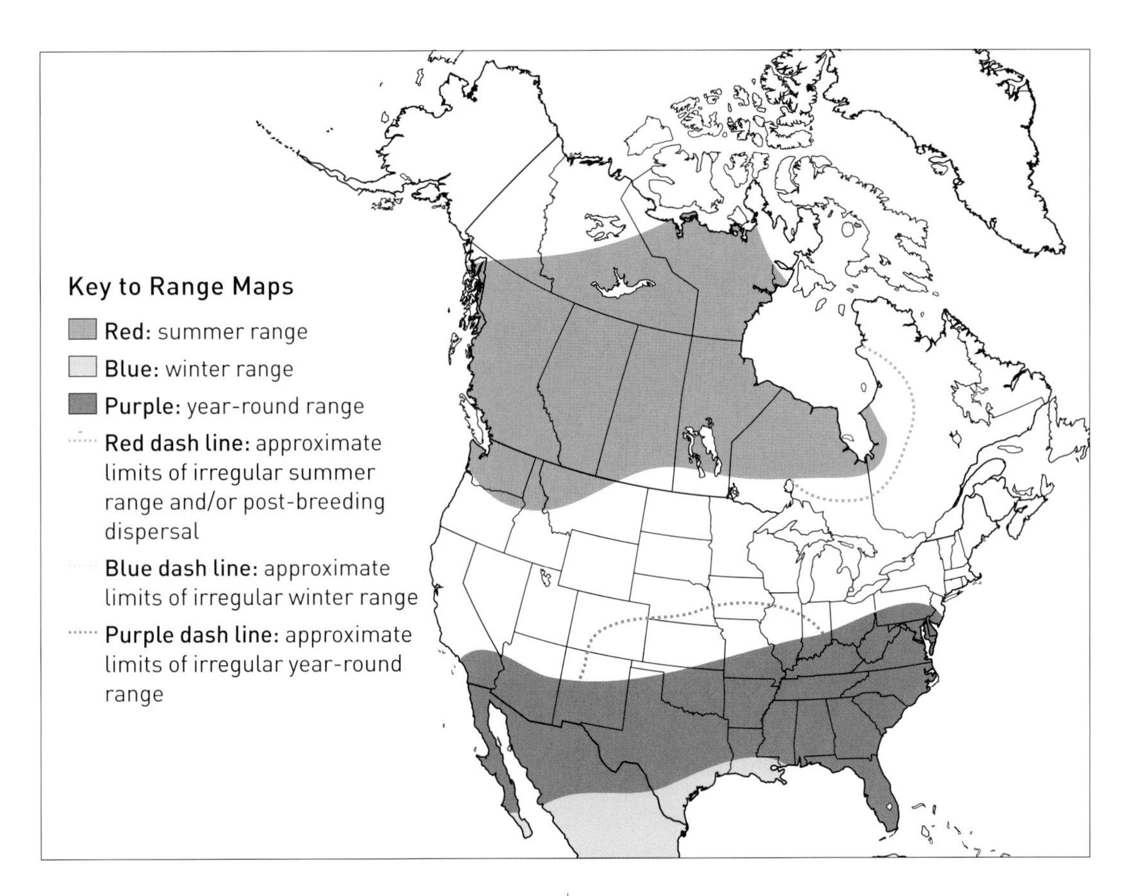


discussion on Neotropical woodpeckers that are either closely related to or sympatric with North American species.

I have made a concerted effort in most chapters to minimize the use of scientific jargon. That said, a familiarity with some specialized terminology will make your reading more enjoyable; a glossary begins on page 262.

At the ends of this chapter and each species account I list the primary references used for those sections. Full citations for all references can be found in the bibliography. The appendices contain a summary of woodpecker systematics, as well as physical measurements of each species and quantitative information on reproductive behaviors, such as nest specifications and breeding periods.

You will probably get more out of the species accounts if you take the time to read here how they are structured and what they include.

## SPECIES INTRODUCTION

The common and scientific species names used in this book (and also the taxonomic order) follow the American Ornithologists' Union's *Check-list of North American Birds*, 7th edition (1998), through the 55th supplement of 2014.

Below each species' name you'll find the ranges of three measurements: length (L), as measured from bill tip to tail tip, with head extended; wingspan (WS); and weight (WT). (See Appendix 1 for more detailed measurements on all 23 species.)

A brief introduction then tells an interesting story for each species.

## DISTRIBUTION

**BREEDING RANGE:** For migratory species, the region where this bird raises its young, from courtship through postbreeding dispersal. For nonmigratory species, this subhead is called "Permanent Range."

**SEASONAL MOVEMENTS:** Descriptions of postbreeding dispersal, migration patterns, seasonal wandering, nomadism, and any other movements outside the breeding season and away from the range of winter residency, where applicable.

**WINTER RANGE:** This subhead applies mostly to species that perform a traditional migration, moving generally southward in fall, wintering in the same region each year, and returning to the same breeding grounds in spring.

**MAP:** The map for each species depicts distribution in greater detail than the abbreviated written sections. Species maps are taken from the *Peterson Field Guide to Birds of North America.*

## HABITAT

Habitat elements may include elevation, forest type, canopy cover, and hydrology, as well as characteristics of urban and rural developed lands, including agricultural areas. Species prone to seasonal movements, migratory or otherwise, may use different habitats in different seasons; these variations are listed here in subsections.

## DETECTION

Woodpeckers betray their presence in several ways, from their characteristic undulating flight style to an array of vocalizations and drumrolls.

**VOCAL SOUNDS AND BEHAVIOR:** Many woodpeckers utter similar types of vocalizations, but most are identifiable to species, with a little practice. Call notes, rattles, and contact calls are common, and a few species possess their own distinctive vocalizations.

Describing bird sounds with words is always problematic. With some exceptions, each birder hears something a little different when the same bird calls. That said, I have attempted to provide basic interpretations for vocalizations wherever they may be helpful.

**NONVOCAL SOUNDS:** The telltale sign of a nearby woodpecker is the archetypal drumroll echoing from a resonant surface nearby. Descriptions here focus on general patterns of sound. Other conspicuous aural identification clues, such as tapping, feeding, flight, excavation, and shuffling, are included where applicable.

## VISUAL IDENTIFICATION

Most woodpeckers do not pose identification problems. However, learning woodpecker plumage variation with attention to some of the subtler characters will enhance your enjoyment of woodpecker watching. This section serves as a guide to woodpecker identification for birders of all interest and ability levels. Detailed woodpecker plumage descriptions, down to the barbs and shafts of each feather, are available in other references. What I provide here is a practical guide to identifying woodpeckers as we typically experience them in the field.

We generally see woodpeckers in two contexts: perched and in flight. Sexual dichromatism and general dimorphism are explained here, along with the basic identifying characters of each species. Adults are described in their freshest basic plumage, juveniles in the plumage held between fledging and the preformative molt.

**INDIVIDUAL VARIATION:** Albinistic, leucistic, or melanistic individuals may rarely be encountered. These are discussed here, along with more typical variation.

**PLUMAGES AND MOLTS:** Identifying molt patterns in birds is an art, best practiced with birds in the hand. However, basic knowledge of seasonal changes may help explain some field-identifiable plumage variation.

**DISTINCTIVE CHARACTERISTICS:** Two species have only three toes, and some have a differently shaped bill or tongue. Some species are built better for flight, and others are built for heavy-duty excavating. This section is absent from the accounts of species that fit neatly into the general woodpecker image.

**SIMILAR SPECIES:** Tips for separating species with similar plumage, voice, or habits.

**GEOGRAPHIC VARIATION:** Plumage and morphological characters that vary across a species' range. This section is absent for species that vary little across their ranges.

**SUBSPECIES:** This section includes recognized subspecies as published in the American Ornithologists' Union's *Check-list of North American Birds*, 5th edition (1957), with changes through the 55th supplement of 2014, along with hypothetical forms described but not necessarily accepted by Short (1982). This section is absent for species that do not have recognized subspecies.

Subspecies occurring north of Mexico are listed in trinomial form with basic variation in plumage characters and geographic ranges; strictly Neotropical subspecies are listed in trinomial form but only with geographic reference. Forms not formally recognized by the AOU but discussed by Short are in italics with quotation marks and without the genus and species reference.

As in other families of birds, subspecies designations have been debated, accepted, and rejected for decades. With modern advances in DNA research, our understanding of relationships at many taxonomic levels will continue to change. Some genetic and evolutionary biologists even go so far as to reject the notion of subspecies altogether. Regardless of your position on their existence, separating currently recognized subspecies in the field adds an engaging, dynamic twist to a normally easy-to-identify group of birds.

**HYBRIDIZATION:** For species that have been documented hybridizing, this section discusses the species crossed, the locations of contact zones, and the basic challenges of identifying hybrid individuals.

Some authors use the term "hybridization" to

refer to interbreeding between two subspecies as well as between two species. I use "introgression" and "intergrade" to refer to interbreeding between subspecies and "hybridization" for interbreeding between species.

## BEHAVIOR

North America's 23 woodpecker species share many common traits, but each also exhibits some distinctive behaviors. The behavioral accounts portray the most characteristic mannerisms of each species based on my own observations and those from two centuries of American naturalists.

**BREEDING BIOLOGY:** This section covers the entire breeding cycle, from nest-site selection to courtship, parenting, and growth and dispersal of the young.

**NONBREEDING BEHAVIOR:** Search and defense of food supplies, feeding styles, social interactions among family groups and neighbors, daily survival, and information on commensalism, competition, and carpentry represent some of the woodpecker behaviors discussed here.

**MISCELLANEOUS BEHAVIORS:** Includes behaviors such as bathing, sleeping, perching, and others. This section may be missing when such behaviors have not been documented.

## CONSERVATION

Historic and current conservation issues, including impacts to the species and its habitat, as well as the species' response to these impacts and conservation priorities for each species.

**HABITAT THREATS:** Known natural and anthropogenic impacts on the species' habitat, as well as historic habitat loss and potential threats of ongoing human activities.

**POPULATION CHANGES:** Reports on episodic changes in local, regional, or rangewide populations, including documented irruptions, as well as known or possible causes of these changes.

**CONSERVATION STATUS AND MANAGEMENT:** State, provincial, and federal designations, including levels of sensitivity. Any level of sensitivity indicated with a state name means the listing is designated by the state wildlife management agency. Other listings include American Bird Conservancy/National Audubon Society "WatchList 2007" (WatchList), which is an analysis of species most susceptible to environmental changes. This section also provides formal habitat prescriptions from various management agencies as well as recommended habitat management activities.

## ERRATA, UPDATES, AND OVERFLOW

Any errata discovered after publication, along with updated or new information and additional material left on the "cutting room floor," will be maintained regularly at www.woodpeckerwonderland.com. Please contact me through this site with any errata or updates you discover that are not already listed there.

### REFERENCES

**ANATOMY AND ADAPTATIONS**

Baumel et al. 1993; Bock 1999; Bock and Miller 1959; Borror 1960; Browning 2003; Burt 1930; Burton 1984; Chapin 1921; Cole 1949; Coues 1884; Delacoeur 1951; Elert 2007; Feduccia 1999; Gardner 1925; Gleason and Gleason, B. and D., pers. comm.; Goodge 1972; Ingold and Weise 1985; Kirby 1980; Kozma, J., pers. comm.; Levin 2007; Lucas 1895, 1896; Mailliard 1900; Manolis 1987; Mayr and Short 1970; McGregor 1900; Miller 1933; Owen 1866; Pettingill 1970; Podulka et al. 2004; Pyle 1997, 2012; Pyle and Howell 1995; Schwab, I., pers. comm.; Schwab 2002; Short 1970; Sibley 1957; Spring 1965; Swierczewski and Raikow 1981; Villard and Cuisin 2004; Walters, E.L., pers. comm.; Wiebe 2000a; Wiebe and Bortolotti 2001, 2002; Wygnanski-Jaffe et al. 2007.

**BEHAVIOR**

Abbott 1929, 1930; American Association for the Advancement of Science 1880; Bock 1970; Brackbill 1953, 1969b; Bull et al. 2001; Bunnell et al. 2002; Cassirer 1993; Chapin 1921; Clevenger 1881; Conner 1975; Cruickshank 1953; Daily 1993; Dodenhoff, D., pers. comm.; Dodenhoff et al. 2001; Eberhardt 1997; Eckstorm 1900; Farris and Zack 2005; Garrett, K., pers. comm.; Gilman 1915; Hailman 1959; Harrison and Loxton 1993; Ingold 1991; Jackson and Hoover 1975; Joy 2004; Kellogg et al. 1953; Kilham 1958b, 1959d, 1959e, 1960, 1961c, 1974d, 1979b; Lawrence 1966; Little 1920; Miller and Bock 1972; Noble 1936; Saab et al. 2004; Skutch 1933; Stark et al. 1998; Tobalske 1996; Wiebe et al. 2006; Wilkins and Ritchison 1999; Winkler and Short 1978; Yom-Tov and Ar 1993.

**ECOLOGY AND CONSERVATION**

Aquilani 2006; Barlow 1900; Batts 1953; Beal 1906; Blewett and Marzluff 2005; Center for Conservation Biology 2008; Conner et al. 1976; Crenshaw 1954; Crockett and Hansley 1978; Erskine et al. 1976; Farris et al. 2004; Farris and Zack 2005; Forbush 1902; Frenzel 2005; Heald and Studhalter 1913; Hill et al. 2006; Hoose 2004; Jackson and Jackson 2004; Kerpez and Smith 1990; Livingston 1995; Morrison and Chapman 2005; Neal 2005–2007; Nicholls and Ostry 2003; Salomon and Payne 1986; Science 1907; Tanner 1942; Yeager 1955.

# SPECIES ACCOUNTS

## LEWIS'S WOODPECKER

### *Melanerpes lewis*

L: 10–10.75 in. (25.3–27.3 cm)
WS: 20–21 in. (50.8–53.3 cm)
WT: 3.0–4.9 oz. (85–138 g)

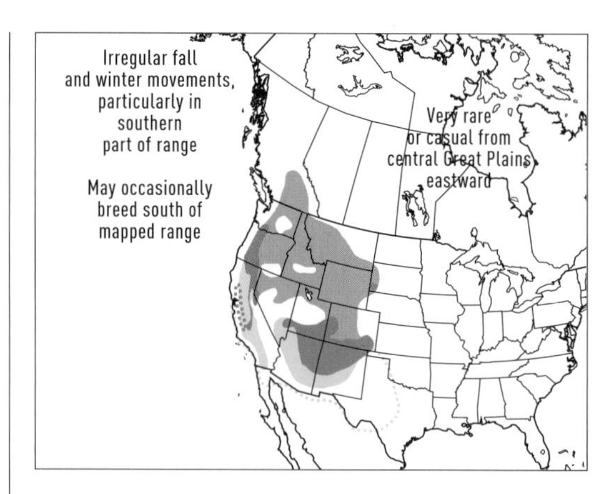


**In May 1806,** snow impeded Lewis and Clark's eastward return journey across the Bitterroot Mountains, forcing the expedition to camp on the Kooskooske River, near present-day Kamiah, Idaho. Here, Meriwether Lewis secured a new avian species he would name the "Black Woodpecker." In 1811, from Lewis's specimens, Alexander Wilson sketched and described the holotype for the bird he named *Picus torquatus*, "woodpecker with a necklace," and he formally named the bird in honor of the venerable explorer. Today, the bird Wilson held in his hands rests in the Harvard Museum of Natural History; it is the only biological specimen remaining intact from the epic journey of Lewis and Clark.

Along with its close relative the Red-headed Woodpecker, Lewis'sWoodpecker is one of only four sexually monochromatic woodpeckers worldwide. Males and females differ only in size and are generally inseparable in the field, unless one learns the male's unique vocalizations. North America's only green-colored woodpecker occasionally breeds in high concentrations, with as many as 10 pairs nesting in a single acre (25 pairs per hectare [ha]). Multiple family groups often gather in the fall and move upslope before moving south to escape from winter snows. Up to 1,000 birds have been tallied in northern California in a half-hour of observation at the peak of fall migration.

Lewis's Woodpecker lacks many of the anatomical adaptations of the proficient excavators, but it possesses its own special adaptatons that make it expert at flycatching. Populations are now declining throughout much of the species' range because of habitat loss, but Meriwether Lewis's "Black Woodpecker" may benefit from recent efforts in central Oregon, where the species has bred successfully in artificial nest boxes.

### DISTRIBUTION

**BREEDING RANGE:** Scattered breeding populations throughout w. N. Am. Generally occupies dry ranges of intermountain West.

**SEASONAL MOVEMENTS:** Northernmost populations may move more than 600 mi. (1,000 km), but birds in southern regions may remain within 12 mi. (20 km) of breeding territories year-round. Multiple family groups may coalesce and wander in fall, occasionally through winter. Opportunistic in migratory timing, route, and distance, depending on food supplies.

Migration is generally diurnal, slow, and silent; observed flying 90 ft. (150 m) or more above ground in direct or circling patterns, occasionally perching in trees to loaf or feed. Generally seen in small, tight groups of 5–15 individuals, occasionally with multiple groups spaced closely together; some large flocks to 150 birds. Migration departure dates may be synchronous at widely scattered locations in a given season.

**WINTER RANGE:** Small numbers overwinter sporadically through much of breeding range, typically among lower-elevation slopes and valleys. Interior CA hosts largest winter concentrations, but these are scattered and erratic.

### HABITAT

**BREEDING RANGE:** Open pine, juniper, and oak woodlands; open cottonwood riparian and aspen woodlands; cutover forest patches, orchards, and agricultural areas. Strong local affiliation with ponderosa pine and recent stand-replacement fires, including burned mixed-conifer forest.

Requires open canopy and brushy or grassy understory; must have tall perches and abundant aerial insects. Found in coniferous forests at 1,000–9,000 ft. (300–2,700 m), in open riparian habitats at lower elevations. May frequent woodland adjacent to open water. In plains areas, may select riparian areas adjacent to fallow and mowed fields rather than streams near grazed fields.

Presence in burned forests depends on years since, and severity of, wildfire; may breed in stand-replacement burned forest in small numbers the summer after burn, increasing in abundance for up to 12 years and declining as snags fall from decay

**OPPOSITE:** An adult Lewis's Woodpecker returns to feed nestlings in a burned ponderosa pine forest. *Steve Brad; June 2010, Mono Co., CA.*

and windthrow. Less common or absent in less intense burns where canopy remains.

**MIGRATION AND WINTER RANGE:** Frequently moves upslope in fall, especially into burns; may linger in winter in areas of abundant food supply, such as burns and orchards, at wide range of elevations. Oak woodlands and agricultural areas are typical winter habitat.

## DETECTION

First cues to presence are typically visual, when a large dark bird swoops and circles above canopy. Least vocal of North American woodpeckers, though it gives frequent and varied chatter on breeding grounds. Nestlings are less vocal than those of other woodpeckers, so nest is typically located by watching adult approach site rather than by hearing nestlings, as with other species.

**VOCAL SOUNDS AND BEHAVIOR:** Compared with other woodpeckers, uses vocalizations far more than drumming during courtship. During mating season, unmated males may call three times more often than paired males. Three primary calls: chatter, screech, and call note.

Chatter is reminiscent of Flipper, the television bottlenose dolphin, often beginning with single or double descending *tseea, tseea* notes followed by descending squeaky chatter. Heard throughout the year, more often during courtship; may be primary territorial call. Given by both sexes but mostly male.

Screech sounds like *skeer, skeer* or *keea, keea*, also descending. Given irregularly and only by male, primarily during courtship.

Call note is like extremely plaintive version of Hairy Woodpecker's *peek*; more like *keeyuk* and given singly or with a few notes in well-spaced succession.

**NONVOCAL SOUNDS:** Lewis's Woodpecker rarely drums, often once or less per hour, and only during courtship. Muffled, plaintive roll is given from atop a snag or rarely from inside cavity. Starts quietly and increases slightly in volume, sometimes followed by single repeated beats, somewhat like a sapsucker's drum. Both sexes drum. Unmated males may drum 10 times more often than paired males.

Tapping may be heard when birds are excavating or pulverizing prey or nuts at anvil site. Flight is not often heard since birds typically fly above canopy.

## VISUAL IDENTIFICATION

**ADULT:** The only North American woodpecker dressed in dark iridescent green. Dark head, back, wings, and tail, with prominent silvery gray collar and upper breast, dark red face, and pinkish or salmon red lower breast and belly. Sexes are identical.

Lewis's Woodpeckers are nothing if not resourceful, often combining their flycatching skills with fruit harvesting and ground foraging to feed hungry nestlings. *Stephen Shunk; July 2008, Deschutes Co., OR.*

In flight, typically appears all black, darkest above, but almost never greenish. Long, broad wings form distinctive aerial profile, much like that of a small corvid.

**JUVENILE:** Fledglings are plain slate black above, barred sooty below. Juveniles are distinct from adults up to several months after fledging. Overall much darker and less glossy, with black parts brownish instead of greenish, and lacking gray col-

A leucistic adult Lewis's Woodpecker. *Stephen Shunk; Aug. 2009, Deschutes Co., OR.*

A juvenile Lewis's Woodpecker. *Jim Burns; Oct. 2006, Pinal Co., AZ.*

lar. Reddish color starts to show through in breast and flanks by late summer, but this is highly variable. Hindneck and mantle may appear light gray because of white subterminal spots. Wide range of variability is seen among juveniles in postbreeding flocks.

**INDIVIDUAL VARIATION:** Males average larger than females, but difference is rarely detectable in the field. Width of silvery collar varies widely among individuals. Belly color varies with age and stage of molt cycle, from pink in fresh plumage to purplish orange when worn. Few records of albinism or leucism.

No known geographic variation, though some NM birds may show buffy tint to chest feathers. No recognized subspecies, and no records of hybridization.

**PLUMAGES AND MOLTS:** First basic plumage is attained in fall and early winter as family units arrive and disperse at wintering grounds. Juvenal flight feathers are retained until the second fall, with second prebasic molt occurring from late summer through Nov. Unlike in most woodpeckers, Lewis's nestlings do not exhibit shortened inner primaries. Adult prebasic molt is often delayed or extended to Dec.–Feb., giving this species the longest molt cycle of all North American woodpeckers.

**DISTINCTIVE CHARACTERISTICS:** Silvery feathers of collar and pinkish red feathers of underparts are coarse and hairlike, lacking typical feather features. Wider gape than in most other woodpeckers is likely an adaptation for aerial feeding. Fibers of pectoral muscles are specially adapted for quick takeoff, frequent flapping, and extended glides required during flycatching. Some evidence of vestigial 11th primary.

**SIMILAR SPECIES:** Could be mistaken for small American Crow or Clark's Nutcracker, especially in poor lighting. Some range overlap with Acorn Woodpecker, which shares dark mantle and upperwing, but Acorn's pied plumage, social behavior, and vocalizations are distinctive.

## BEHAVIOR

### BREEDING BIOLOGY

Occasionally breeds in loose colonies, especially in broad patches of burned forest; up to 10 nests per ac. (25 per ha). As many as three active nests are sometimes found in each of multiple adjacent trees, possibly in areas with low snag density.

#### *Nest Site*

Key components include abundant aerial insects, open canopy, snags, and occasionally dense ground cover. Frequently nests in burned conifer forest, depending on years after fire (see Habitat).

Nest success may be lowest in oak woodland and isolated cottonwood riparian habitats, highest in burned ponderosa pine forest.

Typically uses cavities excavated by other woodpeckers or natural tree cavities. Will nest in same cavity for successive years. Occasionally excavates, usually in trunk or large branch of a large decaying tree. Regionally favored tree species include cottonwood, willow, ponderosa pine, and occasionally juniper and Douglas-fir.

#### *Courtship*

As with other migratory species, courtship ensues upon arrival at breeding grounds. Both sexes may select breeding territory, but male usually becomes established on it first. Female may select nest cavity, then calls and displays to attract mate.

Courtship displays, mostly by male, include wing spreading and bill pointing away from mate; most displays are accompanied by screech and chatter calls, sometimes with faint drumming. Circular flight display ending near mate typically precedes or follows copulation, which is generally in immediate vicinity of nest, occasionally on pole or top of snag. Mates copulate frequently over several days, from mid-excavation through early incubation. Pairs may remain monogamous for up to 4 years, even using same nest cavity, especially in nonmigratory populations.

#### *Parenting*

Both sexes attend nest for roughly similar-length periods. Feeding rates vary with insect availability and weather, not nestling age. When insects are abundant, adults may temporarily cache prey at anvil site, typically in bark furrows or tops of snags, then pulverize prey into boluses to feed young.

Both adults, especially male, defend area around nest cavity during breeding season; interactions may be heightened during incubation, more docile during prenesting and nestling stages.

#### *Young Birds*

Fledging is sometimes drawn out over 2 days. Fledglings sometimes remain perched at nest tree 2–3 days, exploring bark, trunk, and branches but not reentering cavity; they stay near nest site minimum of 10 days. Mates may split up fledglings for feeding.

Initially, juveniles flycatch clumsily, returning to original perch with difficulty. By early Sept., flight and feeding behaviors are indistinguishable from those of adults.

#### *Dispersal*

When breeding in burned forest, families may move to unburned forest after fledging. Multiple family groups may gather in small postbreeding flocks, often moving upslope immediately after fledging. Juveniles are usually not driven off until adults begin accumulating winter mast stores.

### NONBREEDING BEHAVIOR

#### *Feeding*

A diverse, opportunistic feeder, generally preferring locally abundant aerial insect populations or mast sources.

***Preferred Foods:*** Diet varies with season but consists primarily of adult flying insects, nuts, and fruit; rarely larval insects. Favored insects include ants, bees, wasps, butterflies, various beetles, and grasshoppers. Also known to eat snails and rarely bird eggs (possibly its own). Plant foods include acorns, almonds, apples, cherries, peaches, corn, and wide variety of wild berries and seeds.

Grit may constitute more than half of stomach contents in fall and winter, when acorns or corn dominate diet; much less in summer. Simple fall diet may include half acorns and half grit. Generally favors shelled peanuts at feeders, but also eats suet and sunflower seeds.

***Foraging Behaviors:*** Supremely adapted for aerial

An adult Lewis's Woodpecker feeding a recently fledged juvenile. *Tom Grey; July 2007, Sierra Co., CA.*

insect capture; capable of catching multiple insects on a single flight foray. Requires tall perches—snags, utility poles, fenceposts—and open canopy for efficient foraging. Usually sallies from perch, from which it scans constantly while lit, but also engages in longer hawking flights over fields or open water, sometimes with swallows and swifts. Frequently exploits seasonal insect hatches over water; observed gorging on recently hatched salmonflies over riparian corridors. May also snap insects from air while perched. Spends more time flycatching in pine forest than in oak woodland.

Occasionally gleans from shrubs and ground; rarely bores into bark and almost never into cambium. Searches visually rather than listening for grubs as most other woodpeckers do. Outside breeding season, plucks live acorns and nuts out of their hulls. When eating fruit from trees, especially crab apples, clings to underside of fruit or adjacent small branch and pecks from underneath.

***Food Storage:*** On wintering grounds, individuals, or less often pairs, collect acorns, almonds, corn, and other mast and store them in makeshift granaries, using natural cavities, cracks, and crevices in trees. Typically stores nut parts rather than whole nuts. Birds may rotate or relocate acorn pieces, possibly to minimize fungal decay. Occasionally

An adult Lewis's Woodpecker in a relaxed-wing sunbathing posture, taking a break from preening. *Stephen Shunk; June 2008, Elko Co., NV.*

overstores mast, ignoring or abandoning it upon leaving wintering grounds. Frequently observed storing insects in cracks and crevices of tree bark or snag tops between feeding bouts.

### *Territory Defense and Sociality*

Performs continuum of more aggressive displays in single encounter with an intruder, beginning with bill pointing; then wing flashing from perched stance; and finally circular flight display with wings held in dihedral. Chatter call and flashing of bright pinkish underparts accompany most advanced displays.

Nonbreeding territories range from 2.5 to 15 ac. (1–6 ha), depending on food availability. Individuals in postbreeding flocks may quarrel, but they do not defend fall feeding areas. Many individuals may share and defend the same tree as winter mast source, but no single bird or pair of birds controls access to the supply. See Dispersal for group behavior during postbreeding dispersal and migration.

### *Interactions with Other Species*

Lewis's Woodpecker is intensely territorial toward Acorn Woodpecker at winter food stores, and is known to rob stores of other woodpeckers. Exhibits aggressive territoriality toward Red-headed Woodpecker over shared resources in se. CO, where the two species nest in close proximity.

In areas with limited nest cavities, Lewis's is rarely known to usurp nests from Hairy Woodpecker and bluebirds, evicting eggs and young but usually not depredating live birds. Unlike most woodpeckers, Lewis's is strongly dominant over European Starling; its nests are rarely usurped by starlings, and a single Lewis's is able to overcome multiple starlings. American Kestrel may passively force Lewis's from a nest site, but the two species may also nest in the same tree depending on local food and nest-site availability.

### MISCELLANEOUS BEHAVIORS

Lewis's direct, non-undulating flight differs substantially from that of most other North American woodpeckers. When feeding, may glide continuously from high perch to lower food source, returning to perch with constantly flapping ascent.

Roosts singly in former nest cavities in all seasons. Rarely, may roost in natural crevices, such as behind peeling bark or in rotten tree center. Observed sunbathing atop snags, poles, and fenceposts, with wings slightly or fully extended.

## CONSERVATION

**HABITAT THREATS:** Primary threat is loss of open ponderosa pine forests due to fire suppression, selective timber harvesting, and replanting with closely spaced seedlings. Salvage logging in burned stands, especially large snag removal, may preclude future colonization after fire.

Decline in open cottonwood habitat is due to attrition of standing dead trees and lack of regeneration that results from flood control, low water-flow rates, intensive cattle grazing, and clearing for agriculture. Removal of large snags and mature oak woodlands is probable cause of local population declines over past 100 years in w. N. Am.

**POPULATION CHANGES:** Sporadic distribution makes population assessment difficult. Extirpated from Vancouver I., BC, by 1962, and from lower Fraser R. valley, sw. BC, by 1964. Other notable local population decreases or extirpations include MT; Willamette Valley, OR; s. CA; and n.-cen. UT. Range expansion onto plains of se. CO began circa 1910 and has been particularly evident since 1950s; attributed to presence and retention of mature cottonwoods for nesting and cultivated corn as food supply.

**CONSERVATION STATUS AND MANAGEMENT:** National Audubon Society Blue List in 1975, followed by elevated levels of concern in U.S. and Canada through present; also listed by 15 states, plus AB and BC, at varying levels of conservation priority.

Recommended measures in Georgia Depression ecoprovince of sw. BC include controlling invasive species and reducing watering and fertilizing in patchwork developed areas.

Snag-retention specifications recommended for Blue Mts. of OR and WA, to support maximum density of 6.6 pairs per ac. (16.6 per ha), are 100 snags per ac. (249 per ha) that are greater than 12 in. (30.5 cm) in diameter and more than 30 ft. (9.1 m) in height in ponderosa pine, riparian cottonwood, and burned mixed-conifer forest.

Other habitat conservation efforts may include retaining snags in clumps; restoring stream flows in riparian habitats; facilitating periodic floods or mechanical disturbance; managing for open, parklike stands of ponderosa pine with selective thinning or periodic burns; limiting repeated annual grazing, consecutive grazing days, and herd size near riparian habitats; reducing postburn salvage logging; and avoiding high-density replanting of trees after logging or fire.

Successful use of artificial nest boxes pioneered in cen. OR may facilitate reintroductions into regions where the species has been extirpated, as well as provide support for tenuous populations in marginal habitats.

### REFERENCES

Abele et al. 2004; Adams 1941b; American Ornithologists' Union 1903, 1908, 1976; Baldwin and Schneider 1963; Bock 1970; Bock et al. 1971; Bolander 1914, 1930; Brewster 1898; Brown 1902; Bryant 1929; Committee on the Status of Endangered Wildlife in Canada 2001; Constantz 1974; Cooper et al. 1998; Coues 1892; Currier 1928; Dawson 1921; Galen 1989; Garry Oak Ecosystems Recovery Team 2002; Hadow 1973; Hardie 2005; Heintzelman 1981; Hine 1924; Hlady 1990; Kays and Wilson 2002; Law 1929; Linder and Anderson 1998; Linsdale 1936; Marsden 1907; Michael 1926; Mitchell 1915; Newlon 2005; Pemberton 1923; Robertson 1935; Saab and Vierling 2001; Saab et al. 2007; Schwab et al. 2006; Sherwood 1927; Smith 1940; Sousa 1983; Stone 1915; Tobalske 1997; Velland and Connolly 1999; Vierling 1997; Welch 1899, 1900; Williams 1905; Yocom 1960.

# RED-HEADED WOODPECKER

*Melanerpes erythrocephalus*

L: 7.6–9.5 in. (19.4–24 cm)
WS: 16–18 in. (40.6–45.7 cm)
WT: 2.0–3.4 oz. (56–97 g)

**John James Audubon once** described the Red-headed Woodpecker as living its "whole life (as) one of pleasure." Perhaps the most striking of the continent's Picinae, the "Red-head" glows like a burning flame atop its favorite snapped-off tree-tops. Its bold white wing-flashes are unmistakable in flight, even to the naked eye.

One of only four sexually monochromatic woodpeckers in the world, the Red-headed is also one of North America's four woodpecker species that commonly stores food (with Lewis's, Acorn, and Red-bellied Woodpeckers), but it is our only woodpecker known to place pieces of wood or bark over its stores as some sort of protective covering. Like its closest North American relatives—the Acorn and Lewis's Woodpeckers—the Red-headed enjoys a wide variety of forage, from berries to beetles, alternately harvesting, flycatching, and ground gleaning for the most accessible food source.

Red-headed Woodpecker populations have fluctuated wildly over the last 200 years, peaking during successful crop years of the once-abundant northern beech forests and during invasions of the now-extinct Rocky Mountain grasshopper. Despite its relatively large rangewide population, the Red-headed now faces potentially serious declines and is considered a high priority for conservation throughout much of its range.

## DISTRIBUTION

**BREEDING RANGE:** Rocky Mt. states eastward to Atlantic Coast; southern states from e. TX eastward into cen. FL. Rare scattered breeder in Appalachian states.

**SEASONAL MOVEMENTS:** No single population appears to reside permanently in any one area, because of unpredictable fall wandering; occasionally prone to seasonal irruptions. Observed migrating silently, diurnally, and at low altitude in fall, often in loose flocks, probably aiding in search for winter mast supply.

Fall migration on New England coast is mostly juveniles; spring movements in the region are mostly inland; moves along e. Appalachian ridges, rarely to 5,900 ft. (1,800 m). Large spring flocks observed along Great Lakes shorelines. Sept. single-day high count of 880 tallied in 2008 on Missouri R., w. IA, sometimes 50/50 adults and juveniles. Gulf Coast and FL breeders generally move northward in fall.

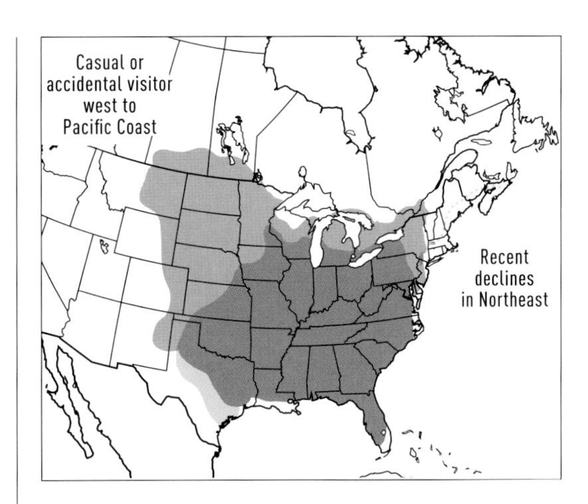


**WINTER RANGE:** Winters from OK and e. KS east to w. and cen. PA and MD, south through remainder of breeding range. Some individuals winter outside breeding range, especially at northeastern and southern limits. Rarely winters in n. Great Plains and Upper Midwest.

## HABITAT

**BREEDING RANGE:** Favors open deciduous woodlands with large-diameter snags for nesting and flycatching; also clear understory for ground foraging. Early in breeding season, moves from denser interior woodlands to edges, occasionally remaining in interiors with open canopy. Also found in open, short-grass meadows or their like (e.g., pastures and golf courses) with broadly dispersed or isolated groves of large deciduous trees and snags.

High concentrations may be found in woodlands where trees have been decimated by herbicides, flooding, fungal disease, or severe winter storms. In prairie habitats, found on utility poles far from nearest trees, and among cottonwood galleries and aspen groves.

**MIGRATION AND WINTER RANGE:** May be abundant or completely absent year to year depending on availability of mast, especially acorn crops. At western limits, frequents tall-grass prairies with abundant oak galleries as well as pecan groves; pine component is more prevalent in southern areas. May use forest edges more in fall than summer.

## DETECTION

Conspicuous; often first observed actively flycatching from treetops or flying over open habitats.

**OPPOSITE:** An adult Red-headed Woodpecker with its "hidden" black bib visible behind the red chest. *Greg Lasley; July 2009, Hansford Co., TX.*

An adult Red-headed Woodpecker in a relaxed posture. *Doug Backlund; May 2008, Hughes Co., SD.*

Vocally distinctive from but generally less vocal than sympatric Red-bellied Woodpecker. Difficult to miss when present.

**VOCAL SOUNDS AND BEHAVIOR:** Vocal repertoire is fairly broad. Gives three basic calls, occasionally similar in style to those of Lewis's and Acorn Woodpeckers: squeal, chatter, and rattle.

Squeal, the typical call note, is a harsh, repeated *rheer, rheer* or *raer, raer*. Chatter—*racka, racka, racka, rak, rak*, with various combinations—is often given by mated pairs when approaching or meeting at cavity tree, before nest exchange, and during copulation. Rattle, the common defense call, is a short, descending, and dry *quirr* or *werrr*, often compared to tree frog (and reported to elicit response from frogs, and vice versa).

Groups call jointly when flycatching. Pairs perform mutual aggression call when defending territory.

**NONVOCAL SOUNDS:** Drum is a steady series of beats in 1-second bursts, consistent in pitch and volume through entire roll. More often given by male than female, mainly in nesting season, and often repeated. Generally not separable in the field from Red-bellied Woodpecker's drum in overlapping range and habitats.

Drumming is elicited by approaching intruders, becoming more emphatic after a successful chase. Role of drumming in courtship is not known.

Like other *Melanerpes*, Red-headed uses tapping during sexual encounters, often between mates, with one inside and one outside cavity. Tree-surface feeding sounds are rarely forceful enough to be audible.

A recently fledged Red-headed Woodpecker. *Alan Murphy; July 2006, Harris Co., TX.*

## VISUAL IDENTIFICATION

**ADULT:** Striking and boldly contrasted black, white, and red; the most flamboyant North American woodpecker. In flight, upper- and underparts display sharp contrast.

**JUVENILE:** Plumage contrast in flight is similar to adult's except for brownish head, dark subterminal band across white secondaries, and brown spots in tertials. Perched, shows thin, tapering white eyeline, brown throat finely streaked with white, and white breast and flanks with dusky streaks.

**INDIVIDUAL VARIATION:** Sexes are indistinguishable in the field by plumage and almost identical in size. Lower belly rarely (possibly on older individuals) washed with dull yellow, orange, or red (see Geographic Variation). Red head is occasionally flecked with orange or yellow. Some albinism and leucism reported.

**PLUMAGES AND MOLTS:** Timing of molt in juveniles varies greatly. First basic plumage is typically acquired Sept.–Mar.; may begin on breeding grounds, especially with some red in head, but some birds retain some brown in head through second fall. Black in wings is typically acquired by first Apr. Ages are generally indistinguishable by May. As in Lewis's Woodpecker, juvenal flight feathers are retained until second fall, with second and subsequent prebasic molts occurring Aug.–Dec.

**SIMILAR SPECIES:** Not likely to be mistaken for any other woodpecker (or any other bird) within its typical range. Boldly contrasting dorsal pattern superficially resembles that of much larger Ivory-billed Woodpecker, the tiny range of which lies within that of Red-headed (some erroneous reports of Ivory-billeds from inexperienced observers have later been determined to be Red-headeds).

Term "red-headed woodpecker" is occasionally ascribed colloquially (and in some literature) to Red-bellied Woodpecker, male of which shows red only on nape and crown.

**GEOGRAPHIC VARIATION:** Generally decreases in size from northwest through southeast, smallest in FL. Some Great Plains and Rocky Mt. populations are larger with more extensive red in belly, once defined as "*caurinus,*" but size and belly color are intermediate in Mississippi R. valley, hence possible clinal variation.

**FIGURE 9.** Variation in Secondaries of the Red-headed Woodpecker. (1) Typical juveniles show the solid subterminal bar across all 11 secondaries; some juveniles show a second bar, as indicated by the dashed lines. (2) The typical adult shows black only on the outer edge of the outermost secondary; also note the black shafts on the outer secondaries. (3) Variation in the outer secondaries of some adults, possibly in the second basic (second fall) molt; note the absence of a solid bar in the subterminal markings. *Redrawn by Christine Elder after Pyle and Howell 1995.*

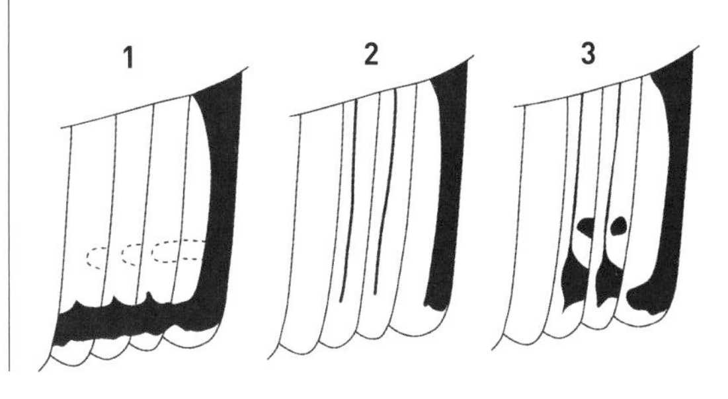

An adult Red-headed Woodpecker returns to the nest with an insect catch. Note the retained juvenal plumage, with black markings in the two outer secondaries (see Figure 7). *Marie Read; Aug. 2013, Seneca Co., NY.*

## BEHAVIOR

### BREEDING BIOLOGY

#### *Nest Site*

Prefers dead trees or dead portions of live trees; wide range of favored species. Will use utility poles and fenceposts, especially in prairies. Prefers barkless snags, possibly to deter snake predation; also tends toward very old trees in areas with clear understory. Well known to nest on golf courses, especially rural courses with plentiful snags.

Relatively weak excavator, often starting with existing crack or fissure in tree; occasionally uses natural cavities, often old cavities of other species, and possibly nest boxes (though evidence is lacking). May reuse same cavity several years or may excavate just below previous year's. May more often nest in dead limbs than trunks of trees.

#### *Courtship*

Monogamous, with pairs occasionally remaining together multiple years. Possible polygyny observed, with two females and one male at same nest.

Male does most of excavating. If female approves of site, mates tap together. Female repeatedly rests, preens, and taps near male as he excavates.

The two may stand parallel to a branch or atop a snag, pointing in same direction, then, often after tapping, will spin 180 degrees to point in opposite direction. Both sexes bob and bow toward each other, with head pointed forward, wings drooped, and tail cocked up. Mates rarely wrestle but frequently chase around surfaces of snags or utility poles, or even in flight from tree to tree.

Copulation is most often within 6.5 ft. (2 m) of cavity entrance. Female flutters onto male's back to perform reverse mounting, the two then switching places.

#### *Parenting*

As in most woodpecker species, full-time incubation begins after last egg is laid. At dawn and throughout morning, male known to call for female from inside cavity. If first brood fails, pair will attempt to renest, usually within 10–12 days, often in same cavity. Up to half of pairs in a given population may double-brood, with broods separated by up to 2 months; unconfirmed reports of three broods in FL. Adults often incubate second brood while feeding first set of fledglings; may drive off first brood after second brood hatches.

#### *Young Birds*

Fledglings remain close to parents or nest cavity, spending most of time resting or waiting for parents to bring food. Capable of flycatching soon after fledging. When young are a little over 3 weeks out of nest, parents may refuse to feed them and begin chasing them off.

#### *Dispersal*

Families may remain around nest site—with immature birds feeding independently—until early Sept. Winter dispersal is highly variable year to year, es-

An immature Red-headed Woodpecker still in preformative molt. Note length of replaced secondaries showing single black bar, contrasting against short secondaries with 2 bars on fledgling in photo on page 55. *Christopher Ciccone; Feb. 2015, Durham, NH.*

pecially in North. Juveniles, especially from second or very late broods, may remain in natal territory until spring. First-season breeders may nest as close as 360 ft. (110 m) from natal site.

## NONBREEDING BEHAVIOR

### *Feeding*

***Preferred Foods:*** Red-headed is most omnivorous North American woodpecker. Diet includes seeds, nuts, berries, insects, and bird eggs and nestlings; occasionally adult birds, various small vertebrates, and dead fish. Summer diet is one-third insects, two-thirds plant foods. Winter diet is mostly hard mast, especially acorns; corn and other crops are eaten if acorn crop is sparse.

Will eat bark in winter, possibly being saturated with sap from the inside. Grit is found in about half of birds, more often in females than males.

Among insects, favors adult beetles and honeybees; strongly affiliated with major insect emergence episodes such as cicadas and midges.

Formerly consumed mass quantities of Rocky Mountain grasshoppers, most notably during NE invasion of 1873–1876. Once fed heavily on expansive beech-nut crops, but exotic chestnut blight decimated old beech forests.

***Foraging Behaviors:*** Nearly equal to Lewis's Woodpecker in flycatching proficiency; more often gleans and ground feeds than does Lewis's. Summer foraging consists of up to 40 percent flycatching, 30 percent stooping from perch to ground. Gleans adult beetles from bark but rarely eats larvae, grubs, or ants. Occasionally chisels at bark in search of wood-boring insects.

Well known to perch on fenceposts, utility poles, and wires to scope paved roads for grain or invertebrates. Rarely known to place pine cones and nuts on road to be crushed by cars, partly accounting for relatively high rate of roadkill mortality.

Feeds at active sapsucker wells and occasionally cuts its own wounds in trees. Will feed at suet feeders, mostly in winter.

***Food Storage:*** Red-headed commonly stores food and is the only woodpecker known to conceal stored nuts and seeds by covering individual caches with pieces of wood or bark, especially after heavy rains; may forcibly pound bark pieces into storage sites, making them difficult to remove. Collects nuts, sometimes in summer but mostly early Sept.–Nov. and into Jan. in some areas. Will exhaust supply of live nuts in trees before collecting from ground.

Uses cracks or crevices as anvil sites to break nuts into smaller pieces, then hammers them tightly into natural openings. Also stores in cavities, under patches of raised bark, and in gateposts, railroad ties, and roof shingles. Frequently stores grasshoppers alive, wedged tightly into crevices. Juveniles reported storing acorns, pine nuts, earthworms, and crickets.

Follows an unusual storage ritual, first stashing foods in single tree or small area (larder-hoarding), then redistributing pieces to scattered storage sites throughout territory (scatter-hoarding).

### *Territory Defense and Sociality*

Aggression toward intruders peaks in early Apr., then again in July when second broods are initiated. Summer territories may be larger than winter, with some overlap among neighboring pairs. Territorial displays are similar to courtship displays.

Guards winter storage sites aggressively, but multiple birds may share same mast source. Winter territories are delineated by locations of evening roost trees.

Usually solitary in winter, but often seen among other Red-headeds if territories are small. Multiple pairs may associate somewhat during breeding season, and family groups may join in loose flocks for fall migration.

### *Interactions with Other Species*

Pugnacious; will chase any other bird at any time of year—even Pileated Woodpecker acquiesces. Known to submit to Lewis's Woodpecker in CO, from which Red-headed frequently departs in winter. Observed successfully defending food stores from Blue Jays. Will defer to Eastern Kingbird when both are flycatching in same vicinity.

Often shares snag with other cavity nesters; Red-headed is most common and dominant primary cavity nester in some FL habitats.

Summer consumption of eggs and young of other birds may be opportunistic consequence of pseudousurpation of nests (tosses eggs and young but does not occupy nest). Known to evict flickers, swallows, orioles, and flycatchers; will also destroy flycatcher nests after intense quarrels with nesting adults. Usurps cavities from Downy, Red-cockaded, and Red-bellied Woodpeckers. Destroys eggs of box-nesting ducks, but not known to nest in boxes that it depredates. Often overcomes European Starling or House Sparrow competition for nest site.

### MISCELLANEOUS BEHAVIORS

May alternate among multiple nightly roost cavities. Juveniles and adults may roost in different habitat types (e.g., pine versus oak), but territory sizes are similar.

Flight style is similar to that of Lewis's Woodpecker; exceptionally acrobatic. Observed anting and sunbathing. Reported drinking water more often than most other woodpeckers; observed chipping and eating ice and drinking from pond in winter.

## CONSERVATION

**HABITAT THREATS:** Removal of snags and dead branches for aesthetics, firewood, agriculture, and so on has precipitated local and widespread declines in both urban and rural areas. Other detrimental impacts include large-scale monoculture farming and habitat conversion for development.

**POPULATION CHANGES:** Red-headed probably exhibits the most dynamic pattern of population change of all North American woodpeckers. Expands dramatically with large-scale tree mortality but crashes when snags come down; thrives with clearing of dense forests and planting in prairies but declines in areas of large-scale reforestation. Also shows short-term changes with acorn crop fluctuations.

An adult Red-headed Woodpecker places a protective bark flake over food stores. *Paul Bannick; Nov. 2013, Lucas Co., OH.*

Local populations peaked in mid-19th century with widespread death of chestnut and elm trees, then increased again in early 1980s following Dutch elm disease outbreak of 1960s–1970s. Expansion of breeding range into NM in mid-1800s correlates with expansion of Santa Fe and Rock Island Railway lines, when telegraph poles along railways became available for nesting sites.

Once bred through much of New England into NL and e. NY; estimated 1,270,000 breeding adults in IL in 1909, compared with 180,000 in 2007.

**CONSERVATION STATUS AND MANAGEMENT:** National Species of Conservation Concern and on Yellow WatchList. Listed as endangered in CT and DE, threatened in NJ; targeted as high conservation priority in 29 states and 3 provinces, more than any other woodpecker.

Conservation measures should include creation and retention of snags, in groups if possible; retention of dead branches on large trees in nonurban areas and selective pruning of hazardous branches in urban areas; and selective thinning of live trees in small woodlots and former grasslands. Prescribed burning and understory thinning create more open forest stands, but care should be taken not to destroy snags.

## REFERENCES

Adkins 1926; Bailey 1912; Bock et al. 1971; Brackbill 1969b; Brackett 1896; Brooks 1934; Bryan 1899; Burr 1926; Cardiff 1972; Conner 1976; Copeland 1926; Crewe and Badzinski 2006; Dill 1926; Doherty et al. 1996; Fenn 1940; Franken and Gillies 2001; Gibbs 1892; Harrington 1924; Hay 1887; Henderson 1931; Hoffman 1928; Holdstein 1899; Ingold 1987, 1989, 1990, 1991; Jackson 1976; Judd 1956; Kilham 1958b, 1958c, 1959a, 1959e, 1977b, 1978; Klaphake 2008; Kohler 1910; Leopold 1918, 1919a, 1919b; MacRoberts 1975; Marqua 1963; Mead 1900; Meanly 1936; Moskovits 1978; Nauman 1932; Nicholson 1997; Pinkowski 1977; Robbins and Blom 1996; Rodewald et al. 2005; Rogers and Jaramillo 2002; Rogers et al. 1979; Roth 1978; Semple 1930; Sherman 1907; Smith 1986; Smith et al. 2000; Southern 1960; Stefferud 1966; Stewart 1931; Stoner 1925; Teachenor 1929; Torrey 1901; Wright 1904.

# ACORN WOODPECKER

***Melanerpes formicivorus***

L: 8.3–9.3 in. (21.1–23.6 cm)
WS: 17–17.5 in. (43.2–44.5 cm)
WT: 2.4–3.1 oz. (67–89 g)

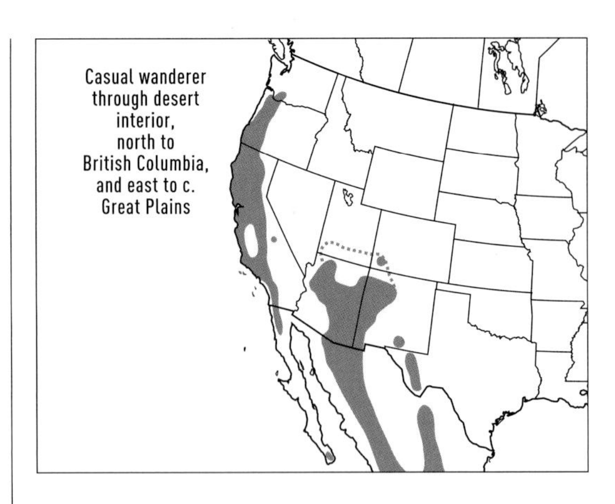


**The well-named** Acorn Woodpecker exhibits some of the most fascinating behaviors among the North American Picinae, from its communal breeding habits to its construction of elaborate storage "granaries." The latter activity of this industrious carpenter is its signature: obsessively pounding acorns into tight-fitting holes it has drilled in the bark of large trees. The largest granary trees have been estimated to hold more than 50,000 acorns.

The Acorn Woodpecker's scientific species name, *formicivorus*, translates as "ant-eating," but this highly omnivorous species could also be called the "flycatching" or "sapsucking" woodpecker. Its distinctive plumage has also earned it the nickname "clown-faced" woodpecker.

Naturalist William Ritter conducted the first in-depth field studies of Acorn Woodpeckers in the 1920s and 1930s, culminating in his delightfully effusive volume *The California Woodpecker and I*. Ritter was the first to prove that this species eats acorns primarily for the acorns themselves rather than for the insects attracted to the stored mast. Opportunistic as it is, the Acorn Woodpecker will gladly consume the larvae of acorn worms, but this could never distract it from its feverish obsession of stocking the pantry. As Ritter so bluntly put it, "The meat is the object of their storing."

## DISTRIBUTION

**PERMANENT RANGE:** Permanent resident except as noted below. Occupies a roughly L-shaped range from w. OR south through CA and east through cen. AZ, NM, and w. TX. Neotropical populations extend through Mex. and Cen. Am. into Colombia.

**SEASONAL MOVEMENTS:** Generally sedentary; postbreeding movements depend on abundance of stored acorns. If granaries are exhausted before winter, groups will wander, distance depending on nearest food supply, typically remaining close enough to return to home breeding territory in spring. Large-scale acorn crop failure can force permanent relocation. Such events, coupled with species' colonial and pioneering nature, may be at least partly responsible for historic range expansions.

Members of at least one population breeding near Huachuca Mts., se. AZ, wander locally after postbreeding dispersal until food supplies are exhausted, then depart independently, presumably to Mexican Sierra Madre Occidental. Individuals return in spring to breed, most occupying former territories. In years of acorn excess, birds may overwinter and breed in small groups the following summer.

## HABITAT

Wide variety of foothill and montane oak woodlands, often mixed with pine or other large conifers. Also riparian woodlands and mixed coniferous forests, as long as oaks are present or nearby. Occasionally colonizes oak-occupied human environments, as long as some substrate exists for acorn storage, such as wooden utility poles or buildings.

Typical woodland characteristics include one or more large granary trees at the center of a well-defined territory, separated from acorn-supply trees by grassland, pasture, or other cleared area. Density of multigroup communities is correlated with oak species diversity for southern (nominate) subspecies but not far western populations. Rarely occurs to sea level, as in s. CA.

## DETECTION

Very gregarious, usually heard before seen; almost constantly vocalizing. Flycatching behavior and tight-knit group activity are easily observed, especially in open oak woodland.

**VOCAL SOUNDS AND BEHAVIOR:** Persistent *waka-waka* or *ricka-ricka* chattering is most commonly identified call, usually repeated several times and performed primarily as a greeting display among group members and in territorial interactions with invaders from other groups.

Dynamic vocal repertoire is correlated with social complexity. Wide variety of calls is associated

**OPPOSITE:** An adult female Acorn Woodpecker, nominate subspecies. Note subtle signs of leucism (brownish tint) in wings and tail. *Alan Murphy; May 2013, Pima Co., AZ.*

An adult male Acorn Woodpecker, *M. f. bairdi* subspecies, delivering an acorn to a palm tree granary. *Steve Brad; Feb. 2012, San Diego Co., CA.*

with territorial defense, predator detection, intrusion on a new territory, attack on a subordinate, and other diverse behaviors.

**NONVOCAL SOUNDS:** Drumming is rare. Rolls include 2–20 beats at even cadence, given by both sexes. Not always but usually associated with extragroup territorial encounters and power struggles. Often heard tapping acorns into granary holes.

## VISUAL IDENTIFICATION

**ADULT:** Generally unmistakable, with distinctive pied facial pattern, patchy black-and-white upperparts, and underparts streaked black on white. In flight, shows strongly contrasting black-and-white pattern above, with crescent-shaped white wing patch and bold white lower back, rump, and uppertail coverts.

**JUVENILE:** Dark-eyed, less glossy above than adult, and with brownish breast band and streaking and

A juvenile male Acorn Woodpecker (note the dark eye), nominate subspecies. *Alan Murphy; June 2006, Pima Co., AZ.*

light tan belly. Black ring does not completely encircle bill. Far more juveniles than adults (70:30) show white spots or bars on tail feathers. Juveniles of both sexes have solid red crown and therefore cannot be identified to sex until fall molt begins.

**INDIVIDUAL VARIATION:** Males are slightly larger and have longer bill than females. Leucism and albinism occur in small fraction of cen. CA *bairdi*. Even more rarely, individuals show golden yellow, rather than red, crown.

**PLUMAGES AND MOLTS:** Eyes become brighter with preformative molt, occurring first May through second summer and into second fall. After juveniles lose all-red crown, no single character reliably separates first-year from adult birds.

**DISTINCTIVE CHARACTERISTICS:** "Clown-faced" pattern is distinctive. One of three North American woodpeckers with a solid, glossy black back and, except for Ivory-billed, the only one with a bright white or yellowish iris. Black "ring" around base of bill is unique among woodpeckers.

**SIMILAR SPECIES:** Nearly impossible to mistake for any other woodpecker, despite some common plumage characters. Bold white wing patch formed by inner half of primaries is found only in Acorn, White-headed, and (much larger) Pileated Woodpeckers.

**GEOGRAPHIC VARIATION:** As many as nine subspecies, divided into seven subspecies groups; *formicivorus* and *bairdi* groups occur in N. Am. Populations vary in amount of yellow in white parts of throat and forehead, pattern of black streaking on breast, and extent or presence of red feathers on upper chest and crown. Bill size and wing length generally decrease from north to south.

## SUBSPECIES

**NORTH AMERICA:** The *formicivorus* group includes *M. f. formicivorus* and debated "*aculeatus*"; occurs from AZ and NM south to s. Mex. Most variable and intermediate plumage of all groups; heavier streaking through belly than *M. f. bairdi*. Separated from *bairdi* group in N. Am. by Mojave Desert. "*Aculeatus*," or "Mearns's" Woodpecker, of AZ described on basis of 15 percent smaller bill (in all dimensions) on average than *formicivorus* or *bairdi*.

The *bairdi* group includes subspecies *M. f. bairdi* and debated "*martirensis*"; ranges through Pacific

An adult male Acorn Woodpecker, nominate subspecies. This color variant shows red in the upper chest, which is far more common in the nominate subspecies than in *M. f. bairdi*. *Copyright Monte M. Taylor; Nov. 2007, Cochise Co., AZ.*

An adult female Acorn Woodpecker, *M. f. striatipectus* subspecies. Note the yellowish color in the throat and lack of solid black in the breast (streaking only), as well as the very narrow black crown stripe. *Paul Bannick; Apr. 2010, San Gerardo de Dota, San Jose, Costa Rica.*

states. Black breast streaking may appear as solid black breast band with only fine streaking below and on flanks. Moderate yellow tinting on white parts of head and throat and little to no red at upper edge of black breast band. Northernmost populations of *bairdi* have largest bill and longest tail and wings of all subspecies.

**NEOTROPICAL SUBSPECIES:**
"*martirensis*": San Pedro de Martir Mts., n. Baja CA
*angustifrons* group: extreme s. Baja CA
*lineatus* group: Isthmus of Tehuantepec to Nicaraguan Depression
*albeolus* group: Belize and n. Honduras
*striatipectus* group: montane Costa Rica and Panama
*flavigula* group: Colombian Andes

## BEHAVIOR

### BREEDING BIOLOGY

One of two North American woodpeckers (with Red-cockaded) to practice cooperative breeding, by which individuals in addition to natal parents cooperate to raise offspring of those parents. Some populations exhibit true communal nesting (polygynandry); multiple females breed with multiple males in the same nest, and multiage family groups all assist in rearing young.

#### *Nest Site*

Prefers large valley oak, sycamore, cottonwood, and ponderosa pine. Prefers snags if available; will use large dead limbs on live trees and will excavate in living wood if necessary. Most cavity entrances are on underside of large branches or leaning trees. Granary trees are favored nest sites. Groups may use any of several roost cavities in their territory as nest site; about half of groups reuse former year's cavity.

Nesting success is largely independent of nest-site preferences. Cavities in live limbs, those farther from granaries, and those in top 20 percent of tree generally fledge more young, despite preferences described above.

#### *Cooperative Breeding Dynamics*

Sedentary and gregarious nature and occupation of habitats accessible to researchers have allowed in-depth study of fascinating breeding behaviors. Mating systems range from predominant monogamy (AZ) and cooperative polygyny (NM) to polygynandry (CA). Records of unusually large clutches and runt eggs from Southwest presumably indicate at least occasional cooperative breeding there.

Typical group size is 4–5 birds. One to 7 breeding males may compete to mate with 1–3 breeding females; also 1–10 nonbreeding helpers, usually offspring from previous years. Maximum recorded group size is 15. Nonbreeding helpers may be up to 5 years old. All group members feed fledglings.

Onset of breeding season varies widely. Groups may renest if first attempt fails, or may double-brood if first brood is successful early in season. In about 20 percent of years, secondary breeding season occurs in fall, following years of good acorn crops and mild winters when granaries are still well stocked into early spring. No fall breeding documented in NM. In AZ, breeding is delayed until midsummer monsoon season, and renesting is rare.

No typical courtship displays or pair bonding. Breeders in resident populations usually remain on home territory throughout their lives, and males are hyperattentive toward breeding females during fertile period. Copulations are infrequent, brief, and rarely observed. Multiple males may attempt to mount a female in succession and often interfere with cobreeders' copulation attempts. Extragroup paternity and brood parasitism are extremely rare.

In groups with two nesting females, each female is the biological parent of half the offspring fledged. Individual females may breed with multiple males in group (cooperative polyandry); this leads to multiple paternity within clutches and parental uncertainty among males, possibly explaining why all males help raise young.

Females engage in a bizarre ritual of destroying each other's eggs until all individuals have begun normal laying sequence. Infanticide is rarely observed. If a lone breeding female or male disappears early in breeding season, the new replacement breeder may destroy eggs or nestlings of its predecessor. Up to 20 percent of all broods may contain a runt egg.

Cobreeding individuals are usually siblings, or father and son for males, mother and daughter for females, but females are unrelated to breeder males. Most successful male in group may sire three times as many young as next most successful male. Father–daughter incest is rare.

Nonbreeding helpers (especially males) often search for breeder vacancies in other groups 9 mi. (15 km) from home territory or farther. Relatively high mortality of breeding females; survivorship of breeding males increases significantly with group size.

Nestlings are fed mostly aerial insects; singly or in boluses (pulverized clumps). As they age, they receive increasing amounts of broken acorn pieces. Eggs and nestlings are lost to a wide variety of factors. Hatching success is greater in smaller groups. Unhatched eggs are removed several days after others hatch.

Chicks hatching more than 1 day after first chick are ignored by adults and usually starve within several days. Early hatching chicks fledge healthier, are dominant over younger siblings, are more likely to survive fledging, and grow larger as adults. Fledging success increases with group size.

Some chicks climb out of nest, but most leave on the wing. When second broods occur, juveniles from first brood may help feed nestlings, but will

An adult female Acorn Woodpecker feeding anestling male at the nest cavity entrance. Note the dark eye and unfeathered orbital ring of the juvenile. *Jim Burns; July 2005, Santa Cruz Co., AZ.*

also reenter nest and intercept food intended for nestlings. Fledglings may also roost together quietly in foliage.

First 2 months after fledging, juveniles frequently follow older individuals throughout territory, gradually expanding home range; by 4–8 months, juveniles' home ranges are larger than those of breeders in group. Helper females typically emigrate by end of third year. Males are much more likely than females to remain in natal territory, often not emigrating until fourth or fifth year. Females also move farther away and are more vigorously competitive than males for breeding vacancies in new group. Both sexes typically disperse in unisexual units of one to three group members.

Less than half of offspring breed in first year, most at 2 years, a few not until 5 or 6 years. About 50–80 percent of adults and 35–60 percent of first-year birds survive to next nesting season, fewer females than males and fewer overall in Southwest.

## NONBREEDING BEHAVIOR

### *Feeding*

Despite insect preference noted below, relatively complex social behavior is probably related to widespread dependence on mast rather than on insect food availability.

***Preferred Foods:*** Insects are preferred at all times of year. Takes a wide variety, especially ants and beetles; wood-boring insects and larvae are almost entirely absent from diet. Also eats some small herptiles, acorns (both immature and stored), sap, fruit, flower nectar, and occasionally grass seeds. Feeds at oak blossoms, most likely for pollen; oak catkins may be large part of spring diet.

Acorns are mostly supplemental, stored in fall and eaten primarily when insects are unavailable. Acorn caches supply less than 20 percent of energy needs in winter; majority comes from flying insects and sap.

***Foraging Behaviors:*** When flycatching, makes single-insect flights, unlike Lewis's Woodpecker, which may catch many insects per flight. Also does not forage for insects on ground or in shrubs like Lewis's. May rarely go to ground to collect grit or retrieve dropped acorns. Flycatching predominates in Apr. and May, until sapsucking becomes more important.

June and July are peak sapsucking period. Each group uses one or two active sap trees. Sap wells are smaller and shallower than acorn storage holes but are spaced the same and scattered, unlike linear sapsucker wells. Individuals take turns "connecting the dots" by successively starting, adding to, and then finishing a series of holes. Will drink from hummingbird feeders.

An Acorn Woodpecker storage granary in a fencepost. *Stephen Shunk; Apr. 2012, Tuolumne Co., CA.*

Collects and eats acorns without storing from late summer through early fall, using anvil site to split nuts. In mid-Sept., birds begin to store mature acorns, but still eat some. Usually remove acorns individually from trees, but rarely may break off twigs containing multiple acorns or eat large acorns off branch.

***Food Storage:*** Groups store acorns communally. Territories center around one or two and up to seven granary trees, with thick-barked conifers preferred. Birds will expand a current granary before exploiting another tree in territory. Storage holes do not penetrate cambium. Some large conifers are filled with acorns 200 ft. (61 m) high.

Occasionally taps a fresh acorn into the hull of a rotten one. As acorns dry, they shrink and must be moved to smaller holes. Opportunistically stores other nuts in drilled holes or natural crevices, occasionally behind peeling bark or inside open pine cones. Also known to use artificial structures such as

**OPPOSITE:** An Acorn Woodpecker family group, *M. f. bairdi* subspecies, placing and/or moving acorns in a storage granary. *Marie Read; Feb. 2006, Orange Co., CA.*

utility poles, fenceposts, roof shingles, and even vehicle radiators, as well as old nest and roost cavities.

Does not favor acorns parasitized by insects, but will eat insects when present. Acorn parasites include various beetles; typically, larvae in acorns pupate and leave acorn as it dries, before all the meat is eaten and before the woodpecker gets to the nut.

### *Territory Defense and Sociality*

Acorn Woodpecker will aggressively defend 5-ac. (2-ha) or larger territory from nongroup members, especially in spring, when intruders search for breeding vacancies. Aggression is highly sex-specific; opposite sexes are known to ignore each other. Nonbreeding helpers and breeders are equally defensive of territories.

Aggression peaks during power struggles over reproductive vacancies. Up to 20 nonbreeding helpers, including same-sex siblings from different groups, may engage in veritable melees involving days or even weeks of constant battle. Largest coalition of siblings generally wins, remaining to share breeding status in new territory. Losers return to natal groups and resume status as nonbreeding helpers.

During encounters, birds call and chase extensively, performing wide array of displays and occasionally grappling in midair, pecking at each other's heads, and sometimes dropping to ground before paring off.

Internal aggression is rare among groups. Dominance is established early among siblings, possibly in nest. Frequency of fights decreases over first several months after fledging and is subsequently low. Males are generally dominant over females.

### *Interactions with Other Species*

Chases other bird species from granaries (most vigorously in fall), sap wells (primarily spring and early summer), cavities, anvil sites, acorn-supply trees, and flycatching perch trees, regardless of whether intruder competes directly for food. Will still defend storage tree even if stores are exhausted. Where ranges overlap, competes extensively with Lewis's Woodpecker for winter acorn supplies.

Known to attack other cavity nesters, especially

Biologist Walt Koenig prepares a trapping device to capture and band an Acorn Woodpecker after it enters its known roost cavity for the evening. *Kristine Falco; Nov. 2006, Monterey Co., CA.*

European Starlings, aggression elevating once nesting has begun. Will hassle Great Horned Owls in home territory and will attack snakes approaching cavity. Anna's Hummingbird is known to visit Acorn Woodpecker sap wells only when woodpeckers are absent.

Intruding birds fly away quickly when chased, but squirrels will withstand continual attack while attempting to extract nuts from granary; squirrels may also team up and eventually usurp storage trees until stores are depleted.

### MISCELLANEOUS BEHAVIORS

As in many other *Melanerpes*, undulating flight is less pronounced than in other woodpeckers. Preens commonly, but only self. Sunbathes with wings partially spread, appearing to sleep for up to several minutes at a time. Observed perching on utility wires. Frequently drinks water from holes or horizontal furrows in trees, also from springs or other groundwater sources and artificial watering stations.

## CONSERVATION

**HABITAT THREATS:** Habitat loss, partly due to overgrazing in riparian and pine-oak habitats, is primary threat to species. Clearing of oaks for development and poor regeneration may eventually have an impact on this species, but there is no current evidence of population declines. Any snag removal eliminates current or potential nesting and storage portals.

**POPULATION CHANGES:** Often the most abundant woodpecker in its range. Most populations are limited in size by local acorn supplies and granary availability; acorns may be underused in absence of adequate storage facilities. Disjunct populations, especially in Southwest, may depend on immigration and population exchange for long-term viability.

Adapts well to suburban conditions; persistent pioneer and colonizer, regularly pushing normal limits of its range, generally within 125 mi. (200 km) of known populations, occasionally resulting in range expansion.

**CONSERVATION STATUS AND MANAGEMENT:** Listed as critically dependent on oaks of oak woodland in Sierra Nevada. Targeted for conservation and monitoring in WA. Generally remains widely distributed and likely occupies most of its presettlement range. Conservation depends on persistence of mature forests with large oaks capable of producing abundant acorn crops and serving as substrate for nesting, roosting, and storage. Retention of trees of all ages, and especially snags and dead limbs, is crucial for survival.

### REFERENCES

Adams 1941a; American Association for the Advancement of Science 1929; American Ornithologists' Union 1916, 1951; Benitez-Diaz 1993; Brown 1987; Bryant 1921; Clabaugh 1928; Corman and Wise-Gervais 2005; Davidson 1999; Du Plessis et al. 1994; Fajer et al. 1987; Fisher 1906; Gignoux 1921a; Grinnell and Swarth 1926; Haydock and Koenig 2002; Henshaw 1921; Hoffman 1931; Hooge et al. 1999; Kattan 1988; Kattan and Murcia 1985; Koenig 1980a, 1991; Koenig and Benedict 2002; Koenig and Mumme 1987; Koenig and Pitelka 1979; Koenig and Walters 1999; Koenig and Williams 1979; Koenig et al. 1983, 1991, 2000; Leach 1925; Linsdale 1936; MacRoberts 1970; MacTague 2004; McKeever and Adams 1960; Mearns 1890a, 1890b; Michael 1926, 1936a, 1936b; Miller 1947, 1955; Mumme et al. 1985, 1990; Myers 1915; Paulson 2007; Peck 1921; Peyton 1917; Ritter 1921, 1922, 1938; Roberts 1979; Utah Bird Records Committee 2008; Verner 1965; Walker 1952.

# GILA WOODPECKER

## *Melanerpes uropygialis*

L: 8–10 in. (20.3–25.4 cm)
WS: 15–18 in. (38.1–45.7 cm)
WT: 1.8–2.9 oz. (51–81 g)

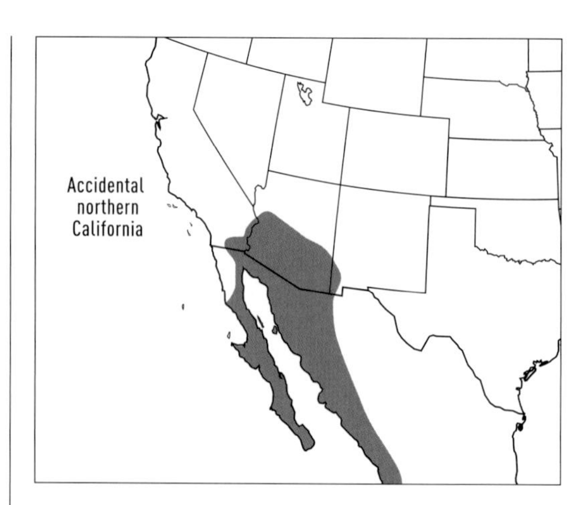


**Described by** Roger Tory Peterson as one of the "master artisans of the desert," the Gila Woodpecker thrives among the towering saguaro forests of the Southwest. Small populations spill into California, Nevada, and New Mexico, with a much larger range in "old" Mexico, but it is on Arizona's Sonoran Desert floor that the Gila dominates our North American avifauna.

Along with its desert companion the Cactus Wren, the Gila Woodpecker is one of the most vocally conspicuous birds in its range. Comical at times—at least to human observers—and always asserting its dominance, it can be quite common in Tucson's suburban neighborhoods, wherever a semblance of native vegetation remains.

This is one of the few woodpeckers known to occasionally raise three broods in the same year. As a result, the Gila may border on abundant in early summer, when family sizes swell with the earlier rounds of still-dependent fledglings. Like its desert relative the Ladder-backed Woodpecker, the Gila's abundance belies our body of knowledge about the species. We know much about some aspects of its behavior, taxonomy, and distribution, but further study is warranted.

A nestling female Gila Woodpecker begging from the entrance of a saguaro cavity. *Laura Stafford; June 2012, Pima Co., AZ.*

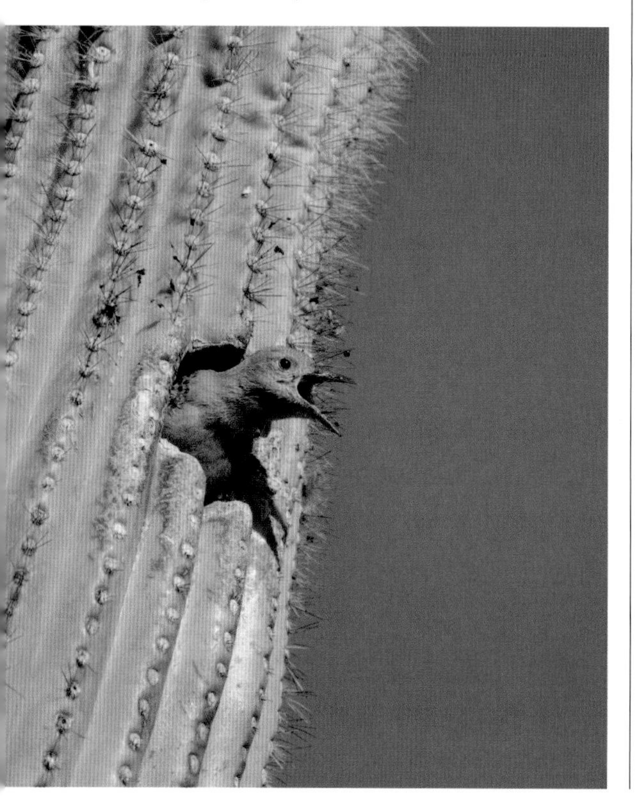

## DISTRIBUTION

**PERMANENT RANGE:** Throughout s. AZ, especially south of Mogollon Rim and west of Graham and Cochise Cos. Locally common in adjacent states, including se. CA, s. NV, and w. NM.

**SEASONAL MOVEMENTS:** Primarily sedentary. Some short-distance movement northward or to higher elevations in winter. Occasionally wanders outside breeding season to areas with abundant food sources.

## HABITAT

Conspicuous in Sonoran Desert habitats of Southwest, especially saguaro forests. Well adapted to suburban landscapes, especially those with date palms and large saguaros, even in dense neighborhoods with small yards. Also found in riparian and dry subtropical woodlands. Observed from below sea level in Imperial Valley, CA, to about 3,300 ft. (1,000 m), rarely above 5,000 ft. (1,524 m).

At edges of range, resident among riparian cottonwood galleries, often favoring easily excavated snags or exotic trees. Can be found in mesquite patches infested with fruiting mistletoe, but generally does not breed there.

## DETECTION

Vocally conspicuous, with one of the most distinctive drums among North American woodpeckers.

**OPPOSITE:** An adult male Gila Woodpecker feeding at—and thereby pollinating—a saguaro cactus blossom. *Paul Bannick; Apr. 2006, Pima Co., AZ.*

**VOCAL SOUNDS AND BEHAVIOR:** Gives two primary calls, a call note and rattle. Call note is a sometimes repetitive series of sharp *eek, eek, eek* or *kee, kee, kee* notes; sequence varies considerably in length and cadence. Loosely resembles slow version of Northern Flicker's "jungle call" when repeated in long series. Given more by female than male. Serves as general alarm call, occasionally with aggression displays.

Rattle is a churring sound like *braay, braay* or *preee, preee* or *whurr, whurr,* similar to that of other *Melanerpes.* Used as communication between mates and as primary territorial display. Given far more often by male than female.

Two additional calls are given less often: a harsh croak and a combination call with parts of both primary calls. Croaking call is given when agitated, combination call in a variety of circumstances. Most vocalizations are independent of aggressive behavior.

An adult female Gila Woodpecker feeding a fledgling male on a cholla skeleton. *Alan Murphy; June 2008, Pima Co., AZ.*

A fledgling female Gila Woodpecker. Note the yellowish gape flange and very faint yellowish wash in the lower belly. *Laura Stafford; July 2011, Pima Co., AZ.*

**NONVOCAL SOUNDS:** Drums infrequently in territorial display, often followed by aggressive behavior. Drum comically resembles sound of a stubborn lawnmower that won't start after each attempt. Gila does most of its foraging and nest excavation in soft substrates, so tapping sounds are rarely heard.

## VISUAL IDENTIFICATION

**ADULT:** Typical "zebra-backed" *Melanerpes* woodpecker, with pale head and underparts. In flight, white crescent-shaped wing patch contrasts against darker outerwing. Flanks and undertail coverts strongly barred with black.

**JUVENILE:** Underparts and head darker and grayer than in adult, dark bars on back and wings grayer and less distinct, with white bars slightly buffy. Fine blackish streaks in crown and breast. Both sexes may show a few red feathers on crown, male usually more than female but less than adult male.

**INDIVIDUAL VARIATION:** Male about 14 percent heavier with 14 percent longer bill than female. Female shows paler yellow belly than male and occasionally has a few red feathers in crown.

**PLUMAGES AND MOLTS:** Preformative molt Apr.–Sept. or Oct. Second prebasic molt occurs July–Oct. Subsequent prebasic molts begin in Sept.

**SIMILAR SPECIES:** No similar woodpecker through most of Gila's North American range. Very similar in plumage to closely related Golden-fronted and Red-bellied Woodpeckers. Gila's barred rump and tail are distinctive, and its solid-colored parts are darker overall than in the other two species.

**GEOGRAPHIC VARIATION:** Complicated, subtle variation throughout range.

### SUBSPECIES

Up to six subspecies recognized, generally one accepted in N. Am. Variation described in back barring, rump pattern, color of underparts and head, bill length, and size.

**NORTH AMERICA:** *M. u. uropygialis* covers most of North American range, south through w. Mex. from Sonora to Aguascalientes and Zacatecas.

Debated *M. u. albescens*, "Colorado River" Gila Woodpecker, of lower Colorado R. valley and Imperial Valley, includes specimens that show broader white than black barring on back; minimal markings on rump; lighter head, neck, and underparts; and to be smaller and shorter-winged than nominate *uropygialis*.

**MEXICO:**
*M. u. brewsteri*: s. Baja CA
*M. u. cardonensis*: n. Baja CA
*"fuscescens"*: s. Sonora; possibly part of *M. u. cardonensis*.
*"sulfuriventer"*: Sinaoa to Aguascalientes; considered part of nominate race.
*"tiburonensis"*: Tiburón I., Sonora; considered part of nominate race.

**HYBRIDIZATION:** Hybridizes infrequently with Golden-fronted Woodpecker at contact zones in s. Zacatecas, Aguascalientes, and e. Jalisco, Mex. One possible hybrid observed near Tucson, AZ, in 2003 may be attributable to individual variation.

## BEHAVIOR

### BREEDING BIOLOGY

#### *Nest Site*

Each year's nest cavity site is selected during previous nesting season. Once site is selected, excavation occurs in fall and winter, allowing several months required for inner cactus pulp to dry before use. Saguaro nests are often excavated entirely within outer cortex (soft tissue, outside woody skeleton).

Favors large, multibranched saguaro cacti in arroyos and on lower alluvial slopes; less often on foothills and open desert valleys. Also nests in mesquite, fan palm, and exotic shade trees, less often in sycamore and ash, rarely in oak and palo verde. In riparian habitats prefers cottonwood and willow.

Cavities are typically lower on saguaro than those of Gilded Flicker; wide variation in entrance diameter, often not circular. Northward-facing cavity entrances moderate interior temperatures by avoiding direct sun. May use individual nest sites for repeated years.

#### *Courtship*

Little information available. Pairs remain together year-round. Courtship may ensue mid-Feb., when pairing and territorial behaviors may be more conspicuous than normal.

**OPPOSITE:** An adult male Gila Woodpecker on an ocotillo inflorescence. *Alan Murphy; May 2012, Pima Co., AZ.*

#### *Parenting*

Occasionally triple-brooded; some pairs may use separate nests for each brood, but first and third clutches are occasionally raised in same cavity. Second and third clutches contain fewer eggs than first.

Female may deliver food more often than male, but male spends more time guarding nest while female gathers food, although these roles may reverse spontaneously.

#### *Young Birds*

Earlier broods remain with adults while later broods are raised. Nest productivity may be extremely high in dense riparian areas, away from European Starlings, with three to five fledglings from some nests.

#### *Dispersal*

Local populations may be very dense in midsummer, after all broods have fledged, and young may continue begging for several months.

### NONBREEDING BEHAVIOR

#### *Feeding*

Gila is probably one of the least discriminate feeders among North American woodpeckers. It feeds 90–100 percent on animal foods, from cicadas to bacon rind to young songbirds, and does it all with an attitude.

***Preferred Foods:*** Favored insects include cicadas, ants, beetles, grasshoppers, termites, and butterflies. Also takes earthworms and small lizards. Plant foods include saguaro and other cacti fruit, berries of mistletoe and lyceum. Will ground glean among cornfields and will raid human-stored corn, crop fruits, and cultivated pecans. During nesting season, known to steal and eat eggs and young of some small songbirds. Also reported to eat chicken eggs.

Common at bird feeders, preferring suet and meats; known to repeatedly visit beef bones and bacon rind. Drinks readily from hummingbird feeders and water containers.

Young are fed primarily insects; also saguaro fruits, pollen, and food products of exotic plantings.

***Foraging Behaviors:*** Gleans and probes in trees, cacti, and shrubs, occasionally moving to ground. Favors saguaro fruits and flowers during cooler hours, spending heat of day among inner foliage.

Sexes occasionally practice niche partitioning, with males found on saguaro trunks and main

Optimal Gila Woodpecker habitat in the saguaro cactus "forest" of Arizona's Tucson Mountains. *Stephen Shunk; Aug. 2013, Pima Co., AZ.*

branches and females on periphery of trees and in diseased areas. When feeding young, males peck into substrate for insect larvae, and females search surfaces for adult insects.

Mates feed together on cacti flowers in late spring and cacti fruit in summer. Late summer through fall, males favor larger branches and tree trunks, and both sexes gradually discontinue cactus foraging. In early winter, both forage mostly on mistletoe berries. When breaking into galls, uses cracks in trunk or fencepost as anvil site, wedging gall inside and hammering at it to split it open.

Observed calling loudly in early morning at unstocked feeding stations, including hummingbird feeders, until feeders are filled. Also known to hammer on windows to gain attention at regularly stocked feeding stations, presumably to get feeders filled.

***Food Storage:*** Infrequently caches acorns. Reported collecting fresh acorns from exotic oaks in urban settings, storing them among fibers of palm fronds.

### *Territory Defense and Sociality*

Highly aggressive, males more than females. Performs a variety of displays to defend nesting and feeding territories, including bill pointing, head bobbing, and head shaking, sometimes chasing and attacking with bill. Observed attempting to usurp other Gila nests, unsuccessfully.

Territory sizes range from 11 to 25 ac. (4.5–10 ha) depending on availability of nest sites and food resources. Dominant individuals will push subordinates to less productive feeding areas in winter.

### *Interactions with Other Species*

Gila is one of the most aggressive birds throughout its range; observed antagonizing a wide variety of bird species. At onset of nesting season, aggressively defends 130- to 165-ft.-diameter (40- to 50-m) roughly circular territory from Gilded Flickers and European Starlings, as well as other Gilas and other cavity nesters. Will nest closer to Gilded Flicker than to closest adjacent Gila Woodpecker, rarely nesting in same cactus with the former. Starlings will successfully usurp Gila Woodpecker nests in saguaro cacti; if early enough in season, woodpeckers can excavate and breed in new cavity.

## CONSERVATION

**HABITAT THREATS:** Urbanization of Sonoran Desert habitats may pose future threat. Increasing occurrence of desert wildfires fueled by exotic grasses can destroy large expanses of saguaro habitat.

Recent population declines in Colorado R. valley and Imperial Valley may be tied in part to clearing of woodlands and possibly to nest-site competition with European Starlings, which are likely greatest threat in rural and urban settings; problem is expected to increase in severity as starling populations spread and thrive with urban sprawl.

**POPULATION CHANGES:** Maximum densities recorded in riparian forest, Yuma Co., AZ, of about 1 bird per 2.5 ac. (1 per ha) during breeding season. Peak densities in developed habitats are found in urban areas dominated by native vegetation.

The species is increasing along riparian corridors in AZ, but has declined substantially over the last century from edges of range. Small population in se. CA has fluctuated drastically. First expanded northward from Colorado R. delta, possibly as a result of planting of shade trees, then declined steeply because of clearing of riparian lands for agriculture and other development.

**CONSERVATION STATUS AND MANAGEMENT:** Endangered in CA; Partners in Flight Focal Species for conservation in Colorado desert of se. CA.; NM Threatened and Level 2 for Biodiversity Conservation.

Habitat conservation for this species requires retention and healthy regeneration of large saguaro cacti. Conservation efforts are currently being practiced on public lands in Sonoran Desert habitat. Retaining large saguaros on suburban lots is a key element of success in developed areas. Local municipalities have the opportunity to enact conservation measures prohibiting saguaro removal. Conversion of native habitat to agriculture may continue to threaten local populations.

Given that Gila Woodpecker is the most abundant primary excavator on the desert floor, its decline could negatively affect the entire saguaro natural community, as well as survival of the saguaro itself, since woodpeckers may be key pollinators of the cacti.

## REFERENCES

American Ornithologists' Union 1903, 1976; Anderson 1934; Antevs 1948; Bradley 2005; Brenowitz 1978a, 1978b; Edwards and Schnell 2000; Gardner 1959; Garrett and Dunn 1981; Gilman 1915; Grinnell 1915, 1927; Hickey 2003; Hoffman 1927; Inouye et al. 1981; Kerpez and Smith 1990a, 1990b; Korol and Hutto 1984; Linsdale 1936; MacRoberts and MacRoberts 1985; Martindale 1984; McAuliffe and Hendricks 1988; McCreedy 2006; Peterson 1948; Speich and Radke 1975; Van Rossem 1933, 1942.

# GOLDEN-FRONTED WOODPECKER

***Melanerpes aurifrons***

L: 8.5–10 in. (21.6–25.4 cm)
WS: 16–18 in. (40.6–45.7 cm)
WT: 2.2–3.5 oz. (63–100 g)

**One of the "zebra-backed" species** in the *Melanerpes* genus, the Golden-fronted Woodpecker is a common permanent resident from southwestern Oklahoma through Texas and another 1,000 miles (1,600 km) into Central America. It is one of the most variable of our 23 woodpecker species, with as many as 14 subspecies recognized, although nearly all of this variability occurs south of the Rio Grande. Up to 10 of the southernmost subspecies, occupying tropical rainforests from southern Mexico to northern Nicaragua, were described in 2009 as belonging to a possible new species, the Velasquez's Woodpecker, *M. santacruzi.*

An adult female Golden-fronted Woodpecker returning to a cavity. *Cissy Beasley; Apr. 2012, Bexar Co., TX.*

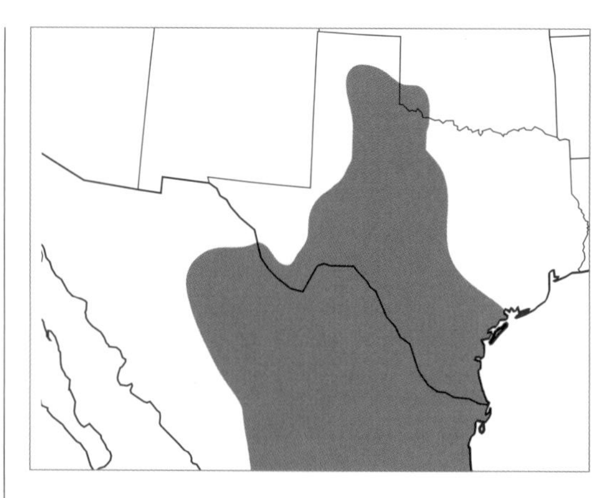

In the southern United States, the Golden-fronted Woodpecker has undergone a notable northward range expansion over about the last half-century, only first being recorded in Oklahoma in 1954 and first confirmed nesting there in 1958. It has since become established in the southwestern corner of the state, where it meets and rarely hybridizes with the closely related Red-bellied Woodpecker; the two also hybridize in central Texas.

Earlier in the twentieth century, this abundant resident of the lower Rio Grande valley inspired the lifting of protective regulations for woodpeckers because of its purported habit of excavating in telephone and telegraph poles. However, in the 1960s ornithologist John Dennis proved that the Ladder-backed Woodpecker initiates the majority of such carpentry, and that Golden-fronteds simply enlarge existing holes for nesting and roosting.

## DISTRIBUTION

**PERMANENT RANGE:** Sw. OK through cen. and s. TX, increasing in abundance from north to south. Often sporadically distributed and localized throughout. Generally sedentary, with some limited seasonal wandering, usually in winter. Also n. Mex. through Belize and into nw. Nicaragua.

## HABITAT

This species is amazingly adaptable, as evidenced in part by the wide diversity of habitats it occupies throughout its range: open woodlands, second-growth forests, and scrublands from sea level up to 8,200 ft. (2,500 m). Favors drier habitats, especially with mesquite, but also occupies riparian corridors, oak-juniper savannas, parks, and suburbs.

In w. TX, favors scrublands and narrow riparian corridors, scrublands dominated by mesquite

**OPPOSITE:** An adult male Golden-fronted Woodpecker. *Doug Backlund; Feb. 2011, Hidalgo Co., TX.*

A fledgling Golden-fronted Woodpecker practicing with its extensible tongue. Note the juvenal plumage trait of dusky streaking in the flanks. *Cissy Beasley; Aug. 2012, Bee Co., TX.*

and associated flora, and riparian woodlands partly dominated by hackberry, pecan, black willow, and bur oak. Confined to riparian cottonwoods along Rio Grande in Big Bend region; downstream on Rio Grande, drawn to dead junipers and other trees around Falcon Reservoir. Favors isolated patches of native palms and associated vegetation conserved in lower Rio Grande valley, but also widespread in suburban landscapes around Corpus Christi and in s. TX.

## DETECTION

Hard to miss in core of its range, especially Rio Grande valley and Brush Country. Often loud when vocalizing, and prominently affixed to urban (and rural) utility poles.

**VOCAL SOUNDS AND BEHAVIOR:** Conspicuously vocal all year, with all calls given by both sexes. Has three basic calls, similar to those of other "zebra-backed" *Melanerpes*: churr, check, and rattle.

Churring is most often heard Feb.–Oct. A repetitive, harsher version likely serves to advertise territory and attract a mate; a softer, more subtle version is given in courtship.

Checking note is given as a warning call, sounding like *chuk, chuk, chuk* or *rek, rek, rek*, repeated on fairly steady cadence but varying slightly in pitch, like a slower, raspier, and less-resonant version of Pileated Woodpecker.

Metallic rattle, *wahhh* or *werhhh*, is similar to that of Red-bellied Woodpecker.

**NONVOCAL SOUNDS:** Drum is a series of rolls followed or preceded by lighter, slower tapping. Used for both territoriality and courtship, most often Feb.–Aug. Roll often tapers (less depth to beats) smoothly toward end. Females rarely drum.

Golden-fronted is not a particularly strong excavator, so feeding sounds are rarely heard; may be heard expanding the opening or inside of a previously excavated cavity.

## VISUAL IDENTIFICATION

**ADULT:** Conspicuous throughout its range, with male showing the most striking plumage among all "zebra-backed" woodpeckers in N. Am. Male has long and variable orangish patch from mantle to rear crown, with bright red crown patch above gray forehead and lemon yellow nasal tufts. Female has yellower nape patch, lacks red crown.

In flight, white wing patch contrasts with darker outerwing. Bold white lower back, rump, and uppertail coverts contrast strongly with nearly all-black tail.

**JUVENILE:** Similar to adult female but duller; crown and breast with fine dusky streaks, nasal tufts and nape lacking distinct color, and barring on back indistinct. Males usually have small patch of red on crown, females very little to no red. Yellow on female nape is fainter than on adult female and weakly demarcated from gray parts. Brown iris turns reddish by first spring.

**INDIVIDUAL VARIATION:** Males are up to 14 percent heavier than females and their bill 9–15 percent longer. Females occasionally show several red or orange-yellow feathers in crown. Albinism and leucism are rarely reported. Reported with purple-stained faces from eating fruit of prickly pear cactus.

**PLUMAGES AND MOLTS:** Molt patterns are similar to those of other woodpeckers. Prebasic molt occurs Jul.–Nov., peaking Sept.–Oct.

**SIMILAR SPECIES:** Closely related to and hybridizes with Red-bellied Woodpecker (see Geographic Variation). Golden-fronted shows far more variability in head color than Red-bellied, and stark contrast between black tail and white rump. Calls of Red-bellied are generally higher pitched, softer, and less raspy than those of Golden-fronted.

Barring on back and contrast between white rump and black tail may be mistaken for Northern Flicker in flight; usually only a problem in winter, when flickers are widespread in cen. TX.

**GEOGRAPHIC VARIATION:** One of the most variable of all woodpeckers. Separation of up to 14 subspecies in 4 subspecies groups, and the complex intergrades among them, is most challenging from cen. Mex. south into Nicaragua. Among all populations, body measurements are greatest in TX and n. Mex.

Some authorities describe the southernmost form of *M. aurifrons* as "Velasquez's Woodpecker (*Melanerpes santacruzi*)," but AOU rejected such a proposal in 2013; as of 2015, AOU had no additional proposals under consideration.

### SUBSPECIES

**NORTH AMERICA:** Twelve to 14 subspecies recognized, with only 1, *M. a. aurifrons*, universally accepted as occurring in N. Am. Subspecies are categorized into four subspecies groups, with only the *aurifrons* group occurring north of Mex.

Debated *M. a. incanescens*, described as lighter overall than *aurifrons*, with white bars on back wider than black ones, and pure white rump; said to occur from Big Bend region northeasterly into n.-cen. TX and north of San Antonio. Variation generally thought to be clinal rather than geographically distinct.

**NEOTROPICS:**

***Dubius* group**

*M. a. veraecrucis: s. Veracruz to ne. Guatemala*
*M. a. dubius: Yucatan to ne. Guatemala*
*M. a. leei: Cozumel I., Mex.*
*M. a. turneffensis: Turneffe I., Belize*
*M. a. canescens: Roatan and Barbaretta Is., Honduras*

***Santacruzi* group**

*M. a. grateloupensis: s. Tamaulipas to cen. Veracruz*
*M. a. santacruzi: se. Chiapas to interior Honduras*
*M. a. hughlandi: cen. Guatemala*
*M. a. pauper: lowlands, n. Honduras*
*M. a. insulanus: Utila I., Honduras*

***Polygrammus* group**

*M. a. polygrammus: Pacific slope, Oaxaca to Chiapas, Mex.*
*"frontalis": interior Chiapas*

**HYBRIDIZATION:** Range overlaps with that of Red-bellied Woodpecker roughly from sw. OK to cen. TX coast. Two primary hybrid zones: upper Red R. at TX–OK border and s.-cen. TX on eastern edge of Edwards Plateau. Hybridization first confirmed genetically in 1987, probably evidence of very recent range expansion of one or both species. More hybrid males than females, with plumage characters broadly intermediate in hybrid individuals. Beware that some Red-bellieds may show yellow to yellow-orange nape patch and may have reduced white in central tail feathers.

In contact zones, Red-bellied and Golden-fronted Woodpeckers may occupy complementary microhabitats, the former preferring moister mixed woodlands, the latter arid mesquite brushland.

Golden-fronted also hybridizes with Gila Woodpecker in nw. Mex. and with Hoffmann's Woodpecker (*M. hoffmannii*) in s. Honduras. Reported Gila x Golden-fronted hybrids from N. Am. may be variant Gilas. One purported Golden-fronted spent nearly a year in Pensacola, FL, siring two young with a female Red-bellied.

## BEHAVIOR

### BREEDING BIOLOGY

Information on breeding biology is limited, partly because of species' relatively small population size and restricted range in N. Am.

#### *Nest Site*

Excavates in limbs or trunks of live or dead trees, including mesquite, pecan, oak, hackberry, and cottonwood. Also uses utility poles and fenceposts, and very rarely nest boxes. Extreme variation in cavity height.

Pairs occasionally use male's roost cavity for nest

An adult male Golden-fronted Woodpecker exiting a cavity to dispose of excavated wood chips. *Cissy Beasley; Apr. 2012, Bexar Co., TX.*

site; typically reuse old cavities from prior years and same cavity for multiple broods in a single season. Nest cleaned of waste and debris between broods, and supplemental excavation may result in increased depth over season.

Golden-fronted has responded to widespread loss of native habitat by nesting in introduced palm trees and other tree species of suburban landscapes.

### *Courtship*

Monogamous; pairs are often observed foraging close together year-round. Pairs may form at any time of year, with pair bonding and renewal observed from late in breeding season through winter.

Territorial and courtship displays are similar. Both sexes perform bill pointing as greeting toward mate, with neck and bill extended below breast, male flashing red crown. Pairs also engage in synchronized tapping at potential nest site.

Copulation sequence is similar to that of some other *Melanerpes*; female may first mount male, then trade places. Bill pointing and mutual head swinging may precede reverse or normal mounting.

### *Parenting and Young Birds*

During nest exchange, incubating or brooding parent waits inside cavity, peering out of entrance and calling as mate approaches. Hatching is asynchronous, occasionally over 2 days. Parents dispose of eggshells within first few days. Young can feed alone within first week after fledging but generally remain dependent on adults for at least the first few weeks. Pairs occasionally double-brood; typically fledge one to four young per brood.

### *Dispersal*

Family groups often stay together into fall; some young are independent by early summer. Juveniles are rarely observed defending roost territories during first summer.

Size of home range fluctuates year to year. Golden-fronted exhibits unusual seasonal range dichotomy among woodpeckers: summer ranges may average 50 ac. (20.5 ha) per pair, winter ranges about half that.

## NONBREEDING BEHAVIOR

### *Feeding*

***Preferred Foods:*** Golden-fronted is omnivorous and opportunistic. Almost half of diet is vegetable matter; includes pecans, acorns, and corn in fall and winter and mesquite beans, fruits of persimmon and cactus, and wide variety of berries in spring and summer. Known to visit suet feeders; also observed eating bananas, oranges, and other fruit at feeders.

Favored animal prey includes wide variety of adult and larval arthropods, especially grasshoppers (up to one-quarter of total food volume), spiders, ants, cicadas, and various beetles. Also takes small vertebrates, especially lizards. Rarely observed depredating passerine nests.

***Foraging Behaviors:*** Almost never excavates for wood-boring insects; instead, gleans from surfaces of trunks, cracks in poles or trees, and old cavities. Also probes into cracks and crevices and occasionally flycatches. During nesting season, mates exhibit similar foraging behaviors. In late summer and fall, however, males prefer to peck at bark surfaces, whereas females begin to forage more on ground. By late winter, foraging behaviors reverse.

When foraging on tree trunks and limbs, usually found below 20 ft. (6 m); clings to small branches when feeding on fruits and nuts. Commonly feeds in open, grassy areas or on bare ground; rarely forages beneath brush. Known to join mixed-species flocks (occasionally including flickers) while ground foraging.

***Food Storage:*** Rarely stores fruits and nuts in natural or excavated crevices or holes, or in artificial cavities such as open-ended pipes. Observed hoarding corn and driving away other birds from game feeders.

### *Territory Defense and Sociality*

Energetically defends breeding territory. Both sexes will engage intruder of either sex. Interactions typically end with intruder escaping, often being chased to edge of residents' territory. Acrobatic aerial chase may culminate in midair contact.

Most displays are given by both sexes; include head swinging in downward arc, head bobbing with bill pointed at intruder, and full threat display with wings stretched out and up and nape and crown feathers raised. Calling and drumming may accompany any displays.

After nesting is completed, territorial vigor declines, sometimes to nil by winter, when boundaries may be nonexistent depending on food distribution; may feed among other Golden-fronteds in nut orchards during this time.

### *Interactions with Other Species*

In TX, Golden-fronted and Red-bellied Woodpeckers maintain mutually exclusive breeding territories. In limited range of winter overlap, Red-headed Woodpecker dominates Golden-fronted.

May act aggressively toward other species while foraging, performing any of displays described above; other times, may forage in close proximity. May compete with Eastern Bluebird and House Sparrow.

An adult male Golden-fronted Woodpecker bathing in an artificial water feature. *Cissy Beasley; May 2013, Bee Co., TX.*

**MISCELLANEOUS BEHAVIORS**

Drinks from just about any water source, including dripping faucets and fountains. Frequently hops on ground and along tree limbs. Sunbathes from tall perches. Bathes in puddles and drinking fountains. Mud-bathing, during which breast and wings are lowered into mud and bird hops forward, is rarely observed.

## CONSERVATION

**HABITAT THREATS:** Intensive cultivation and range management may involve clearing mesquite chaparral, thereby eliminating prime nesting and foraging habitats and causing wholesale displacement of local populations. However, northward expansion of mesquite woodlands due to fire suppression may have facilitated species' range expansion.

**POPULATION CHANGES:** Notable population declines in TX in early 20th century were due to shooting and agricultural development. Since about 1950, populations have rebounded, possibly due to return of mesquite scrublands in unsuccessfully managed rangeland. Species expanded northward and westward during 20th century, especially into sw. OK.

**CONSERVATION STATUS AND MANAGEMENT:** Considered a Species of Greatest Conservation Need in mixed-grass prairie region of OK because of small population size and unknown trend; conservation efforts are focused primarily on stream and associated riparian forests. Species of Concern in TX.

**OPPOSITE:** An adult male Golden-fronted Woodpecker in aggressive and vocal display posture. *Steven Holt; June 2006, Starr Co., TX.*

**REFERENCES**

American Ornithologists' Union 1903, 1976; Attwater 1892; Clements 2007; Dennis 1964, 1967; Eubanks 2005; Garcia-Trejo et al. 2009; Gehlbach 1993; Gill and Donsker 2015; Gorman 2014; Greenway 1978; Hickey 2003; Husak 2000; Husak and Husak 2002, 2005; Husak and Maxwell 1998, 2000; Lockwood 2001; Martin and Kroll 1975; Norton 1896; Oklahoma Department of Wildlife Conservation 2005; Ragsdale 1890; Rupert and Brush 2006; Selander and Giller 1959, 1963; Sennett 1878; Smith 1912; Smith 1987; Styrsky and Styrsky 2003; Wauer 1973; Wauer and Elwonger 1998; Wetmore 1948.

## RED-BELLIED WOODPECKER

***Melanerpes carolinus***

L: 9–10.5 in. (22.9–26.7 cm)
WS: 15–18 in. (38.1–45.7 cm)
WT: 2.0–3.2 oz. (56–91 g)

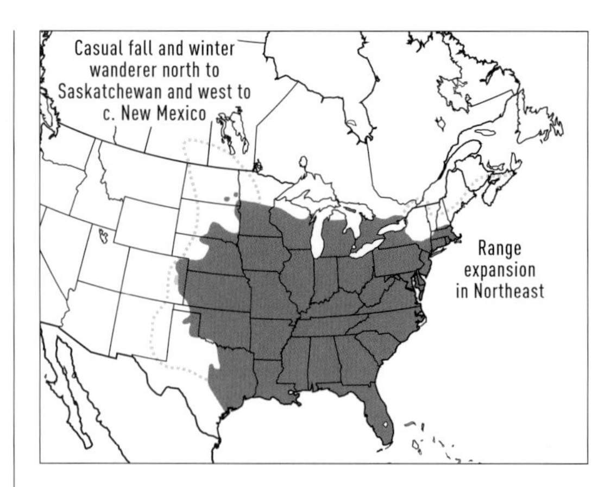


**The chirping and whirring** of the Red-bellied Woodpecker fill mixed woodlands and suburban neighborhoods across the eastern United States. As one of the most prominent of the continent's Picinae, the Red-bellied has entered the daily lives of many observers, giving us an excellent record of this woodpecker's dynamic behavioral repertoire.

An adult Red-bellied was once seen feeding a recently fledged Tufted Titmouse, and another carried up to seven sunflower seeds away from a feeding station lined up in its bill "puffin style." Predatory behaviors include the theft and consumption of nestling songbirds. Red-bellieds have also been observed stealing and eating the eggs of Eastern Bluebird, Indigo Bunting, and Acadian Flycatcher. Courtship behaviors include mutual tapping between mates, with both birds tapping on opposite sides of the outer wall of the nest cavity—one mate inside the cavity and the other on the outside.

An adult male Red-bellied Woodpecker calls to its mate from inside a nest cavity. Note the backward-facing barbs on the tongue tip. *Scott Carpenter; Mar. 2010, Montgomery Co., TX.*

Extremely adaptable and opportunistic, the Red-bellied Woodpecker has undergone a remarkable range expansion in the early twenty-first century. Perhaps the greatest example of its adaptability is the fact that it was the only woodpecker species successfully bred in captivity by twentieth-century virologist, naturalist, and woodpecker behaviorist Lawrence Kilham.

### DISTRIBUTION

**PERMANENT RANGE:** One of the most conspicuous and abundant woodpeckers in e. N. Am. Permanent resident throughout range, east of Rockies and south of boreal forest; rare and irregular at northernmost limits, increasing steeply in abundance southward.

**SEASONAL MOVEMENTS:** Not truly migratory, but a regular postbreeding wanderer and pioneer, often exploring beyond current limits in fall and winter and occasionally expanding breeding range in successive years. Great Lakes birds and other northern populations may retreat southward during harsh winters.

### HABITAT

This species is hugely adaptable and opportunistic. Found in dry or wet sites with relatively mature hardwoods or mixed pine-hardwoods. Also oak-hickory and pine-oak forests, heavily timbered bottomlands and floodplains, wooded swamps, ri-

**OPPOSITE:** An adult female Red-bellied Woodpecker prepares to lodge an acorn into an anvil site for easier feeding. *Marie Read; Oct. 2012, Polk Co., FL.*

An adult male Red-bellied Woodpecker. *Gene McGarry; Sept. 2010, Woodstock, NY.*

parian forests, wooded suburban areas with mature trees and snags, and longleaf pine savannah. May occupy multistory woodlands with denser ground vegetation than other woodpeckers. Typically occurs below about 2,000 ft. (600 m), but up to about 3,000 ft. (900 m) in Appalachians, and in fall up to 6,000 ft. (1,830 m) in CO.

## DETECTION

One of the most easily detected woodpeckers in N. Am.

**VOCAL SOUNDS AND BEHAVIOR:** Extremely vocal year-round, throughout the day, perched or in flight. Gives two primary calls, a rattle and chirp note, as well as a territorial call. Rattle resembles *kwirrr* or *whirr*. Slightly scratchy chirp note is usually repeated, with second and later repetitions at slightly lower pitch than first; resembles *chew, chew, chew* or *cher, cher, cher*; may sound like some squirrel calls. Also known to give what has been described as a *chee-wuck* call during territorial bouts and possibly courtship.

**NONVOCAL SOUNDS:** Drum is a loud steady roll with fairly even cadence and pitch throughout. Both sexes drum, female very infrequently. Drum generally is not separable in the field from that of sympatric Red-headed Woodpecker. Tapping and drumming are performed against resonant surfaces, including utility pole transformers, metal flashing, streetlamps, and other artificial surfaces, often in association with rattle or chirp note.

## VISUAL IDENTIFICATION

**ADULT:** Typical "zebra-backed" *Melanerpes*, with grayish body parts. Pinkish or reddish belly patch with diffuse yellowish edges is difficult to see at any angle. Male shows wide swath of red from upper mantle through forehead; female has red restricted to nape. In flight, faint white wing patch contrasts against dark outerwing. Mostly white rump and central tail contrast against grayish wing.

**JUVENILE:** Duller overall than adult. Light parts of head, throat, and chest dusky, lower nape and nasal tufts pale yellowish. Few red feathers on male crown (fewest in female), typically not visible in the field. Barring on back appears mottled gray; fine dusky streaking on breast; belly faint yellow-orange, brighter in male than in female.

**INDIVIDUAL VARIATION:** Sexes similar in size. Male's longer bill and larger tongue tip may be an adaptation for complementary foraging niches. Various color aberrations include orange belly, pink tinge below throat, orange or yellowish nape, and yellow iris. Female rarely shows small amount of black at forward edge of red nape, a few reddish-tipped (rarely all-red) feathers in crown, or orangey or pinkish nasal tufts.

**PLUMAGES AND MOLTS:** Juvenal plumage retained through Sept. or Oct. Adult fall molt occurs July–Oct.

**SIMILAR SPECIES:** Red-cockaded Woodpecker also shows pale face and "zebra" back but is far more contrasting black and white overall. Northern Flicker has similar back pattern but different colors on back, and underparts are heavily marked.

Red-bellied is closely related to other "zebra-backed" *Melanerpes*, including Golden-fronted Woodpecker, with which it hybridizes in limited contact zone (see below). Yellowish-plumaged Red-bellieds may be mistaken for Golden-fronteds; in questionable individuals, uppertail pattern (and nape color in females) is best distinction.

**GEOGRAPHIC VARIATION:** Morphological variation may be attributed to subtle geographic tendencies or broad individual variation within regional populations. Northeastern birds tend to have longer bill; FL birds have slightly shorter wings; western birds may show broader white barring on back and deeper red in belly than eastern birds.

### SUBSPECIES

Four subspecies recognized by AOU from at least 1945 to 1957, but later studies and assessment by Short (1982) asserted all variation to be individual and/or geographically clinal. Generally smaller s. FL population, once designated "*perplexus*," may show greatest degree of character divergence of the four previously recognized forms, averaging lighter back, tail, and nasal tufts than typical *M. carolinus*.

**HYBRIDIZATION:** Interbreeding with Golden-fronted Woodpecker first genetically confirmed in 1987; ranges meet from OK to cen. TX coast. See Golden-fronted Woodpecker, Hybridization, page 81.

One female Red-bellied fledged 2 young with a purported vagrant male Golden-fronted Woodpecker in Pensacola, FL.

## BEHAVIOR

### BREEDING BIOLOGY

#### *Nest Site*

Forested habitats with dense ground vegetation. In most areas, excavates into underside and near top of leaning snag or dead limb of live tree; commonly uses fenceposts and utility poles, especially in prairies and grasslands. Selects wide variety of tree species. Often returns to same stub or limb used in previous years, but usually excavates fresh cavity.

A fledgling male Red-bellied Woodpecker vocalizing. *Paul Bannick; Oct. 2012, Houston, TX.*

Cavities frequently have bark near entrance. Male's roost cavity is typically selected as nest site.

### *Courtship*

Seasonally monogamous, bonding for 7 months or more. No confirmed cooperative breeding, but single report of two females and one male tending two nests.

Male may begin excavation, while calling, drumming, and tapping adjacent to incomplete cavity to attract female. Female then flies in beside male and the two tap synchronously. Female assists in finishing excavation. If no female responds, male may move elsewhere and repeat ritual.

Male may call prior to copulation, after which female climbs on his back in reverse mounting. They then switch places, male sliding down female's left side, culminating in cloacal contact. Copulation is most frequent just before egg laying, but may occur more than 2 months before nesting. Nest exchanges are often accompanied by chattering or tapping.

### *Parenting, Young Birds, and Dispersal*

Foods fed to older young include wide range of arthropods and fruits. Adults often pulverize insects before feeding to hatchlings. Young often skulk near nest for first 2 days after fledging, then split up between parents. Adults continue to mash large arthropods before feeding them to fledglings, and may bring fledglings to artificial feeding stations. Following a 2- to 10-week dependency period, adults, mostly male, chase young from territory. Very rare for nestlings to remain within natal site.

## NONBREEDING BEHAVIOR

### *Feeding*

***Preferred Foods:*** Highly flexible and opportunistic feeder; eats a wide range of fruits, mast, seeds, arboreal arthropods, and other invertebrates. Averaged over range and among all seasons, diet is about 69 percent vegetable and 31 percent animal by volume. Vegetable matter is strongly preferred in fall and winter, with mostly animal food in breeding season.

Vegetable matter is mostly nut mast and fruits; animal matter nearly one-third weevils and ground beetles, followed by ants and other arthropods. Known to take small vertebrate prey or young, including herptiles and small fish; documented eating eggs and nestlings of various small passerines and other woodpeckers.

Observed feeding at sapsucker wells and eating flower nectar or rotting fruits. Eats from wide variety of bird feeders.

***Foraging Behaviors:*** On average, feeds most often by gleaning, followed by fruit and nut collection and surface probing, only rarely excavating, bark scaling, ground foraging, or flycatching. May hang underneath or hover-glean at clusters of small fruits, berries, or seeds, also hanging directly from large fruits. Prefers unfallen fruits and nuts, but occasionally forages on ground.

Swallows most arthropods whole; pecks small vertebrates with bill or slams them against tree, then tears into pieces before swallowing. Uses crevice in tree or post as anvil to remove heavy nut shells and break mast into small pieces. Some observations of birds catching dropped food by pressing body against tree and cupping wings, like a baseball catcher blocking a wild pitch.

Males may forage more on trunks and females more on limbs, females higher on trees and in taller trees than males, but there is much overlap. Males may peck, pound, and gather mast and fruits more than females, females gleaning and probing more than males.

***Food Storage:*** Observed storing nuts, acorns, corn, grapes, seeds, berries, and insects. One individual reported storing cow dung. Observed carrying several sunflower seeds at once, lined up perpendicular to bill, similar to the way puffins carry fish.

May store food year-round, but mostly in fall. Stores small items whole, breaks larger objects into pieces. Typically uses natural crevices and cavities rather than excavating. May be able to remember locations of stored items. Does not consistently defend food stores.

### *Territory Defense and Sociality*

Red-bellieds call, drum, display, or chase in progressive intensity to expel intruders, rarely ending in aggressive physical contact. Males, and occasionally females, raise crown and nape feathers to flash red colors at aggressor; may also raise feathers of mantle and arch back. Some interactions lead to full display, with tail spread and wings stretched and raised. Flight display involves slow, floating descent to high perch, with wings in same position as standing display. Will press spread wings against trunk when defending nest cavity.

Males exhibit strong site fidelity all year and throughout lifetime. After pair bonds dissolve temporarily in early fall, mates may chase each other to defend individual feeding territories. On rare occasions, wintering birds may roost or forage together, but birds are generally solitary or paired loosely with mate outside breeding season.

**OPPOSITE:** An adult male Red-bellied Woodpecker with an extensive amount of red in its chest and belly. *Jan Wegener; Mar. 2010, Sarasota Co., FL.*

A pair of adult Red-bellied Woodpeckers in mutual courtship/pair-bonding display. *Marie Read; Mar. 2012, Polk Co., FL.*

Aggressively defends radius of about 30 ft. (9 m) around nest tree from competing individuals; proportionately less defensive with distance from nest. Maintains general territory of 4–40 ac. (1.6–16 ha), only half of which is typically defended, with some sectors rarely visited by resident pair. Edges of territory are often defined by natural features such as waterways or ecotones, but there are no abrupt edges within contiguous habitat.

### *Interactions with Other Species*

Often aggressively usurps nest and roost cavities from Red-cockaded Woodpecker; may also evict Downy Woodpecker. Observed stealing and carrying away, but not eating, eggs from House Sparrow nest in mutual nest tree.

Frequently documented losing its own cavity to European Starling, causing delayed breeding and limiting pair to one brood for that season. In some regions, starlings will usurp up to 50 percent of local Red-bellied cavities, and may kill adult woodpecker in the process. Red-bellied is known to carry its own eggs and young from nest, possibly in response to starling encroachment and eventual usurpation. Also successfully usurped by other woodpeckers and flying squirrels. However, when multiple cavities are available, will also nest in same tree as other cavity nesters.

Female Red-bellied almost never interacts with Red-headed Woodpecker, but splits starling conflicts with male about 50-50. Where ranges overlap, both sexes aggressively defend territory from Golden-fronted Woodpecker, and vice versa. Red-bellied is aggressively attacked by passerines when approaching their nests. Often dominant at backyard feeding stations.

#### MISCELLANEOUS BEHAVIORS

Observed bathing in dust and water, as well as anting. Sunbathes from any high perch with crown erect, often fanning tail or wings. May alternately preen, stretch, and call while sunning, and observed in a sort of warm-weather torpor while sunbathing. Individuals often alternate among different nightly roost cavities.

### CONSERVATION

**HABITAT THREATS:** Extreme adaptability and opportunism allow this species to exploit broad diversity of habitats, making it one of our woodpeckers that is least vulnerable to habitat changes. As with other woodpeckers, loss of cavity trees could affect local populations.

**POPULATION CHANGES:** Rapid northward expansion continues in early 21st century. Northeasterly range increase is attributed to maturing forests and cultivation of exotic flora in developed areas. Expansion into Great Plains follows wooded riparian bottoms with maturation of planted trees. Major 2004–2005 winter incursion into ne. U.S. and Maritime Canada (over 230 birds recorded in ME, from every county except one) resulted in immediate expansion of breeding range into previously unrecorded sections of ME; first breeding records from NS in 2005 and ME in 2006. Upper Midwest birds may be expanding up northern rivers into SK, and post-breeding wanderers from CO may be drifting into w. NM.

**CONSERVATION STATUS AND MANAGEMENT:** Representative nesting species, and therefore important to monitor, in coastal plain of e. Gulf of Mexico.

Opportunistic adaptability likely prevents any long-term population declines. However, typically requires snags and mast-producing trees, and wholesale clearing of either would affect local populations. Expansion of European Starling also may have significant impact on local breeding productivity.

### REFERENCES

Abbott 1929; Abbott 1937; American Ornithologists' Union 1903, 1976; Brackbill 1969a; Brewster 1889; Chamberlain 1884; Conner 1974; Cronenweth, S., pers. comm.; Curry 1969; Fisher 1903; Hauser 1957; Hazler et al. 2004; Hickman 1970; Kilham 1958a, 1961b, 1963; Lawrence 1896; Lewis 1890; McGuire 1932; McNair 1999; Miller 1892; Mueller 1971; Neill and Harper 1990; Peck 1890; Petersen 2007; Poirier 2008; Saul 1983; Shackleford 2000; Shackelford et al. 2000; Smith 1987; Smith 1986; Stickel 1962, 1963, 1964, 1965; Watt 1980; Whitney 1917; Wilkens and Ritchison 1999; Zeranski and Baptist 1990.

# WILLIAMSON'S SAPSUCKER
## *Sphyrapicus thyroideus*

L: 8.5–9 in. (21.6–22.9 cm)
WS: 17 in. (43.2 cm)
WT: 1.6–2.3 oz. (44–64 g)

**Breeding across montane forests** of western North America, Williamson's Sapsucker is among the most sexually dichromatic woodpeckers in the world. In fact, the Williamson's male and female are so different in plumage that they were once thought to be two distinct species.

Williamson's is the most conifer-dependent of the sapsuckers, feeding primarily among mixed coniferous forests throughout its range. It will also nest in conifers, but it prefers to excavate nest cavities in mature aspens softened by heart-rot fungi or in aspen snags. Outside the breeding season, Williamson's can be found tapping juniper, pine, or exotic conifers and sometimes feeding at high elevations among the twisted pines at timberline.

Two subspecies of Williamson's Sapsucker can be distinguished only by their bill sizes—and barely at that. Western populations, primarily of the Sierra Nevada and Cascade Mountains, have a slightly larger bill than the Rocky Mountain and Great Basin birds, and they are less migratory than their more easterly relatives.

Williamson's stands apart genetically from the other three sapsuckers—which comprise the *varius* superspecies—and DNA testing has shown it to be the most ancestral of the *Sphyrapicus* genus. Further DNA analyses may one day show sufficient separation within the *thyroideus* + *varius* ancestry to warrant a new genus for Williamson's Sapsucker.

### DISTRIBUTION

Fairly evenly distributed across montane interior West, but generally restricted to elevations above 3,000 ft. (914 m).

**BREEDING RANGE:** Rocky Mts., south to Mogollon Rim. Cascade Mts., from s.-cen. BC south along eastern slope to CA; south through cen. Sierra Nevada. Isolated mountain populations on fringes of primary breeding range.

**SEASONAL MOVEMENTS:** Nature of migration is somewhat mysterious in this species; some populations are known to be resident, others to migrate. Observed foraging in fall with small groups of Red-breasted Sapsuckers, but little is known about

**OPPOSITE:** An adult male Williamson's Sapsucker with a billful of ants for its nestlings. *Paul Bannick; Yakima Co., WA.*

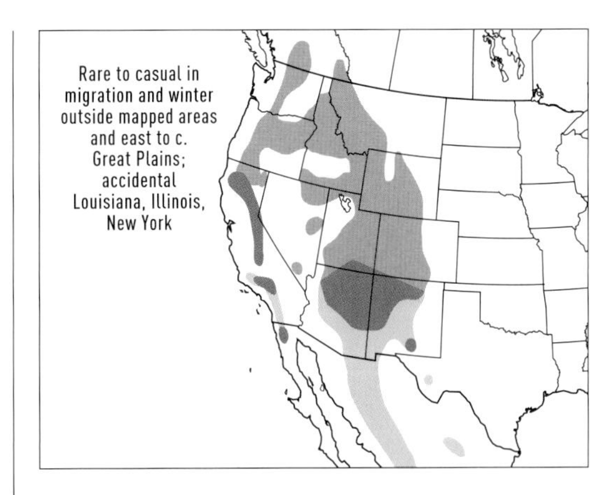


origin of birds observed outside nesting regions. Throughout range, birds engage in elevational migration, with downslope postbreeding dispersal evident in many areas. Southernmost winter records are typically females, implying sexual variance in distance covered.

**WINTER RANGE:** Lower-elevation and southern populations are generally nonmigratory. Far western migratory populations move as far south as s. CA and n. Baja CA. Rockies and Great Basin populations likely move into mountains of e. AZ and w. NM, south through s.-cen. and e. Mex. Rarely recorded in winter in northern parts of breeding range, but snowbound breeding habitats are difficult to survey in winter, hence more birds may be sedentary than is currently known.

### HABITAT

**BREEDING RANGE:** Mixed-conifer forests with various dominant tree species depending on region and elevation, often with an aspen component. Rarely forages above timberline in summer. Regardless of tree species, prefers snags or aspen trees with heart rot for nesting, often in or adjacent to fairly open canopy. Where available, aspen snags are typically preferred over conifer snags. Generally limited to higher elevations, 4,900 to 10,500 ft. (1,500–3,200 m), but as low as 2,800 to 4,265 ft. (850–1,300 m) at higher latitudes.

**MIGRATION AND WINTER RANGE:** Low- to mid-elevation forests with mix of oak, juniper, and pine, occasionally deciduous riparian woodlands. Females typically overwinter in wider diversity of habitats and at lower elevation than males.

### DETECTION

Drum and calls are similar to those of other sapsuckers but fairly easy to separate where ranges overlap. May be more vocal in breeding season than other sapsuckers.

An extremely dark variant adult female Williamson's Sapsucker delivers ants to young nestlings (not visible at the cavity entrance). *Tom Grey; June 2013, Sierra Co., CA.*

**VOCAL SOUNDS AND BEHAVIOR:** Gives three primary calls, with variations of each: churr, squeal, and growl.

Churr is deeper and less shrill than that of other sapsuckers; slightly higher pitched in female than in male. Used mostly by male in territory defense, by both sexes in pair communication. Frequency decreases after pairing and increases just before fledging.

Squeal is a raptorlike call, *raer, raer, raer*, descending in pitch on second syllable. Given at beginning of territorial or courtship interaction; also used as alarm call.

Growl, a hoarse guttural roll, is given in response to intruders near nest or sap tree.

**NONVOCAL SOUNDS:** Drumming, similar to that of other sapsuckers, is a stuttered series of rolls that decrease in number of beats per phrase toward end, often ending with single or double tap. Beats are slower and phrases more evenly spaced than those of other sapsuckers in its range. Mostly given by male, most often in early morning and during pair formation. Females may drum only during encounters with other females, and typically more softly than males. Drumming usually performed more than 100 ft. (30 m) from nest; territory may contain up to four favored drumming sites.

Feeding sounds often heard as rapid bursts in repeated sets, like repetitious pulsing of a sewing machine.

## VISUAL IDENTIFICATION

**ADULT:** Male is striking and unmistakable, with solid black back, white wing patches, and pied facial pattern, with bright red patch on chin and throat. Female superficially resembles a small, dark Gila Woodpecker, and is the only *Sphyrapicus* lacking white wing patches. Female throat and chin may

**OPPOSITE:** A fledgling female Williamson's Sapsucker. Note the very faint creamy wash in the belly and very dull blackish parts. *Alan Murphy; Aug. 2006, Deschutes Co., OR.*

show some red (see Individual Variation). Yellow belly of either sex often difficult to see when pressed against tree trunk.

In flight, male shows nearly solid black wings and back broken by sharply contrasting white wing patches. Female appears mottled dark gray. Both sexes show bold white rump.

**JUVENILE:** Male closely resembles adult male. Black parts duller, and less extensive on chest; throat white instead of red; whitish nape sometimes connects two eyebrow stripes; belly with little to no yellow. Female like adult female, but light parts much warmer brown overall, barring on underparts more extensive, face markings less distinct, and no black on breast.

**INDIVIDUAL VARIATION:** Relative length of female tail exceeds that of male, but difficult to discern in the field. Males are strikingly consistent in plumage, but female underparts vary considerably. Chin of some females may look orange or pale red; breast may show solid black patch or only barring blending with upperparts.

**PLUMAGES AND MOLTS:** Preformative molt begins July–Aug., complete by late Aug. or early Sept. Adult basic plumage is acquired mid-July–late Sept. Flight feathers molt before migration, but body molt continues into migration.

**DISTINCTIVE CHARACTERISTICS:** Among the most sexually dichromatic of all woodpeckers worldwide.

An adult male Williamson's Sapsucker tends both xylem and phloem wells in the same tree. White markings across the back are unusual in this species, and documented hybrids with Williamson's parentage are even rarer than this back pattern. Lacking any other signs of hybridization, this individual should simply be considered a plumage variant. *Kris Kristovich; May 2007, Yellowstone National Park, WY.*

Both male and female plumages are unlike those of the other three *Sphyrapicus* species.

**SIMILAR SPECIES:** Male shares some plumage characters with different members of *Melanerpes*. Female closely resembles "zebra-backed" *Melanerpes*, possibly owing to a genetic link to *Melanerpes* genus. Call notes and drumming patterns are similar to those of other sapsuckers.

**GEOGRAPHIC VARIATION:** Little detectable difference throughout range, especially when observed in the field, as opposed to in the hand.

#### SUBSPECIES

Two barely differentiated subspecies, varying minutely in bill size: western *S. t. thyroideus* with larger bill than eastern *S. t. nataliae*. *S. t. thyroideus* generally breeds in Cascades and Sierra Nevada; less migratory subspecies, wintering at lower elevations throughout range. *S. t. nataliae* occurs throughout Rockies and Great Basin; migrates farther south in winter. Subspecies intergrade in Blue Mts. of ne. OR and se. WA, possibly into Ochoco Mts. of cen. OR, and once in se. BC.

**HYBRIDIZATION:** Despite shared range, only three hybrid Williamson's x Red-naped (*thyroideus* x *nuchalis*) hybrids identified, each collected in winter. Specimens most closely resemble Williamson's.

### BEHAVIOR

#### BREEDING BIOLOGY

*Nest Site*

May prefer drainage bottoms over ridge tops, and sites with mix of live and dead aspen and green ground cover. Where available, strongly prefers to excavate in aspen, about 50–85 percent in snags; regional preference for western larch and other conifers. Regardless of species, selects trees with heart rot more than 50 percent of the time, and those with snapped-off tops 64 percent of the time. Will nest in burned forest with broad matrix of burn severity, often in later years than wood-boring woodpeckers, foraging in adjacent live or lightly burned stands. Prefers 30–60 percent canopy cover.

*Courtship*

Males set up territories upon arrival at breeding grounds. Females arrive 1–2 weeks later, and pair formation follows. Excavation usually begins within 3 weeks of female arrival. Some unmated males excavate multiple cavities and defend them as territories throughout breeding season.

Monogamous, bond lasting only through fledging of young. Rarely, female may reject mate for choosing unsuitable nest site. Some far western populations (those that migrate shorter distances) may remain near nesting territory all year, with some pairs remating in subsequent years.

Copulation is preceded by male performing various displays in response to female solicitation. Flight displays include male fluttering down toward mate from above while chattering, then landing beside her. Pair then engages in series of head-bobbing and body-swaying displays, followed by copulation. Copulation rarely occurs in nest tree but always nearby; lasts about 15 seconds. Female crouches on branch and droops wings; male flutters down, chattering and flipping tail repeatedly, then mounts from left side. Preening often follows.

Adults engage in highly unusual postfledging promiscuity with members of other nesting pairs. Individuals may perform various displays and may excavate new cavities, as well as inspect previously excavated cavities, but no actual nesting has been observed.

*Parenting*

Except for early morning calling and drumming, adults are generally quiet during laying period. Females may limit daily activities to immediate vicinity of nest site. Unmated males may rarely feed neighboring pair's nestlings, to disapproval of natal parents; will also play stepfather if male parent abandons before fledging. Adult male Williamson's observed feeding Red-breasted Sapsucker young, along with natal parents, and more frequently than Red-breasted male.

*Young Birds*

Young begin gleaning insects within 1–2 days of fledging, but generally remain totally dependent on parents for first few days. Adult males often abandon mate and young up to 2 days before fledging. Siblings may disperse short distances as soon as 1 week after fledging, but may remain in home range for up to several weeks.

*Dispersal*

Small mixed flocks of adults and juveniles occur in migration, with no observed conflicts. Generally overwinters singly. Individuals frequently return to previous nesting territory, and often same nest tree (but new cavity), for successive years.

#### NONBREEDING BEHAVIOR

*Feeding*

***Preferred Foods:*** Generally omnivorous, with simple seasonal variety in food preferences: in early breeding period (before young hatch) eats almost

An adult female Williamson's Sapsucker feeds a nestling male in a mature aspen tree. Note the fungal conch above the cavity entrance. *Tom Lawler; June 2014, Deschutes Co., OR.*

exclusively conifer sap and phloem; after young hatch and often through Sept., eats mostly ants; after young become independent, returns to sap and phloem. Winter diet often supplemented by madrone, piñon, and juniper fruits, as well as cultivated and other exotic berries.

Preferred ants include carpenter and wood ants; during breeding season, also eats small amounts of beetles, crane flies, and aphids. Feeds nestlings mostly ants. Ingests small amounts of cambium when drilling and feeding at sap wells.

***Foraging Behaviors:*** Feeds primarily among conifers, with more than 90 percent of year-round foraging on trunks of live trees. Sap-tree species preference depends on regional availability but includes a wide range of conifers; rarely feeds in aspen. Uses individual sap trees multiple years. In some areas, sexes differ in foraging location, with females preferring trunks and males favoring ground and limbs. May collect ants below chosen sap trees.

Gleans insects from bark surface; occasionally probes beneath bark; also removes sections of bark to facilitate greater sap drainage and/or to directly eat underlying layer of phloem fibers. Observed gleaning aphids from underside of maple leaves. Flycatches infrequently.

### *Territory Defense and Sociality*

Mean densities reported from ne. OR of about 19–27 birds per 247 ac. (100 ha), highest in grand fir forest; much lower (approximately 3 birds per 247 ac.) in mature ponderosa pine forest. Defends breeding territory of 10–22 ac. (4–9 ha), ranging up to 100 ac. (40 ha). Male takes on primary role of defining territory, and individuals often return to same territory annually.

Both sexes are aggressive toward other Williamson's. Intruding male and resident male will alternate drumming and calling to each other until intruder leaves on its own or is chased by resident.

Variety of displays performed, ranging from vocal-only to aggressive physical attacks. Resident bird often exhibits combination of bobbing, fluttering, and various flight displays. Following flight display, male may land with wings up and puff out throat and belly to flash red and yellow colors of underparts. In bobbing display, one individual chatters, raises crest, puffs out chest, and points bill up, waving head up and down or side to side. Retreating male may depart with non-undulating mothlike flight. Females known to vocally spar with neighboring females during nesting season.

### *Interactions with Other Species*

Often nests in close proximity to Red-naped or Red-breasted Sapsuckers, with as little as 50 ft. (15 m) separation reported, but interactions may be minimal because of different foraging microhabitats. Observed nesting in same snags as Northern Flicker and American Kestrel, with distances between cavities less than 20 ft. (6 m) in both instances. May interact aggressively with Hairy Woodpecker, also with cavity-nesting passerines such as Pygmy Nuthatch and Violet-green Swallow.

## MISCELLANEOUS BEHAVIORS

Typically approaches active nest by landing above entrance and shuffling downward to deliver food. When foraging, works toward top of tree, then glides down to base of next tree. Birds occasionally roost under conifer boughs.

Stretching and preening observed after copulation and following nest exchange. Unusual anting behavior performed by holding a single ant in bill tip and stroking feathers, changing ants every 2–3 minutes. Bathes in and drinks from water puddles. Sunbathes by facing away from sun with tail spread, wings extended, and crest raised.

Like other sapsuckers, when a Williamson's finds a tree that produces ample sap, it may exploit the food source to the fullest extent. Williamson's favors conifers more than other sapsuckers, and this mature ponderosa pine has likely been tapped for many years. *Stephen Shunk; Mar. 2008, Deschutes Co., OR.*

## CONSERVATION

**HABITAT THREATS:** Requires live conifers; therefore, large-scale stand-replacement fires and large-scale logging in mixed-conifer forests may be greatest threats to habitat.

**POPULATION CHANGES:** Species' preference for montane and often rugged habitats limits ability of standard surveys to adequately sample for distribution.

**CONSERVATION STATUS AND MANAGEMENT:** Listed as Yellow on WatchList and at varying levels of conservation concern in CO, MT, NM, NV, and WA. Greatest threat is currently in BC, where Williamson's was first listed as Sensitive in 1980 because of potential logging threat in favored larch habitat. In 1994, placed on BC's Blue List for species of special concern; *S. t. thyroideus* remains on Blue List, but *S. t. nataliae* was designated as Endangered in 2005 and subsequently upgraded to Red List. In 2013, BC population estimated at 837 breeding adults in three distinct populations, only 39 individuals of which may have represented *nataliae*. High priority for conservation in most breeding states in U.S.

Optimal habitat management would retain large snags and maintain natural fire regime. Because it depends on live conifers for sap production, Williamson's Sapsucker does not benefit from fire like other woodpeckers; notable population declines have been documented in areas of large-scale stand-replacement fire. However, it needs snags for nesting, so mixed-severity fire may foster optimal habitat. Fire also has detrimental effect, though, by hardening snags that may fall before they are soft enough to be used for nesting by this species.

Conserving dense patches of large snags is preferable to conserving individual trees of various sizes and may be most advantageous in drainage bottoms or other low-lying habitats. Conserving large live conifers as sap trees is also important.

## REFERENCES

Bendire 1888; Bezener et al. 2003; Bock and Block 2005a, 2005b; Bock and Larson 1986; Bock and Lynch 1970; Cicero and Johnson 1995; Committee on the Status of Endangered Wildlife in Canada 2005; Cooper 1995; Cowan 1938; Crockett 1975; Dobbs et al. 1997; Environment Canada 2013; Johnson and Zink 1983; Kennedy 1999; Li and Martin 1991; Machmer and Steeger 2004; Michael 1930, 1935; Nielsen-Pincus 2005; Oliver 1970; Pilliod et al. 2006; Raitt 1960; Raphael et al. 1987; Ritter 2000; Russell 1947; Saab et al. 2005; Short 1970; Smith 1890; Stark 1998; Swarth 1917; Trombino 2000; Wilson 1976.

# YELLOW-BELLIED SAPSUCKER

***Sphyrapicus varius***

L: 7.5–9 in. (19.0–22.9 cm)
WS: 16–18 in. (40.6–45.7 cm)
WT: 1.4–2.2 oz. (40–62 g)

**The Yellow-bellied Sapsucker** heads one of the most closely knit woodpecker taxa on the planet—the *varius* superspecies, comprising the Yellow-bellied (*S. varius*), Red-naped (*S. nuchalis*), and Red-breasted (*S. ruber*) Sapsuckers. Modern ornithology still struggles to define this triad, which has undergone numerous "splits" and "lumps" in the last 150 years. Recent genetic research has helped clarify relationships within the *varius* superspecies, but birders will face the challenges of hybridization and field identification of mixed-parent progeny for centuries to come.

In the annals of American folk lingo, "Yellow-bellied Sapsucker" joins the "snipe" as one of the birds believed to be fictitious by many Americans. Calling someone a "yellow-bellied sapsucker" may connote cowardice in some circles, but the derogatory nature of this epithet could have originated from this species' proclivity for damaging valuable timber in the eastern United States. The Yellow-bellied Sapsucker earned its reputation as a nuisance by drilling its sap wells in the trunks of trees, which in turn caused defects in the wood and introduced the trees to various pathogens. In 1911 the U.S. Department of Agriculture estimated that sapsuckers caused $1.2 million a year in economic loss to the timber industry.

## DISTRIBUTION

Yellow-bellied Sapsucker is the most migratory woodpecker in the world, breeding as far north as e. AK and the northern boreal forest, and wintering as far south as Panama and the Caribbean.

**BREEDING RANGE:** E. AK across n. Canada, increasing southward and throughout boreal forest. Generally decreases in abundance southward, often patchy in distribution, east of Rocky Mts. into Upper Midwest. Widespread in Atlantic provinces south through interior New England. Common in Adirondacks; disjunct populations in Appalachians of e. WV and TN–NC border.

**MIGRATION AND WINTER RANGE:** From central plains and Lower Midwest southward, west of Appalachians; Long Island, NY, through Mid-Atlantic, east of Appalachians and south throughout se. U.S. All but extreme w. TX and w. Mex., south to Costa Rican highlands; rare and irregular to Panama and Colombia. Regular in West Indies, mainly on larger

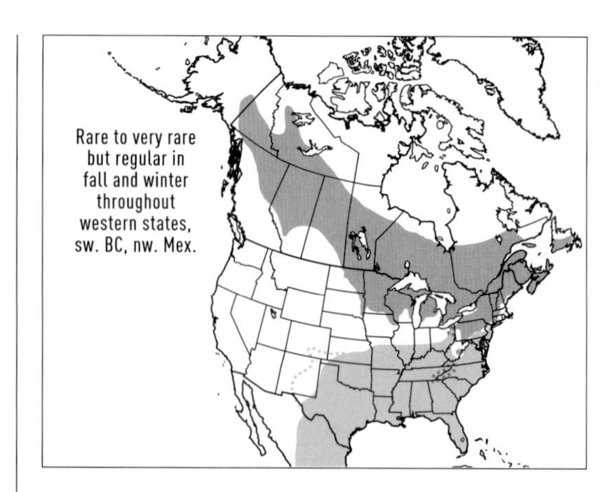


islands, rarely to Netherlands Antilles. Rarely winters north of these limits.

## HABITAT

**BREEDING RANGE:** Medium-aged deciduous woods, mixed-conifer forests, and riparian zones, generally below about 6,500 ft. (2,000 m). Favored tree species include aspen, birch, and maple. Unlike most woodpeckers, Yellow-bellied thrives in cutover conifer forest where maple and birch remain in relatively open stands; also similar forests with succession of young conifers. Also favors swampy habitats at edges of such forests, likely because of concentration of willows used as sap trees.

**MIGRATION AND WINTER RANGE:** Found in habitats with broad mix of tree composition. Inland habitats include forest edges and other relatively open woodlands at wide range of elevations. Locally, favors rolling hills dominated by oaks and bottomland hardwoods, preferring areas rich with hickory. Uses a wide range of developed areas, including orchards, parks, and suburban neighborhoods, as well as wet and dry hammocks and swamps.

In Neotropics, especially Caribbean, often more common in exotic plantings and developed areas than native forests; most common native Caribbean habitat is probably tropical pine-oak forest.

## DETECTION

Fairly vocal in breeding season; typically quiet in winter. Cryptic plumage makes visual location difficult if bird is silent, especially in juvenal plumage.

**VOCAL SOUNDS AND BEHAVIOR:** Gives three basic calls, each one variable depending on context, and

**OPPOSITE:** An adult male Yellow-bellied Sapsucker and its fledged female offspring at a series of fresh phloem wells. *Marie Read; May 2007, Tompkins Co., NY.*

all somewhat similar to those of other sapsuckers in *varius* superspecies: whine, chatter, and squeal.

Whine is the most recognizable call: squeaky, nasal *meehhr* or *meeahh*, descending in pitch and commonly repeated. Has broad contextual use.

Variable dry chatter is given when approaching other individuals: harsh, repeated, and descending *jeea, jeea, jeea* or *yeeh, yeeh, yeeh*, often eliciting a similar but deeper and more raspy *shh-ya, shh-ya, shh-ya*.

Squeal is given mostly by males, often from high perch or drumming post to proclaim territory early in breeding season: *reer, reer, reer, reer* or *rhhea, rhhea, rhhea*, typically audible from some distance.

Variations of these calls are often given in succession during prolonged or intense interactions. Unmated males are presumably responsible for drumming and squeal calls after incubation begins among breeding pairs. Most sounds are heard only in breeding season, though whine can be heard in winter and migration. Some calls are given aerially.

**NONVOCAL SOUNDS:** Drumming is typical of sapsuckers: a series of brief punctuated rolls, with strikes and pauses in irregular cadence. Specific pattern of drum is highly variable, even among individual birds.

Most drumming is by males. Female drum is softer and shorter in duration. Like squeal call, drum of male serves as both attractant to incoming migrant females and territorial announcement to competing males; it is most often given as long-distance signal early in breeding season and recurs if male loses mate. Males avoid drumming on trees with nest cavity or sap wells, heavily favoring one or two highly resonant sites. Counter-drumming occurs between competing males and between interacting male and female.

Both sexes perform ritual tapping in stiff, exaggerated manner during excavation and as prelude to nest exchange during incubation or brooding; bird either outside or inside cavity taps upon return of mate. Tapping may also be heard when adult is maintaining nest cavity from inside, during confrontation with intruders, and after arrival at feeding station. Drilling of sap wells resembles quiet sewing machine, in rapid bursts.

## VISUAL IDENTIFICATION

**ADULT:** Pied facial pattern of both sexes is superficially similar to that of most *Picoides* species. Solid black malar stripe completely borders red throat in male (solid white throat in female). Bright red forehead and forecrown and bold white nape in both sexes.

Broad swath of whitish mottling down back, tapering toward white rump, often tinted buffy throughout. Bold vertical white stripe on forewing. Belly is buffy or pale yellow below black breast patch, but difficult to see when perched.

In flight, appears speckled white through black parts, with bold white wing stripes and whitish rump.

**JUVENILE:** Body plumage is drabber overall and buffier above than in adult. Dark parts of flanks, chest, and head are olive brownish; light parts on back and tail are darker than in adult, often with more extensive mottling; and rump is barred rather than solid white. Belly is pale grayish gold, lower throat and breast scalloped brown. Crown is grayish, finely streaked with buff; light facial parts buffy and not cleanly demarcated; upper throat pale. Males may show some red feathers in chin, forehead, and crown, but this varies greatly.

**INDIVIDUAL VARIATION:** Females occasionally show some red-tipped feathers on chin and throat. Some females show reduced red or solid black on forehead; frequency of solid black crown increases eastward—from approximately 1 in 18 individuals in AB to about 1 in 6 in eastern provinces. Small number of Yellow-bellieds show some red-tipped white feathers in nape.

**PLUMAGES AND MOLTS:** Preformative molt is protracted from natal summer through May of first spring; solid black breast patch is often the last character to be acquired. All primaries and tail feathers are usually replaced by Aug., whereas body feathers and wing coverts are molted on wintering grounds. Second basic and subsequent plumages are acquired June–Oct.

**DISTINCTIVE CHARACTERISTICS:** Subterminal primaries (6–8) are notably longer than inner and terminal primaries, giving a long-winged appearance. Most sexually dimorphic of the three sapsuckers in *varius* superspecies.

**SIMILAR SPECIES:** Yellow-bellied is most similar to Red-naped in *varius* superspecies, differing in subtle characters, all of which can vary in both species. Variation may or may not indicate signs of hybridization, so care should be taken to recognize and compare as many characters as possible when faced with a questionable identification. Chief characters to observe when separating Yellow-bellied and Red-naped include extent and color of back markings;

**OPPOSITE:** A pair of Yellow-bellied Sapsuckers performs courtship display with synchronous bobbing and swaying, erect postures, and crests raised high. *Marie Read; Apr. 2009, Tompkins Co., NY.*

**OPPOSITE:** An immature male Yellow-bellied Sapsucker that has yet to complete its preformative molt, lacking the full black breast shield. *Christopher Ciccone; Feb. 2011, Naples, FL.*

completeness of malar extension around red throat in male; throat color in female; and extent of red in nape.

Most summer identification problems occur in hybrid contact zones; away from these zones, plumages are fairly crisp and distinct.

**GEOGRAPHIC VARIATION:** Little detectable variation in "pure" individuals, though hybrids show broadly variable shared plumage traits.

**SUBSPECIES**

Additional subspecies to nominate form described from s. Appalachia, "*appalachiensis*," is not universally recognized as separate from nominate form. Smaller and darker than northern individuals, and formerly described as *S. v. atrothorax*.

**HYBRIDIZATION:** Yellow-bellied contacts and hybridizes with the other two *varius* superspecies sapsuckers; Red-naped in s-cen. AB and, to a lesser extent, Red-breasted in cen. BC. Hybridization with Red-naped on east slope of AB Rockies is subject of ongoing studies (see Figure 10).

## BEHAVIOR

### BREEDING BIOLOGY

#### *Nest Site*

Yellow-bellieds nearly always nest in deciduous trees, about 60–68 percent aspen, less often birch, maple, beech, or elm. More than 85 percent of nests are in live trees with heart rot or other fungal decay. In northeastern limit of range, may more often select snapped-off snags than live trees. Generally selects large trees in a stand of mixed sizes.

Usually starts cavity on a smooth trunk surface, and often excavates multiple "starts" in a single season before selecting one for nest cavity. As in other sapsuckers, entrance is barely big enough to fit adults, causing heavy progressive feather wear through season. May use same nest tree for up to 7 years, but rarely the same cavity. Clutch sizes average 30 percent larger in reused cavities, but fledging success is more than 80 percent higher in fresh cavities. Inclement weather may delay onset or prolong excavation time.

#### *Courtship*

Male arrives on territory about 1 week before female, selects nest tree, and establishes drumming

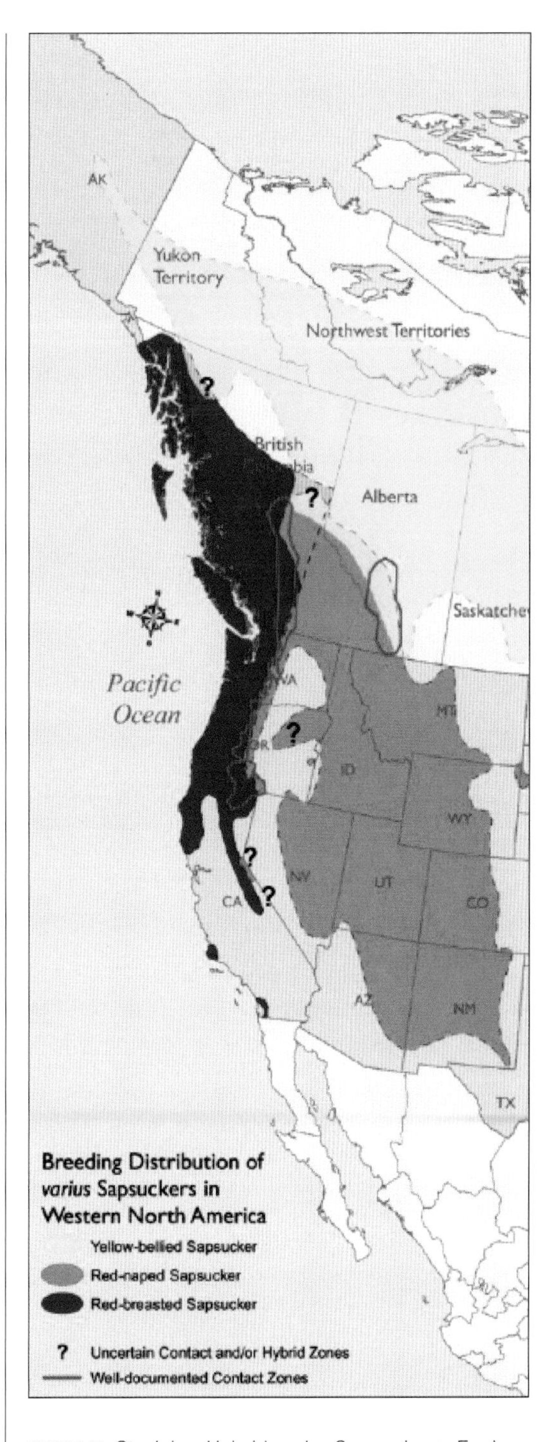


**FIGURE 10.** Studying Hybrid *varius* Sapsuckers. Each of the three sapsuckers in the *varius* superspecies is known to hybridize with each of the other members of the superspecies. Identifying adults and recent fledglings in the contact zones—as well as on the wintering grounds—can be extremely problematic. *Kei Sochi, courtesy of American Birding Association.*

An adult female sapsucker that looks superficially like a typical Yellow-bellied, with a faint hint of red in the nape. *Jocelyn Hudon, courtesy of Royal Alberta Museum; July 2008, Clearwater Co., AB.*

The same individual as the bird in image to the left shows red flecking in the throat as it exits the nest cavity, possibly indicating mixed ancestry between Red-naped and Yellow-bellied Sapsuckers. *Jocelyn Hudon, courtesy of Royal Alberta Museum; July 2008, Clearwater Co., AB.*

posts before female arrives. Pair then shuffles around trunk, tapping in search of excavation site. Rarely, two females may actively pursue an unpaired male.

Monogamous; one female rarely seen at two nests, with two different males. Courtship and territorial displays are identical. Copulation occurs late in excavation period; male may tramp a few times on female's back before sliding down her side; act lasts up to 10 seconds, sometimes followed by mutual tapping. Pair bond is maintained through fledging period and usually reestablished in successive years if mate survives. Mate fidelity is correlated with general site or even specific tree.

### *Parenting*

Female typically lays entire clutch within 8 days of nest completion; inclement weather may delay laying. Incubation begins after third or fourth egg is laid. Nest exchange is generally silent, except for first exchange of morning; male exits, flies to specific perch, and calls, then female enters nest. During typical exchange, relieving mate lands near entrance, taps, calls quietly, and moves aside to allow mate to exit. Both mates may raise and lower crest and fluff throat feathers during exchange.

Adults remove eggshells and unhatched eggs within first day of hatching. Parents deliver small insects to recent hatchlings; stoneflies favored in some regions. Adult may drag insects through sap before delivery, but not observed feeding sap to young directly. Usually removes insect wings and pulverizes remains of large insects before feeding. As nestlings develop, commonly feeds carpenter and other ants, also serviceberry and currant fruit.

Parents that lose mate may still successfully fledge young with longer nestling period, or surviving parent may occasionally pair up with a new adult that shares feeding duties. Three adults rarely observed feeding young at one nest, possibly because of nest failure of third adult.

### *Young Birds*

Nestlings fledge over 2–3 days and remain totally dependent 2–3 days after fledging, then able to feed on sap and insects attracted to sap wells. Begin drill-

An adult male sapsucker with multiple plumage traits that may indicate mixed Yellow-bellied (YB) × Red-naped (RN) ancestry: a red throat with a thick black border (YB); minimal whitish back markings in two fairly neat stripes (RN); and far less red in the nape than in the typical RN but a bit more than in the usual YB variant. *Jocelyn Hudon, courtesy of Royal Alberta Museum; June 2009, Clearwater Co., AB.*

A very worn adult sapsucker with mixed YB × RN plumage traits: the red throat appears not to be entirely bound in black (RN), though it's severely worn; the red in the nape is also very worn but appears too slight for RN and within the range of variant YB; whitish markings on the back are far too extensive for the typical RN and more typical of YB. Perhaps even more problematic is the fact that this bird bred in central Oregon, far from any known RN × YB contact zones (although several possible hybrid records exist from this same region). *Kris Kristovich; July 2014, Deschutes Co., OR.*

ing rudimentary sap wells within a few weeks of fledging, but often in dead trees or limbs and in tree species not normally used by adults.

### *Dispersal*

Family sociality centers around sap wells and is maintained through mutual whine calls. Young seem to actively associate with one another. Young are fed insects for 1–2 weeks after fledging; female often abandons young earlier than male. At about 6 weeks, young are able to effectively drill wells in suitable trees. They remain in natal territory up to 8 weeks, feeding mainly at sap wells. Juveniles are observed associating with adults into Oct.

Juveniles and adult females leave breeding grounds first and migrate farther on average; female records south of U.S. outnumber male records approximately 3.5:1. North of Mex. and Caribbean, males are more abundant, constituting 75–100 percent of local wintering populations, highest farther north.

Nocturnal migrant; may travel up to 25 mi. (40 km) per day. Birds may begin migration in large loose flocks, stringing out as they move south. Mostly observed alone on wintering grounds, although some regions can support high concentrations.

## NONBREEDING BEHAVIOR

### *Feeding*

***Preferred Foods:*** General year-round food consumption is about half plant, half animal matter; seasonally omnivorous, with simple diet of mostly

An adult female Yellow-bellied Sapsucker suspended while feeding on winter fruits; sapsuckers of the *varius* superspecies are known to occasionally practice adroit acrobatics while feeding. *Robert Royse; Jan. 2004, Franklin Co., OH.*

sap and various arthropods. Ants make up about 70 percent of foods May–Aug.; caterpillars, moths, dragonflies, spiders, grasshoppers, and wasps seasonally. Eats small fruits, berries, and seeds outside breeding season. In spring, plucks fresh buds of aspen trees from outermost branches.

In process of drilling and tending sap wells, also consumes bast (inner tree layers). About half of spring diet is sap and cambium; sap averages about 20 percent of annual diet. Sap wells probably also constitute general fluid source.

Fall plant foods are about 71 percent fruits, supplemented by sap and some insects. At northern edge of winter range, forages mainly on arthropods, frozen fruit, and frozen sap; will tear bark from crab apple and maple in winter to tap sap. Farther south, individuals persistently attend xylem wells. Will eat suet and nuts at feeders, preferring nuts if both are available.

***Foraging Behaviors:*** Most feeding time is devoted to developing and maintaining sap wells; one pair of sapsuckers may drill more than 3,000 wells per year for up to 3 years. May feed at one tree for up to 2 hours; wintering birds may spend up to 70 percent of time drilling wells, possibly because of relatively low sugar content of winter sap.

Generally selects wounded or weakened trees as sap source, especially those already oozing sap because of damage from disease, weather, or other disturbance; may choose older trees with fungal infection, or trees with old sapsucker scars. Sap-tree species vary widely; about 1,000 perennial plants have been recorded with sap wells or scars.

Gleans insects mostly from tree bark; also probes into bark furrows or pries scales, especially in colder climates. Fond of flycatching from large tree branches or snag tops. Observed using bark crevice as anvil site to extract meats from fruits and nuts.

### *Territory Defense and Sociality*

Both sexes defend nest, each other, and their sap wells, most aggressively in early nesting period. Territorial displays are identical to those of courtship, characterized by complex vocalizations and physical posturing.

At onset of breeding period, individuals advertise presence and defend space around nest tree, as well as sap wells. Further into season, they actively chase competing individuals from immediate vicinity; the most aggressive displays occur inside about a 40-ft. (12-m) circle around nest.

Territory size typically ranges from about 1.5 to 7.5 ac. (0.6–3.1 ha). Breeding-season dominance ranks adults over juveniles, and male juveniles over female juveniles. In winter, adult males are occasionally aggressive toward approaching juveniles.

### *Interactions with Other Species*

Direct conflicts with other species are usually over nest cavity or sap wells. Yellow-bellied is most aggressive toward other woodpeckers, including Pileated, that approach or land on nest tree. However, also known to nest in the same tree with Pileated. Known to abandon nest after carpenter ants tunnel into cavity.

Will chase insects away from sap wells; observed alternately chasing and being chased from sap wells by Ruby-throated Hummingbird, with both later feeding in close proximity.

Adult female observed feeding young Red-breasted Nuthatch even in close presence of her own begging fledglings. Red-cockaded Woodpecker observed chasing sapsucker from resin flows around Red-cockaded cavity entrance.

Sap wells are important food source for many bird, insect, and other species, including Bananaquits and lizards in Virgin Is.

### MISCELLANEOUS BEHAVIORS

Sapsuckers are more adept at flight within dense forested canopies than other woodpeckers. Yellow-bellied rarely observed stretching both legs at once with belly resting on horizontal branch. Rarely observed bathing in bird bath or shallow pond. May practice anting on hot days. Typically does not roost in cavities outside breeding season. Sunbathing performed with dorsal feathers fluffed, back toward sun, and wings and tail partly spread.

## CONSERVATION

**HABITAT THREATS:** Any clearing of sap-producing trees may compromise local food supply.

**POPULATION CHANGES:** First confirmed breeding in AK in 1983 but likely present there much earlier. Current rangewide populations may exceed those in presettlement times; sapsuckers thrive in early-successional habitats and exotic plantings, both of which flourished following the clearing of old-growth forests. Habitat loss in southern Blue Ridge Mts. has precipitated a steep decline in s. Appalachian populations since mid-1900s.

**CONSERVATION STATUS AND MANAGEMENT:** Listed as priority for conservation in ON, PA, TN, NC, and WV, and as endangered as a breeding species in OH, owing to extremely limited habitat.

## REFERENCES

American Ornithologists' Union 1985; Bailey 1953; Bolles 1891, 1892a, 1893; Bond 1957; Conner and Kroll 1979; Davis and Howell 1950; Eberhardt 1997, 2000; Erwin et al. 2004; Franzreb and Higgins 1975; Freer 1933; Freer and Murray 1935; Frisch 1987; Gibson and Kessel 1992; Hebard 1949; Howell 1952, 1953; Johnson 1947; Kennard 1895; Kessel 1986; Kilham 1953a, 1953b, 1962a, 1962b, 1964, 1971, 1977a, 1977c; Longcore et al. 2005; Loughman 2006; MacArthur and MacArthur 1974; MapArt 1999; McClelland 1977; Miller 1923; C. Morris 1905; E. Morris 1905; Nye 1918; Ohio Department of Natural Resources 2006; Reed 1929; Rissler et al. 1995; Robinson and Rosenberg 2003; Rudolph et al. 1991; Seaman 1954; Semo and Leukering 2004; Shelley 1934; Shire et al. 2000; Short and Morony 1970; Shunk 2005; Smith 1954; Taggart 1912; U.S. Fish and Wildlife Service 2007b; Van Vleit, G., pers. comm.; Varner et al. 2006; Williams 1980b; Wood 1891, 1905b.

# RED-NAPED SAPSUCKER

## *Sphyrapicus nuchalis*

L: 7.5–9 in. (19.0–22.9 cm)
WS: 16–18 in. (40.6–45.7 cm)
WT: 1.3–2.2 oz. (37–61 g)

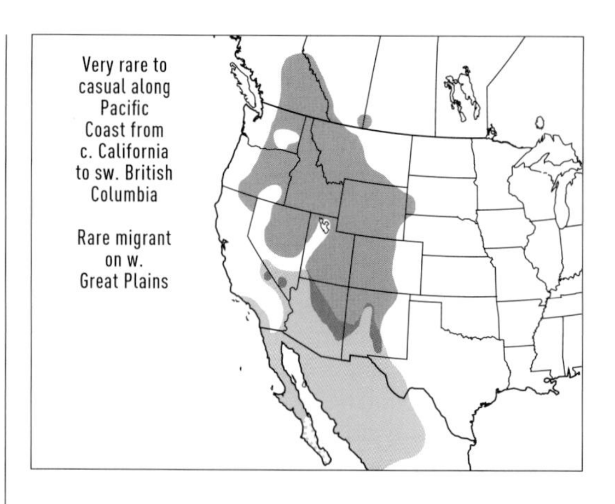


**A punctuated drumroll** echoing through aspen woodlands of the intermountain West is often the first sign of a breeding Red-naped Sapsucker. Once considered a single species with the Yellow-bellied and Red-breasted Sapsuckers, the Red-naped generally resides in montane aspen habitats throughout its range.

Like its *Sphyrapicus* relatives, the Red-naped Sapsucker plays a keystone role in forest ecology by excavating cavities that are later used by numerous secondary cavity nesters and by drilling sap wells on which many small birds feed. Breeding among the "sky islands" of the Great Basin and surrounding states, the Red-naped Sapsucker is one of our most migratory woodpeckers, occasionally reaching the southern end of central Mexico's Sierra Madre Occidental and rarely occurring as far south as Honduras.

South-central Alberta's Kananaskis Country serves as a fertile study ground for observing hybridization between Red-naped and Yellow-bellied Sapsuckers, whereas the east slope of the Cascades and the western edge of the Great Basin bring the Red-breasted and Red-naped together. Most of what little research exists on Red-naped Sapsuckers has occurred in and since the latter part of the 20th century, and DNA evidence has shown us that these three species—under the *varius* superspecies—are some of the most closely related woodpecker taxa on Earth.

### DISTRIBUTION

**BREEDING RANGE:** Generally breeds in mid-elevation forests of intermountain West, from Rocky Mts. westward to eastern slopes of WA and OR Cascades. Local ranges may extend away from mountains along riparian zones. Isolated populations in Cypress Hills, Black Hills, and Lake Tahoe region.

**MIGRATION AND WINTER RANGE:** Most individuals migrate short distances; females migrate farther than males on average. Northern birds move south into w. and sw. U.S., Baja CA, and interior Mex., rarely to Guatemala and Honduras. A few overwinter in AZ and NM breeding ranges.

### HABITAT

**BREEDING RANGE:** Strongly affiliated with aspen groves, especially in montane riparian areas and other wetlands, often with cottonwood, willow, sumac, or birch, surrounded by a wide range of associated forest types. May be found at higher-elevation forest edges, including stands of live trees surrounded by high-severity burns. Typically found up to about 10,000 ft. (3,048 m).

**MIGRATION AND WINTER RANGE:** Adapted to diverse range of habitats, from orchards and parks to riparian zones, and a wide diversity of oak habitats, occasionally into mesquite woodlands. Often found moving through foothills and valleys adjacent to Douglas-fir and western hemlock forests. Typically below 5,600 ft. (1,700 m); up to 8,200 ft. (2,500 m) in Mex.

### DETECTION

Drums frequently and is quite vocal during breeding season, conspicuously belying its presence. Fairly quiet outside breeding season.

**VOCAL SOUNDS AND BEHAVIOR:** Gives three basic calls, similar to those of other sapsuckers in *varius* superspecies: whine, squeal, and chatter, each somewhat variable.

Descending, nasal whine, a high-pitched *mehhh*, often repeated, given in a variety of circumstances. A harsher version is uttered in takeoff and flight, and most often around fledging time. This is most commonly heard call outside breeding season.

Territorial squeal, *reaer, reaer, reaer*, is repeated two to four times, often from a high perch; may be mistaken for a small raptor. Mainly given by males in territory defense and mate attraction, and frequently alternated with drumming.

Harsh chatter, commonly accompanied by softer interaction calls, is given between mates during nest exchange and other close contact situations. Also given with mothlike flight display during territorial encounters.

**OPPOSITE:** An adult female Red-naped Sapsucker. *Stephen Shunk; June 2008, Summit Co., UT.*

An adult male Red-naped Sapsucker returns to the nest cavity with an insect meal that has been draped through sap. *Kristine Falco; June 2009, Jefferson Co., OR.*

Combinations of calls are given when chasing intruder or during aggressive social displays.

**NONVOCAL SOUNDS:** Drum, similar to that of other sapsuckers, is a punctuated series of mini-rolls that typically includes an introductory roll followed by couplets or triplets. Each episode can vary considerably, even from the same individual. Where Red-naped and Red-breasted Sapsuckers nest in close proximity, drums of former may include twice as many doublets as those of latter, but they are difficult to separate.

Drumming is performed mainly by males. About half of all episodes are spontaneous, about one-third in response to drumming by a competitor or during other territorial encounters. About half of all drumming is given less than 98 ft. (30 m) from nest tree. Mates are known to perform "duets."

Ritualized tapping is given by both sexes during excavation, especially when one bird is excavating inside cavity; also performed at nest exchange and during other direct interactions between mates.

Like other sapsuckers, Red-naped can often be heard drilling sap wells, with distinctive rapid bursts of tapping that sound something like a sewing machine.

## VISUAL IDENTIFICATION

**ADULT:** Facial pattern is superficially similar to that of most *Picoides* species. Thin black malar stripe borders bright red throat patch with bold black breast shield below; red throat (with white chin in female) typically invades the black border on sides of neck. White nape feathers are tipped with variable amount of red, rarely all red, and are usually present in both sexes, along with bright red forehead and forecrown.

As in other sapsuckers, white sapsucker wing patch may be difficult to see if worn or obscured by fluffed body feathers. Belly is buffy or pale yellow.

In flight, upperwing appears mottled black, broken by bold white wing stripes and whitish rump.

**JUVENILE:** Dark parts of flanks, chest, and head olive brownish; light parts on back and tail darker than in adult, often with more extensive mottling; rump barred rather than solid white; wing stripes distinctive. Belly grayish gold, lower throat and breast scalloped brown. Crown dark brown, light facial stripes buffy, upper throat pale. Males may show some red-tipped feathers in chin, forehead, and crown, but this varies widely.

**INDIVIDUAL VARIATION:** White in female chin is highly variable, possibly geographically (see Geographic Variation); occasionally mottled with red-tipped feathers or showing minimal white, rarely showing completely red chin and throat. Some females show reduced or no red on forehead. Back pattern in both sexes may also be variable, and individuals may not show red nape.

**PLUMAGES AND MOLTS:** Preformative and prebasic molts occur mostly on summer grounds, Jun–Oct. Late summer plumage of adults can be confusing,

A juvenile female Red-naped Sapsucker, with its preformative molt nearly complete and a few black feathers coming in on the breast. *Robert Royse; Nov. 2005, Socorro Co., NM.*

especially throat, wing patches, chest feathers, and head plumage, because of feather wear after repeated passes through tight cavity entrance. Juvenile acquires black breast patch in late fall or winter.

**DISTINCTIVE CHARACTERISTICS:** Except for male Williamson's Sapsucker, the only woodpecker in its breeding range with vertical white wing patches.

**SIMILAR SPECIES:** Yellow-bellied, Red-naped, and Red-breasted Sapsuckers comprise the *varius* superspecies. Fresh fall plumages vary in two distinct ways: sexual dichromatism, which is greatest in eastern Yellow-bellied, moderate in Red-naped, and nonexistent in "Northern" Red-breasted (*S. r. ruber*); and extent of red in head and chest, which is least in Yellow-bellied, moderate in Red-naped, and extensive in "Northern" Red-breasted.

Most summer identification problems occur in hybrid contact zones; at cores of the three species' ranges, plumages are fairly crisp and distinct.

**GEOGRAPHIC VARIATION AND HYBRIDIZATION:** Geographic variation is generally attributed to hybridization, although females in far West may show less white in chin than those in Rockies. Young Red-naped x Yellow-bellied Sapsucker hybrids may retain some juvenal body feathers and may not acquire black breast shield until late winter.

## BEHAVIOR

Decades of ambiguity in the taxonomy of the *varius* superspecies have led to ambiguity when discussing behavioral elements of any one species. For example, early studies did not distinguish between Red-naped and Yellow-bellied Sapsuckers, since they were once considered the same species, so few diet studies exist solely for Red-naped Sapsucker. Also, because of widespread range of Yellow-bellied Sapsucker and extremely close relationships among members of the *varius* superspecies, many behaviors may have originally been observed in Yellow-bellied, with descriptions for the others postulated to account for geographic variation.

### BREEDING BIOLOGY

#### *Nest Site*

Nest location is typically selected for easy access to foraging areas rather than for individual tree or stand characteristics. Adults often return to previous year's breeding site, frequently in same tree, but only occasionally in same cavity. Often start multiple cavities ("starts") before selecting final excavation site.

Aspen is preferred in all regions, but nests are also reported in cottonwood and wide variety of conifers. More than 80 percent of nests are excavated in live trees infected with heart rot, often easily detectable by presence of fungal conchs on bark surface; in conifer-dominated forest, dead trees are more often selected than live trees.

In aspens, heart rot generally enters tree at base and works upward; sapsuckers occasionally (but not always) follow suit, excavating each year's cavity above last year's; higher cavities may also limit nest predation. Conversely, heart rot enters larch and other conifer snags through snapped-off top, progressing downward, occasionally leading sapsuckers to excavate progressively lower each year in these species.

May nest in logged and unlogged stands with equal frequency, but adult depends on adjacent unlogged stands for feeding, especially in early season when conifer sap is flowing but aspen buds have yet to emerge.

#### *Courtship*

Generally monogamous; very rare reports of polygyny. Pairs are usually mated within 3 weeks of arrival at nest site and begin excavating almost immediately; cold or inclement weather delays excavation. Most courtship displays are similar to territorial displays; bobbing and swaying are often evident during nest exchange and later during feeding encounters. Rarely performs snipelike winnowing flight.

Copulation ensues when excavation is nearly complete and peaks during laying, rarely extending into incubation. Male initiates by calling from horizontal branch near nest; both birds utter soft churring calls as female approaches. Act is similar to what it is in other woodpeckers; may last 10 seconds, and either or both individuals may follow with light tapping.

Pair bond remains through fledgling care period, with strong mate and site fidelity through following season, when mates rejoin on nesting grounds. Spontaneous mate separation is rarely observed after initial mate selection.

#### *Parenting*

Clutches average larger when adults select prior years' cavities; smaller in newly excavated cavities. Some studies show only about half of eggs hatch. Nests of older females may be more productive. Observed carrying and eating eggs but unclear

**OPPOSITE:** A variant adult male Red-naped Sapsucker showing a red patch across the auriculars, a trait exhibited more frequently in juvenile birds than in adults, and possibly more frequently in western adults than in Rocky Mountain populations. *Paul Bannick; May 2005, Yakima Co., WA.*

A nestling female Red-naped Sapsucker begs from its cavity entrance. *Stephen Shunk; June 2009, Deschutes Co., OR..*

whether eggs were their own or stolen. Male may select a single tree as sanitation drop site for entire nestling period, and possibly for successive years.

### *Young Birds*

Young at most nests fledge over 2 days. Postfledging parental attentiveness varies among individual parents and pairs, but family group generally remains together around nest site for at least 1 week after fledging. Parents then may attempt to lead juveniles upslope or to active sap wells. After fledging, all family members roost alone on tree trunks and do not reenter nest cavity.

Fledglings are usually able to forage soon after leaving nest; often observed at sap wells 2 weeks after fledging, but still being fed insects by adults. Adults occasionally drag insects through sap before feeding young; also observed crushing insect prey before feeding young.

**DISPERSAL:** Some first-spring birds breed within 1 mi. (1.6 km) of natal site, but difficult to measure since as few as 5 percent may survive migration. One study found that about 98 percent of adults that did survive to next breeding season nested within about 650 ft. (200 m) of previous year's nest site; about one-quarter to one-third used same nest tree, and one-third to one-half nested within 165 ft. (50 m) of previous tree.

Individuals usually breed in first spring, some not until following year. Fertility may be reduced or lost after fifth year. Older females may nest earlier in season than first-year breeders.

## NONBREEDING BEHAVIOR

### *Feeding*

***Preferred Foods:*** Seasonally omnivorous, with a simple diet of mostly sap and arthropods. Year-round food consumption is about half plant, half

**OPPOSITE:** An adult male Red-naped Sapsucker removes fecal material from an active nest cavity. *Kristine Falco; June 2006, Deschutes Co., OR.*

animal matter. Sap averages about 20 percent of annual diet. Carpenter and *Formica* ants make up about 70 percent of foods May–Aug. Other arthropods include caterpillars, mayflies, beetles, moths, dragonflies, spiders, and grasshoppers. Nestlings are fed mostly ants, caterpillars, mayflies, and spiders; also observed being fed ungulate (probably elk) bone fragments.

Fall plant foods are about 71 percent fruits; supplemented by sap and some insects. Will eat small fruits, berries, and seeds outside breeding season.

Broad mix of deciduous and coniferous trees, including willow, alder, aspen, cottonwood, Douglas-fir, larch, and juniper, serve as sap sources. In process of drilling and tending sap wells, also consumes bast (inner tree layers), hence about half of spring diet is sap and cambium.

***Foraging Behaviors:*** Foraging techniques include drilling and feeding at sap wells, plucking aspen buds, gleaning, and flycatching. When feeding at sap wells, opportunistically retrieves insects trapped in sap.

### *Territory Defense and Sociality*

Both sexes defend nest site and sap wells, with greatest intensity early in breeding season. Most extreme interactions between rival males include birds grasping each other's bill and tumbling to ground. Otherwise, both sexes perform diverse array of displays, usually accompanied by vocalizations. These may include posturing body horizontally and pointing bill directly at intruder; swinging bill side to side; raising bill upward to flash red throat and yellow belly; puffing out throat feathers while erecting crest; flicking one or both wings in and out rapidly; drooping forewings with wingtips crossing at rump; or any combination of these.

Display flights during breeding season include bouncing in tight undulations, exaggerated swooping up to perch, and mothlike flight with head dropped and wings bowed and flapped in shallow beats.

Territory size ranges widely, from 1.7 to 112 ac. (0.7–45.2 ha); individuals typically remain within 1,760 ft. (500 m) of nest site, rarely farther if accessing particularly productive food source. Territories are typically bounded by natural features rather than circular distance; largest sites often include ponds, small lakes, meadows adjacent to nest area, or large stands of old-growth forest.

Territories are often maintained long term, with regularly maintained sap wells defining boundaries. Pairs maintain mutually exclusive territory boundaries with adjacent breeding pairs, with minimum distance between nests of 328 ft. (100 m).

Some northbound migrants defend feeding territories at lower elevations in spring before moving upslope to breeding grounds.

### *Interactions with Other Species*

Engages in most interactions to defend nest or sap wells. Williamson's Sapsucker antagonizes and dominates Red-naped; Red-naped may not respond to Williamson's drumming or calling, but latter may react aggressively to those of Red-naped. Red-naped may nest in close proximity to Red-breasted Sapsucker, Downy and Hairy Woodpeckers, and Northern Flicker.

Sap wells attract numerous species, some of which receive significant caloric intake from sap. Especially important to Rufous, Broad-tailed, and Calliope Hummingbirds, all of which may follow sapsuckers north in spring before flowers bloom. Red-breasted Nuthatches are known to place sap from sapsucker wells around their own cavity entrances.

Red-naped Sapsucker may rarely usurp nests from other cavity nesters.

#### MISCELLANEOUS BEHAVIORS

Male typically roosts in cavity, female on tree trunk under base of limb. Sunbathes with back to sun and with nape, mantle, and scapular feathers fully fluffed. All sapsuckers excel at flying through forest vegetation.

## CONSERVATION

**HABITAT THREATS:** Fire suppression may be one of biggest threats, for three reasons: (1) aspens depend on disturbance such as fire for regeneration and development of diverse age classes; (2) stand-replacement fires may wipe out larger stands of trees than normal fire regimes do; and (3) lack of fire allows conifer intrusion into aspen habitat, further impeding aspen regeneration. Seedlings in fire-suppressed habitats are eaten mostly by livestock, elk, and deer. All of the above factors often lead to large stands of decadent trees that will not sustain sapsucker populations in long term.

Absent in Toiyabe Range of NV 1940–1950 when aspens were logged for mining purposes and road development. Aspen harvesting for pulp-wood products, paneling, flooring, and furniture parts may also degrade habitat.

**POPULATION CHANGES:** Red-naped Sapsucker was second-most abundant cavity-excavating species surveyed in 1996 in n.-cen. BC; now second-most abundant woodpecker in NV's Toiyabe Range. Possibly the most common non-passerine in aspen-birch habitats in Black Hills of SD. May completely withdraw from Great Basin and drier interior mountains during periods of extreme drought.

**CONSERVATION STATUS AND MANAGEMENT:** Targeted for conservation management at various levels in most breeding regions, including CO, ID, AZ, NM, NV, WY, Blue Mts. of OR and WA, and Canada high desert.

Conservation of berry-producing shrubs, willows, and aspens is important for long-term viability; especially linked to healthy aspen riparian habitats. Local prescriptions include retention of mature trees and snags over 10 in. (25.4 cm) diameter at breast height (approximately 4 ft. [1.2 m] aboveground), with mature stands containing 15 snags per 10 ac. (4 ha).

## REFERENCES

American Ornithologists' Union 1985; Bayer 1995a; Beidleman 2000; Browning 1977, 2004; Butcher et al. 2002; Cicero and Johnson 1995; Covell 1936; Daily 1993; Daily et al. 1993; Davis and Davis 1959; Dobkin et al. 1995; Elzroth 1987; Grinnell 1901; Hall 1938; Howell 1952, 1953; Hudon, J., pers. comm.; Hudon 2000, 2001; Johnson and Johnson 1985; Johnson and Zink 1983; Losin et al. 2006; McCelland and McClelland 2000; Mills et al. 2000; Nehls 1985; Nessom 2002; Peterson 1946; Rothenbach and Opio 2005; Scott et al. 1976; Short 1969b; Short and Morony 1970; Shunk 2005; Tobalske 1992; Trombino 2000; Vasquez 2005; Walters 1996; Walters et al. 2002a; Woodbury 1938.

# RED-BREASTED SAPSUCKER

***Sphyrapicus ruber***

L: 8–9 in. (20.3–22.9 cm)
WS: 16–18 in. (40.6–45.7 cm)
WT: 1.4–1.9 oz. (40–55 g)

**Like a warning beacon** in our far western forests, the Red-breasted Sapsucker glows brightly under the coniferous canopies of the Pacific states and provinces. With the most limited range of our four sapsucker species, the Red-breasted can be a prize find, especially for birders from eastern North America. While it may often be found nesting or feeding among aspen galleries, this woodpecker can also be found etching its rectangular phloem wells into the bark of small-diameter willows and alders. Its nesting sites may also be rather diverse, opportunistically excavated in any tree species that is locally available.

The Red-breasted Sapsucker is so closely related to the Red-naped and Yellow-bellied Sapsuckers that the three were once considered to be the same species. Hybridization between the Red-breasted and Red-naped poses interesting identification challenges, especially in central and southern Oregon, where the two subspecies of Red-breasted Sapsucker also interbreed.

The Red-breasted Sapsucker is one of only four sexually monochromatic woodpeckers worldwide, and thus the sexes are generally impossible to distinguish in the field, unless one is lucky enough to observe copulation between a mated pair. This species is also the least migratory of the sapsucker genus, with individuals often wandering into developed suburban areas in winter, where ornamental evergreens offer a year-round sap supply.

## DISTRIBUTION

**BREEDING RANGE:** Generally restricted to Pacific states and provinces; most common from se. AK south through coast ranges and w. Cascades of WA and OR; also n. CA coast ranges, Cascades, and Sierra Nevada. Rare and scattered between e. Sierra Nevada and w. NV ranges; disjunct mountain populations in Blue Mts. of cen. and ne. OR and mountain ranges of sw. CA.

**SEASONAL MOVEMENTS:** Least migratory of all sapsuckers. Generally slow to leave breeding habitat, frequently remaining downslope into late fall. Some

**OPPOSITE:** A pair of adult Red-breasted Sapsuckers, southern subspecies (*S. r. daggetti*), performs nest exchange, with one adult leaving its brooding duty to allow its mate to feed the nestlings. *Tom Grey; June 2005, Sierra Co., CA.*

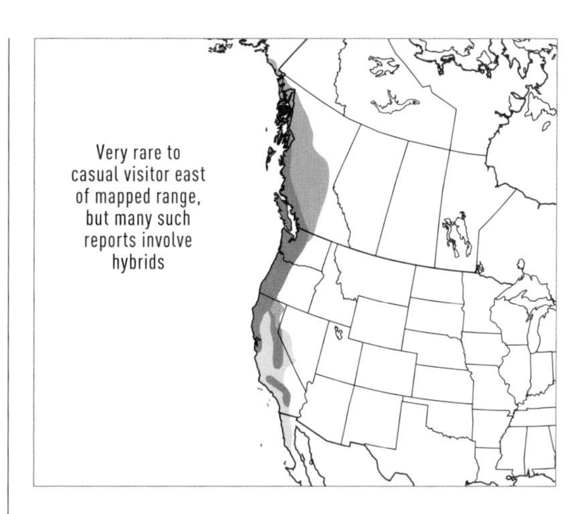


w. Sierra Nevada populations move upslope to subalpine habitats postbreeding, some down into montane valleys; s. CA populations disperse through mountain valleys. Red-breasted Sapsuckers may return to breeding territories 2–3 weeks before Red-naped in OR and WA contact zones.

**WINTER RANGE:** Northern populations may remain on breeding grounds or may move downslope in winter, with inland breeders often moving to coastal areas and foothills. Southern birds may migrate farther than northern birds in latitude but generally overwinter throughout lower elevations of CA. Winter reports of hybrid and intergrade individuals are widespread, often in areas outside breeding range and habitats.

## HABITAT

**BREEDING RANGE:** Breeds to 9,500 ft. (2,900 m) in mixed-conifer forests, often with aspen component, from elevation of western hemlock and Sitka Spruce along the coast up to that of mountain hemlock and Engelmann spruce in interior mountains. In far north, may favor open Douglas-fir and closed spruce-hemlock stands, but conifer associations vary widely by latitude and elevation. In some regions, favors riparian bottomlands dominated by aspen and cottonwood, also second-growth red alder with various conifers. Often associated with old-growth Douglas-fir adjacent to patchy cutover stands and edges of lakes and burned forest.

**MIGRATION AND WINTER RANGE:** Northernmost populations often remain in breeding range and continue to associate with breeding habitat. Southern populations and southernmost northern populations may move into aspen, birch, maple, and other riparian associates at wide range of elevations up to 9,800 ft. (3,000 m). Migrants overwinter in broad alluvial valleys with abundant and diverse cultivated tree species, especially suburban habitats

An adult Red-breasted Sapsucker, northern subspecies (*S. r. ruber*). *Lois Miller; Mar. 2010, Curry Co., OR.*

and orchards. Also found in juniper habitats east of Cascade Mountains.

## DETECTION

Like other sapsuckers, most vocal in breeding season, but usually less vocal than other sapsuckers, especially where ranges overlap. Described by Joseph Grinnell and Tracy Irwin Storer in the 1920s as being "well-nigh voiceless." Bright plumage increases likelihood of visual detection.

**VOCAL SOUNDS AND BEHAVIOR:** Generally gives same basic calls as other sapsuckers in *varius* superspecies, but few authorities describe this species' vocalizations separately from those of its closest relatives. More study is needed of vocal distinction among the three species. Calls are difficult to assign to sex because of monochromatism.

Characteristic nasal whine, a high-pitched descending *mehhh,* often repeated; may be lower pitched than call of Red-naped Sapsucker where ranges overlap. Occasionally heard outside breeding season.

Territorial squeal usually given in doublets or triplets, frequently associated with drumming. Basically indistinguishable from Red-naped and Yellow-bellied squeal; reported as similar to calls of Red-shouldered Hawk and Red-headed Woodpecker.

Courtship or territorial interactions may be accompanied by squeaky, hurried contact calls, similar to those of other woodpeckers: *rittah, rittah, rhut, rhut.*

**NONVOCAL SOUNDS:** Drum, similar to that of other sapsuckers, consists of short, punctuated rolls joined in an uneven cadence, each mini-roll generally decreasing in length (fewer taps per roll) but increasing in frequency (more taps per second). May rarely be distinguished from drum of Red-naped Sapsucker by addition of shorter second phrase. However, when one species is drumming outside presence of the other, drum may not be identifiable to species.

Rolls are highly variable in duration and pattern

**OPPOSITE:** A fledgling Red-breasted Sapsucker, southern subspecies (*S. r. daggetti*). *Alan Murphy; July 2004, Mono Co., CA.*

and believed to be individually distinctive. Mates are known to counter-drum, with dueting observed between Red-naped and Red-breasted Sapsuckers and hybrids breeding in close proximity.

## VISUAL IDENTIFICATION

Description here applies to nominate *S. r. ruber*, the northern subspecies. See Geographic Variation for details on separation from southern *S. r. daggetti*.

**ADULT:** Head is entirely red except for black lores and whitish nasal tufts; red extends to below nape and through entire breast. White patch at base of bill may extend as variable faint mustache toward hindneck; rear portion of this stripe may contain feathers with white bases and red tips, giving appearance of a hidden white stripe. Upper back is generally black with two narrow vertical stripes mottled with off-white or light buff. Blackish wings show white "sapsucker slashes" at forward edge of wing. Center of belly may be quite yellow, but this is usually obscured by dark sides and flanks.

Sexes are alike, with rare sexual variation in tail-feather spotting not distinguishable in the field; see Individual Variation.

In flight, mottled black and white above, with head red and wing patches and rump bold white.

**JUVENILE:** Dark chocolate or olive brown replaces red of adult. Sides are dark gray, faintly barred with darker gray; center of belly is faintly tinted yellow or gold. Wing patches are prominent; back very dark, with markings dull and mostly in single swath of grayish mottling down center; rump barred. Juveniles in hybrid contact zones are often impossible to identify to species unless both parents are observed feeding young bird.

**INDIVIDUAL VARIATION:** Sexes are generally alike in both subspecies, and usually impossible to separate in the field. However, female tail is variable, especially in *S. r. daggetti* and rarely in *S. r. ruber*; pale markings average more extensive in females than in males, though sexes overlap; and in females with palest tail markings, central pair is mostly white, narrowly barred black. Some females may show almost entirely black outer tail; others have white mottling at center of outer tail feathers.

Males on Haida Gwaii (formerly Queen Charlotte Is.), BC, may have 6 percent longer bill than females, on average, but this is not likely detectable in the field. Albinism and leucism rarely reported.

**PLUMAGES AND MOLTS:** Juveniles acquire first fall plumage on breeding grounds or during dispersal, occasionally into Oct. Adult fall molt occurs June–Sept.

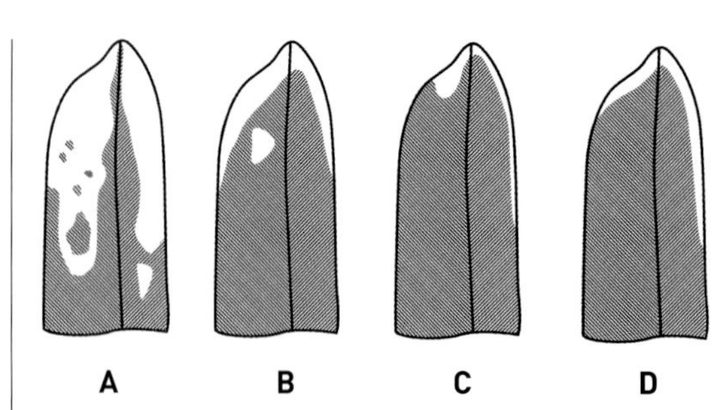


**FIGURE 11.** Sexing Red-breasted Sapsuckers. Sexes of this species are not reliably separable in the field and hence are considered monochromatic. (A and B) While practically impossible to identify in the field, some females of *S. r. daggetti* (and possibly some *S. r. ruber*) may show some white mottling at the center of the outer rectrices; however, birds with all-black centers to these feathers could be either males or females. (C and D) The same feathers of both sexes of *S. r. ruber* typically show fine pale edging when fresh; in some instances, this may be useful for separating subspecies in the field. *Redrawn by Christine Elder after S.N.G. Howell, in Pyle 1997.*

**DISTINCTIVE CHARACTERISTICS:** Shows the most red of any western woodpecker. The least migratory of the sapsuckers, often making it the latest one seen in the fall and through winter, especially in northern regions.

**SIMILAR SPECIES:** Adult is unlikely to be confused with other woodpeckers occurring in same range. Only Red-headed Woodpecker has similar extent of red on head, and its black-and-white patches show far sharper contrast than sapsucker's dull mottling; chance of seeing the two together in the field is astronomically low.

Juvenile Red-breasted Sapsucker shows darker head and breast than juvenile Red-naped and Yellow-bellied Sapsuckers; the latter two also show more prominent facial markings, whereas Red-breasted barely shows a hint of white malar stripe.

**GEOGRAPHIC VARIATION:** Most geographic variation is attributable to subspecies; in general, amount of white, especially in facial and back plumages, increases from north to south.

## SUBSPECIES

Two subspecies are recognized. *S. r. ruber* is mostly a permanent resident of temperate rainforests from se. AK to coastal Oregon and east to western slope of WA and OR Cascades; to date, no confirmed breeding in CA. This is larger race, with more red, and red brighter and more ruby colored (occasion-

Two juvenile Red-breasted Sapsuckers in the final stages of preformative molt. Note the grayish patches in the breast and the subspecific variation in the red parts of the breast: (A) northern subspecies (*S. r. ruber*), with greater extent of red; (B) southern subspecies (*S. r. daggetti*), with far less red in the breast and an exposed white malar stripe. Also note the "hidden" white eye stripe and dark patch at the base of the auriculars in both individuals—traits that do not help identify individuals to subspecies. *A, Stephen Shunk; Oct. 2008, Deschutes Co., OR; B, Tom Grey; Jan. 2008, Santa Clara Co., CA.*

ally closer to purplish red) than in *S. r. daggetti*. Back is faintly marked, often with two narrow vertical stripes.

*S. r. daggetti* breeds from cen. OR south through s. CA and w. NV. Red is paler than in *S. r. ruber* and less extensive; head pattern approaches that of Red-naped Sapsucker, sometimes with blackish ear patch and partial white supercilium, prominent malar stripe, and red chest weakly demarcated from paler yellow belly. Back is more heavily marked than that of *S. r. ruber*, approaching that of Red-naped.

Intergrade zone between the two subspecies stretches from e. Cascade Mts. of cen. OR south to Klamath and Siskiyou Mts. across CA–OR border; possibly includes Ochoco Mts. of cen. OR and Cascade foothills east of Upper Klamath Basin.

Plumage characters of intergrade individuals cover a broad continuum between parental subspecies; it is often impossible to identify them to subspecies in the field, especially juveniles of mixed-subspecies pairs. In winter, identification problem likely extends farther south.

**HYBRIDIZATION:** Hybridizes with Red-naped Sapsucker from s.-cen. BC south along east slope of Cascades, east across Cascades foothills and into Warner Mts. Isolated hybrid populations on e. slope of Sierra Nevada, from Lake Tahoe south to Inyo Co., very rarely into riparian habitats of White Mtns. Rarely hybridizes with Yellow-bellied Sapsucker in nw. BC, possibly into AK.

Especially in southern parts of range and in

An adult Red-breasted Sapsucker, northern subspecies (*S. r. ruber*), tends a series of xylem wells. *Ollie Oliver; Oct. 2012, King Co., WA.*

winter, potential challenges of distinguishing pure Red-breasted Sapsuckers from hybrids should not be underestimated. In addition to plumage characters, season and location must be considered, based on tendencies of parental types and their influences on different hybrid combinations.

## BEHAVIOR

Because Red-breasted and Red-naped Sapsuckers were long considered the same species, early studies did not distinguish between their behaviors. Some studies in regions where Red-naped Sapsucker does not regularly occur likely also apply to Red-breasted. Elsewhere, anecdotal and scientific observations support the notion that various behavioral elements are similar.

### BREEDING BIOLOGY

#### *Nest Site*

Generally selects dead tree for nest cavity, in a variety of favored tree species, including local preference for madrone in CA; on average, fewer than 25 percent of nest cavities are live substrate. Prefers large-diameter snags in advanced decay.

Nest-site characteristics are mostly indistinguishable between Red-breasted and Red-naped Sapsuckers in contact zones; mixed-species pairs may select shorter trees than pure-species pairs.

An adult Red-breasted Sapsucker, southern subspecies (*S. r. daggetti*), tends a series of active xylem wells. *Steve Kaye; Dec. 2010, Orange Co., CA.*

An adult hybrid Red-breasted x Red-naped Sapsucker excavating phloem wells. The facial pattern is nearly perfectly intermediate between those of the two parental forms. *Stephen Shunk; Sept. 2006, Deschutes Co., OR.*

About half of Red-breasted Sapsuckers may return to previous year's nest site, but only about 20 percent of hybrid individuals do so.

In coastal BC, nests relatively high in large trees; farther south, may nest lower. Cavity entrances are almost perfectly round. Does not reuse old nests, but often excavates new cavity in trees with old cavities.

### *Courtship and Parenting*

Difficult to assign display behavior to sex because of monochromatism.

Smacking sound is occasionally heard from adults in last week before fledging, produced by rapid opening and closing of beak. Juveniles often fledge over a few days. Parents feed them irregularly after fledging; young attempt independent feeding almost immediately.

### *Dispersal*

Small groups, possibly family groups or sibling pairs, observed dispersing in early fall, but after about October, individuals difficult to age in the field.

## NONBREEDING BEHAVIOR

### *Feeding*

***Preferred Foods:*** Eats tree sap and cambium, various arthropods, and small fruits. Late fall–early winter diet is mostly ants and other arthropods and about one-third plant material, mostly fruit and cambium.

***Foraging Behaviors:*** In breeding season, forages mostly by gleaning insects from trunks of live trees, occasionally on snags and live branches; also flycatches occasionally and spends about equal time

An adult hybrid Red-breasted x Red-naped Sapsucker. Note the red-tipped feathers showing through the black breast shield and black mustache, the red extending below the black breast, and the thin white eye stripe with no red between the crown and moustache. Also note that among individuals of this hybrid combination, the subspecies of the parental Red-breasted is impossible to determine in the field with certainty. *Jack Booth; Dec. 2012, Mendocino, CA.*

drilling sap wells. Selects most mature forested stands available, preferring old-growth, likely because of increased bark-surface area.

Selects wide variety of coniferous and deciduous trees for sap wells, including coastal redwoods and introduced eucalyptus. May use 10–15 percent of trees in a given stand for sap-well production. May favor trees with previous sap wells or injuries over unblemished trees. Documented tapping up to 2,000 wells in a single tree; many more likely in large trees. Taps lower portions of trunk in winter, where sap is less likely to freeze.

### *Territory Defense and Sociality*

Little information exists describing territoriality outside breeding season. Begins advertising around breeding territory, including sap wells, early in season using both drums and calls. As season progresses, becomes more aggressive and will chase competing individuals from vicinity of nest or sap wells.

Defense of nesting territories may extend in radius of 165–490 ft. (50–150 m) from nest site, farther in open areas or in small dense stands surrounded by openings; territories are smaller in dense forest. Adults will go considerable distance from nest, up to 1,475 ft. (450 m), for productive food source. Resident home ranges average 14.6 ac. (5.9 ha).

### *Interactions with Other Species*

Mutual territory boundaries are defended equally by Red-breasted and Williamson's Sapsuckers. Red-breasted is known to defend nest and sap wells from other species; however, when fully engaged in maintaining a circuit of wells, may not bother chasing off interlopers. Observed interacting aggressively with House Wrens.

Other species observed feeding on sap at Red-breasted Sapsucker wells include hummingbirds, warblers, kinglets, and butterflies.

**MISCELLANEOUS BEHAVIORS**

Observed bathing in and drinking from small stream. May rest vertically on external bark surface for extended periods in vicinity of sap wells.

## CONSERVATION

Partly owing to its limited range and partly to its taxonomic history, little attention has been paid to conservation issues for this species.

**HABITAT THREATS:** Loss of old-growth to logging and of wide swaths of forest to large-scale stand-replacement fire are the most significant threats. Loss of wintering habitat to development may indirectly affect breeding populations. Temperate rainforest populations are highest in old-growth, lowest in heavily managed stands. Fire suppression leads to inordinately large fires that likely push populations into competing territories with other Red-breasteds or pack them deeper into contact zones with Red-naped, where in either situation carrying capacity may already be limited.

**POPULATION CHANGES:** AK is estimated to support 800,000 Red-breasted Sapsuckers, or 35 percent of species' population. Presettlement population in nw. CA Douglas-fir forests estimated at 49,400 birds; estimated to have declined 15 percent by 1988, and expected to continue declining with increased timber harvest. Maximum breeding-season density has been recorded in w. OR in managed riparian woodland of about 22 birds per 100 ac. (40 ha).

Red-breasted Sapsuckers may perish at very low winter temperatures; populations are reduced following extremely cold winters in BC, when conifer sap freezes.

**CONSERVATION STATUS AND MANAGEMENT:** Ranked as a priority species for conservation in w. NV and se. AK; critically dependent on montane meadows in Sierra Nevada.

A fledgling sapsucker of mixed Red-breasted x Red-naped parentage (though it would not be identifiable as such without clear identification of both parents) feeds at fresh phloem wells excavated and tended by adults. *Gerard Gorman; July 2014, Bend, OR.*

## REFERENCES

Alaska Department of Fish and Game 2006; Allen and Wright 1918; American Ornithologists' Union 1902; Andres 1999; Askins 2000; Bendire 1889; Blow et al. 2000; Browning 1977; Browning, M.R., pers. comm 2004; California Automobile Association 1991; Cicero and Johnson 1995; Coppa 1960; Danforth 1938; Fritz 1937; Great Basin Bird Observatory 2005; Hamilton and Dunn 2002; Harris 1983; Howell 1952, 1953; Johnson 2007; Johnson and Johnson 1985; Johnson and Zink 1983; Joy 2000; MacIntosh, R., pers. comm.; Martin and Ogle 1998; Melchior 2006; Miller et al. 1999; Neel 1999; Nehls 1985; Oliver 1970; Raphael 1985; Rosenberg and Witzeman 1998; Rosenberg et al. 2007; Scott et al. 1976; Shuford 1985, 1986; Shunk 2005; Siegel and De Sante 1999; Taylor 1920; Trombino 2000; Van Vleit 2006; Walters et al. 2002a; Weisser 1973.

# LADDER-BACKED WOODPECKER

## *Picoides scalaris*

L: 6–8 in. (15.2–20.3 cm)
WS: 11–13 in. (27.9–33.0 cm)
WT: 0.7–1.6 oz. (21–48 g)

**Nicknamed "Specklecheck"** by El Paso naturalist Elsie McElroy Slater in 1945, the sprightly little Ladder-backed Woodpecker curiously explores the scrub and cactus lands of the desert Southwest. Equally at home on mesquite or cholla, the Ladder-backed also earned the nickname "Cactus Woodpecker," which persists today in the name of the widespread North American subspecies, *P. s. cactophilus*.

The compact Ladder-backed can raise its young in a wide variety of substrates, from agave stalks to fenceposts. Its persistent chatter closely resembles that of the smaller Downy Woodpecker, and its plumage that of its closest relative, the Nuttall's. Hybridization occurs with Nuttall's Woodpecker at a handful of sites where the desert meets the mountains in southern California and northern Baja California.

The breeding biology of the Ladder-backed Woodpecker remains something of a mystery, although its conspicuous presence allows us to learn much about its distribution and ecology. Breeding biology aside, a visit to the Joshua tree forests of California or the mesquite lands of west Texas will certainly expose the conspicuous wanderings and chatter of the Specklecheck.

### DISTRIBUTION

Widespread throughout sw. U.S., from s. CA to cen. TX, and southward through most of Mex.; often rare and local around fringes of primary range.

**PERMANENT RANGE:** Se. CA deserts east of Transverse and Peninsular Ranges, west through AZ and NM, mostly south of Mogollon Rim, and into w. and cen. TX. Also north into se. CO, sw. KS, w. OK, s. NV, and sw. UT.

South of U.S., through most of Mex. and very localized in Belize, Guatemala, w. Honduras, and ne. Nicaragua.

**SEASONAL MOVEMENTS** Generally sedentary, but may withdraw in winter from CO; also some historic indication of migration in lower Arkansas R. valley. Fall upslope movements in Big Bend region and eastward on TX plains.

**OPPOSITE:** A juvenile male Ladder-backed Woodpecker hanging from a cholla cactus skeleton. *Alan Murphy; May 2014, Pima Co., AZ.*

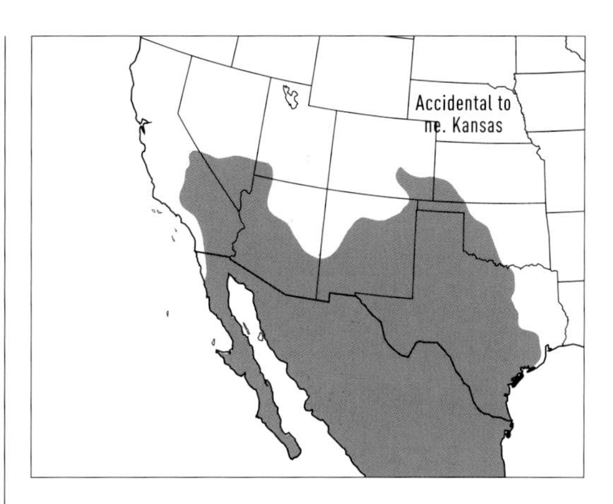


### HABITAT

Arid scrublands of Mojave, Chihuahuan, and Sonoran Deserts, from sea level (Baja CA) to 7,500 ft. (2,300 m). Especially common and widespread in mesquite scrublands; also dry oak-juniper and piñon-juniper slopes and foothills. Often associated with palo verde, catclaw, hackberry, and cholla cacti. Common in Joshua tree forests and through agave belt. In otherwise treeless deserts, concentrated around desert riparian areas with cottonwood and willow; also replaces Nuttall's Woodpecker in riparian areas outside Nuttall's range. Sparse in cleared agricultural lands.

### DETECTION

Vocally conspicuous all year and throughout its range. Often the most prominent woodpecker in suitable habitats.

**VOCAL SOUNDS AND BEHAVIOR:** Vocalizations are much like those of Nuttall's and Downy Woodpeckers. Gives three primary calls: call note, rattle, and churring.

Call note is a sharp *cheet* or *peet*, usually given singly but often repeated. Occasionally given as softer *pewt*. Occasionally gives quick double-call, like *chee-pit* or *kee-pit*, with second note lower in pitch. Quality of call is similar to that of Hairy Woodpecker but thinner; slightly burrier than those of Nuttall's and Downy. Used as low-intensity alarm call or simple location announcement. Often precedes rattle and may evoke drumming or rattle from other Ladder-backeds. Used year-round.

Rattle is similar to Downy's descending whinny, but raspier, and with shallower descent at end notes. Often starts with single, loud call note, as in Hairy Woodpecker. Used year-round; shorter-sequence rattle used Mar.–May.

Churring comes primarily in two forms, ter-

An adult male Ladder-backed Woodpecker in blooming acacia. Note the bird's raised crest. *Alan Murphy; May 2012, Pima Co., AZ.*

ritorial and courtship. During courtship and aggression, a raspy *chewit, chewit* or *chewt, chewt* or *churit, churit*, often linked to drumming. Less often a harsh *kreek* or *kweek*, in territorial defense (or offense), coupled with attacks, wing fluttering, or bill pointing; used Feb.–May.

Often gives sequence of calls during peak of territorial or courtship encounters, starting with call note, then double-call, then rattle and drumming, sometimes ending with repeated churring calls.

**NONVOCAL SOUNDS:** Rarely drums, with drumming notably faster than Nuttall's. Roll sometimes increases in depth and volume after first few beats, occasionally broken with a few less-forceful beats, then finishing with typical beats. Drumming bouts are shorter than in Nuttall's, and not associated with fixed site. Only noted drumming Feb.–Apr.; will respond to drumming by Nuttall's.

Feeding sounds, heard only sporadically, are characteristic of a probing bark stripper rather than wood-boring species.

## VISUAL IDENTIFICATION

**ADULT:** "Zebra-backed" *Picoides* with pied facial plumage. Male has vermilion crown, darkest and most solid at rear, fading toward front; female has all-black crown. Facial striping generally appears as black stripes on white. White parts on head may show variable buffy to dusky brown tint. Chest and belly are often buffy with black streaks or spots. Black and white bars on back are about equal in width.

In flight, head appears very light; back and upperwing mottled light gray, and flight feathers dark gray with long parallel white stripes.

**JUVENILE:** Generally resembles adult male. Both sexes have red crown, but red does not extend through rear crown; female possibly with reduced amount of red. Black parts are duller overall than those of adult.

**INDIVIDUAL VARIATION:** Amount of white in face can vary, with some individuals showing very narrow black facial stripes. Darkness and tone of underparts also quite variable.

Male larger than female, occasionally noticeable in field.

**OPPOSITE:** An adult female Ladder-backed Woodpecker on a cholla skeleton. *E. J. Peiker; Mar. 2010; Pima Co., AZ.*

**PLUMAGES AND MOLTS:** Preformative and prebasic molts occur June–Oct. Little additional information available.

**SIMILAR SPECIES:** Closely related and very similar in plumage to Nuttall's Woodpecker, with which Ladder-backed constitutes the *scalaris* superspecies. Ladder-backed is generally lighter overall than Nuttall's with lighter face and upperparts, more white on back and wings than Nuttall's, and mantle barred rather than solid black. Male Ladder-backed shows more extensive red on crown than Nuttall's (red may extend in front of eye in Ladder-backed but not in Nuttall's). Underparts and nasal tufts of Ladder-backed are tinted buffy or grayish, Nuttall's whiter in fresh plumage. Outer tail feathers are barred black in Ladder-backed, barely spotted in Nuttall's. Ladder-backed bill proportionately longer than Nuttall's.

Ladder-backed is more likely to occur in desert habitats than Nuttall's. Overlaps in range with Gila and Golden-fronted Woodpeckers, both of which also have "zebra-backed" dorsal pattern, though both are larger than Ladder-backed and unmarked below. Also occurs within 100 mi. (160 km) of very similar Red-cockaded Woodpecker in e.-cen. TX.

**GEOGRAPHIC VARIATION:**
Variation in N. Am. is negligible; most notable plumage diversity occurs in the Neotropics.

## SUBSPECIES

Four to 15 subspecies described by various sources. Eight subspecies generally accepted throughout range, only one in N. Am.

**NORTH AMERICA:** *P. s. cactophilus* occurs in desert areas across sw. and s.-cen. U.S., south to w.-cen. Mex.. White bars on back are broadest of all subspecies, showing as lighter upperparts; also generally shorter tail and bill than other subspecies. Group

A possible hybrid Downy (DO) x Ladder-backed (LB) Woodpecker from a narrow, enigmatic contact zone on the upper Texas coast; this may be the first individual identified as this hybrid combination by a photograph from the field. Note the spotting in the wing coverts and primaries and the nearly unmarked breast (DO), and the broad, bold striping down the center of the back (LB). *Greg Page; Mar. 2014, Galveston Co., TX.*

includes "*symplectus*," described from se. CO and w. OK, south to Tamaulipas, Mex.; "*mohavensis*" of upper Col. River valley; and "*yumanensis*" of lower Col. River valley. See below for other debated Mexican forms within the range of this subspecies.

**NEOTROPICS:**
*P. s. cactophilus.* In addition to above, includes: "*centrophilus,*" "*giraudi,*" and "*bairdi.*"
*P. s. eremicus.* N. Baja Cal., Mex. (see Hybridization, below).
*P. s. lucasanus.* S. Baja Cal., Mex. Includes "*soulei.*"
*P. s. graysoni.* Tres Marias Islands, Mex.
*P. s. sinaloensis.* S. Sonora south to central volcanic belt of Mex. Includes "*azelus,*" "*lambi,*" *and* "*agnus.*"
*P. s. scalaris.* Veracruz and Chipas, Mex. Includes "*ridgwayi*" and "*percus.*"
*P. s. parvus.* Yucatan and Cozumel I.
*P. s. leucoptilurus.* Belize and Guatemala south to ne. Nicaragua.

**HYBRIDIZATION:** Hybridizes with Nuttall's Woodpecker in small contact zones of so. CA and n. Baja CA. Hybrid individuals account for fewer than 10 percent of both species affected in CA contact zones; 12 percent of Nuttall's and 30 percent of Ladder-backed in Baja.

One possible Ladder-backed x Hairy hybrid collected from Coahuila, Mex. (see Hairy Woodpecker, Hybridization, p. 165). One possible hybrid with Downy Woodpecker photographed from very narrow contact zone on upper TX coast.

Most hybrid identity is based on museum specimens with plumage and size intermediate in morphology to those of parental species, although some Nuttall's hybrids are observed in the field, with intermediate vocalizations reported. See Nuttall's Woodpecker, Hybridization, page 143, for a discussion of hybrid plumage characters.

## BEHAVIOR

### BREEDING BIOLOGY

Little formally documented but general breeding behavior believed to be similar to that of closely related Nuttall's Woodpecker.

#### *Nest Site*

Favors dead branches of live mesquite trees; able to occupy less-wooded deserts than larger desert-dwelling woodpeckers because small size allows nesting in agave and other small-diameter substrates. Also excavates in live trunks and stalks, including those of Joshua tree, willow, cottonwood, walnut, oak, hackberry, pine, mesquite, agave, palo verde, and various yuccas. Frequently excavates in fenceposts and utility poles.

A Ladder-backed Woodpecker nest cavity in an agave "snag," located in upland Colorado Desert habitat where no trees are present. *Stephen Shunk; Mar. 2008, San Diego Co., CA.*

#### *Courtship*

Apparently monogamous. Courtship displays are similar to territorial defense postures; include wing fluttering and bill pointing, accompanied by harsh *kreek* calls. Gives a raspy, repeated *chewt* or *chewit* call in association with drumming and close-contact mating advances.

#### *Young Birds and Dispersal*

Individuals likely breed in their first summer. Little else is known about nestling and fledgling behavior, but believed to be similar to that of Nuttall's Woodpecker.

### NONBREEDING BEHAVIOR

#### *Feeding*

***Preferred Foods:*** Takes a variety of arthropods, mostly insects; less than 10 percent plant matter. Favors wood-boring beetle larvae, cotton leafworms, and *Formica* ants.

An adult female Ladder-backed Woodpecker (left) and an adult male Golden-fronted Woodpecker "bathe" in an artificial water feature. These two species evolved together as the only resident woodpeckers in the southern brushlands and lower Rio Grande valley of Texas. *Cissy Beasley; Sept. 2012, Bee Co., TX.*

***Foraging Behaviors:*** Occasionally excavates but more often performs acrobatic array of picking, probing, and flaking activities. Males and females may occupy complementary feeding niches; in CA, male is mostly in larger substrates, such as Joshua trees, female more often in smaller cacti and shrubs. In AZ, male strongly favors cholla cacti Oct.–June and mesquite July–Sept.; female prefers mesquite all year, and typically feeds higher than male. When both feed in mesquite, male prefers main trunk and larger branches, female smaller branches and twigs.

### *Territory Defense and Sociality*

May defend a range of about 16.5 ac. (6.7 ha). Display repertoire is almost identical to that of Nuttall's Woodpecker.

### *Interactions with Other Species*

Conflicts with other woodpeckers are typically related to defense of food supply or foraging area. Interactions with songbirds are often with other cavity nesters and center around nest or nest tree. Responds to Nuttall's vocalizations and drumming.

#### MISCELLANEOUS BEHAVIORS

Drinks water from crevices on trees or other plants.

## CONSERVATION

**HABITAT THREATS:** Greatest threat is clearing of mesquite lands for grazing and development. Loss of desert woodland and riparian habitat in se. CA has likely affected local populations. Ladder-backed mostly occurs in habitats not favored by humans, so probably less likely than other woodpeckers to be adversely affected by development.

**POPULATION CHANGES:** Populations in n. TX have increased where mesquite has invaded native grasslands. Grassland restoration efforts require some mesquite removal, and project managers continue to monitor and minimize unnecessary negative impacts on woodpeckers.

**CONSERVATION STATUS AND MANAGEMENT:** Described as sensitive to urbanization in AZ; focal species for conservation in CA, KS, and Mexican highlands.

Species plays a keystone role for numerous cavity nesters in its range; Lucy's Warbler highly dependent on this woodpecker where ranges overlap. Fire suppression in this species' range has allowed mesquite to encroach on native grasslands, facilitating expansion of woodpecker populations.

### REFERENCES

American Ornithologists' Union 1947, 1976; Austin 1976; Corben 1999; Gadd 1941; Kaufman 2005; Lincoln 1917; Linsdale 1936; Lowther 2001; Miller 1894; Miller 1955; Montgomery 1905; Morlan 2002; Oberholser 1911; Sennett 1878; Slater 1945; Steigman 2009; U.S. Fish and Wildlife Service 2006; Van Rossem 1942; Wauer 2001.

# NUTTALL'S WOODPECKER

***Picoides nuttallii***

L: 6.7–7.5 in. (17–19 cm)
WS: 13–13.5 in. (33.0–34.3 cm)
WT: 1.0–1.7 oz. (28–47 g)

**The familiar metallic rattle** of Nuttall's Woodpecker rings across California's oaken landscape, echoing up from wooded ravines as a signature of the Golden State's diverse avifauna. All but endemic to California, the Nuttall's has rarely been reported outside the state. Three recent photographic records exist for Nevada; the only Arizona record, from 1901, is probably a mislabeled specimen; and the only four confirmed records for Oregon are from birds that now rest in scientific collections.

A small population of Nuttall's Woodpeckers resides permanently in northern Baja California, where the species occasionally hybridizes with the Ladder-backed Woodpecker. Limited hybridization also occurs in central California with the Downy Woodpecker, and farther south with both the Downy and Ladder-backed Woodpeckers.

Once classified in the genus *Dryobates,* Nuttall's may be the best representative of that now-obsolete genus. Translation of the name's Greek roots can be interpreted as "one who walks on oaks," and Nuttall's Woodpecker spends much of its life adroitly exploring the surfaces of California's oak trees. And while the oak woodlands are a critical part of this species' feeding ecology, it also requires the decaying wood of riparian willows and sycamores for the excavation of its nest cavities.

## DISTRIBUTION

**PERMANENT RANGE:** Essentially restricted to CA, from far n.-cen. CA south throughout interior valleys and foothills to San Francisco Bay area; southward along coast, generally west of San Joaquin Valley, to n. Baja CA. Rare on coast north to Humboldt Co.; extension of southern populations into s. Sierra Nevada, and isolated population east of Sierras along Owens R., Inyo Co. Very rare in adjacent states.

**SEASONAL MOVEMENTS:** Not migratory, but some fall upslope movement occurs out of foothills and canyons of higher mountain ranges; may overwinter at these higher elevations.

## HABITAT

Typically requires combination of oak and riparian woodlands, with increased dependency on the latter in southern range, where oaks decrease in abundance. Oaks primarily favored as feeding substrate, due to extensive surface area of furrowed bark; riparian snags favored for cavity excavation, especially willow, sycamore, maple, and cottonwood. Rarely in coniferous forests. Northern habitats include hillsides, riparian areas, and canyons.

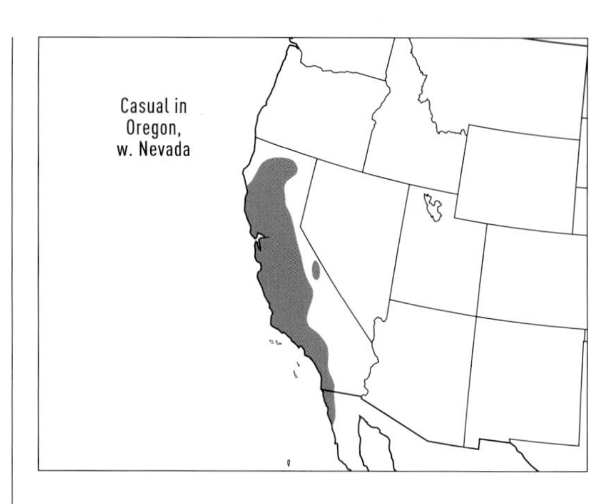


Wide range of regional oak preferences mixed with local combinations of willow, sycamore, maple, California bay laurel, gray pine, and California buckeye. Also found in eucalyptus and cypress groves near riparian areas. Multiple records above 5,000 ft. (1,524 m) in Sierra Nevada (very rarely over 11,000 ft. [3,350 m]). Fairly common to 6,000 ft. (1,829 m) in s. CA, occasionally to 7,000 ft. (2,134 m), among mixed oak, pine, and incense cedar among piñon pines in San Bernardino Mts.

## DETECTION

Often vocally conspicuous, chattering while feeding. Vocalizations and active shuffling around bark surfaces easily betray this species' presence.

**VOCAL SOUNDS AND BEHAVIOR:** Familiar rattle comprises a rapid series of call notes, somewhat crisper or less burry and with much faster cadence than rattle of Ladder-backed Woodpecker; typically given Oct.–May to establish territories, as contact call, and while feeding. Alternative to rattle may be any combination of the various call notes.

Mates utter short, high-pitched, single call notes when foraging out of sight of each other. Typical call notes are more liquid in quality and more often repeated than Ladder-backed's, resembling double-note call of White-headed Woodpecker. Both paired and single notes or a *whut, whut* call are given year-round and in general interactions with other individuals. During prolonged interactions, will give increasingly intense calls beginning with simple call note, followed by double note, then rattle and drumming.

**OPPOSITE:** An adult male Nuttall's Woodpecker shows its acrobatic skill. *Paul Bannick; Nov. 2010, Contra Costa Co., CA.*

An adult female Nuttall's Woodpecker. *Steve Kaye; Dec. 2011, Orange Co., CA.*

**NONVOCAL SOUNDS:** Male drums more frequently than female, up to 20 times per hour in Mar., but no more than 5 times per hour in other seasons. Drumming is most frequent Dec.–June. Ruffling wing sound associated with aerial displays is produced in flight or upon landing near antagonist. Feeding behavior produces only faint sounds associated with flaking and minor excavation into bark, although active shuffling around bark surface can frequently be heard.

## VISUAL IDENTIFICATION

**ADULT:** Small "zebra-backed" woodpecker with pied facial pattern. Head appears to be black marked with white (as opposed to opposite in Ladder-backed Woodpecker). Male has black forehead, streaked with white on center of crown, and bright red rear crown; female lacks red.

Solid black nape and upper back connect to crown by black stripe up hindneck, forming black hourglass shape when viewed from rear. Back barred black and white, black bars wider than white bars. Chest and belly are white surrounded by black spotting.

In flight, appears mottled dark gray. Blackish rump and tail contrast against mottled back.

**JUVENILE:** Lighter than adult above; male typically with small red patch centered on crown, female with fewer, more scattered, red-tipped feathers; red crown feathers in both sexes mixed with white spots and streaks.

**INDIVIDUAL VARIATION:** Little variation noted; possibly variable width of black on upper back. Leucism very rare. Male and female are very close in size, unlike in Ladder-backed Woodpecker.

**PLUMAGES AND MOLTS:** All ages acquire basic plumage June–Oct. Following preformative molt, juveniles appear similar to adults, except for brownish juvenal feathers contrasting with replaced feathers.

**DISTINCTIVE CHARACTERISTICS:** Darkest of North American "zebra-backed" *Picoides*. Except for tiny contact zones with Ladder-backed, Nuttall's is the only zebra-backed woodpecker in its range.

**SIMILAR SPECIES:** Closely related to Ladder-backed Woodpecker and very similar in plumage. Nuttall's is darker overall, with more black on face, solid black upper back, and narrower white markings on back and wings. Male Nuttall's shows less red on crown, not extending forward of eye; underparts and nasal tufts whiter; and outer tail feathers spotted black rather than barred black as in Ladder-backed. Nuttall's is less likely to occur in desert habitats. Female Williamson's Sapsucker also shows barred back, but habitat overlap is minimal.

**GEOGRAPHIC VARIATION:** Extremely restricted range, and hence no variation noted.

**HYBRIDIZATION:** Hybridizes rarely with Ladder-backed Woodpecker in s. CA and n. Baja CA and with Downy Woodpecker in cen. and s. CA.

***Nuttall's Woodpecker* x *Ladder-Backed Woodpecker***

Primary contact zones at northwest corner and eastern edge of San Bernardino Mts, sw. San Bernardino Co.; northwest contact along Mojave R., eastern contact at Morongo Canyon. Two other contact zones closer to Sierra Nevada, near Walker Pass in ne. Kern Co. and Olancha in Owens R. valley, sw. Inyo Co.

Observers typically report first hearing a call note or rattle that "isn't quite right." Some of characters shared by hybrid individuals may include broad

A fledgling male Nuttall's Woodpecker. *Ron LeValley; May 2008, Mendocino Co., CA.*

An adult male Nuttall's x Downy Woodpecker, a rarely encountered—and even more rarely photographed—hybrid combination. *Kimball Garrett; Nov. 2007, Los Angeles Co., CA.*

variation in black and white parts of face and back or in position and extent of red crown patch.

***Nuttall's Woodpecker* x *Downy Woodpecker***
Nuttall's is more distantly related to Downy than to Ladder-backed, and reports of Nuttall's x Downy hybrids are rare. Most hybrid identities are based on specimens intermediate in plumage and size to parental species. See Downy Woodpecker, Hybridization, page 154, for additional details.

Hybrid specimens observed and collected from San Diego, Los Angeles, Santa Barbara, and Orange Cos., s. CA. Multiple observations of a hybrid male from Sonoma Co., n. CA, may be same individual.

## BEHAVIOR

### BREEDING BIOLOGY

#### *Nest Site*

Nearly always excavates in dead trunks or limbs, occasionally fenceposts. Favors willow and cottonwood, also sycamore, alder, maple, and others; rarely in oak.

Male may excavate alone to start and may roost in nest site near completion; female is known to assist once laying begins. New nest cavity is excavated each year, sometimes in same tree as previous years.

#### *Courtship*

Probably monogamous, remaining in home territory year-round. Mates may forage independently (late summer through winter) or almost constantly together (just before excavation). Vocalizations between mates increase during courtship.

Pairs copulate two or three times per morning, usually preceded by mutual call notes (see photo on p. 15). Varied rituals may include either mate approaching the other; female lying parallel or perpendicular to limb; female lifting tail or moving it to one side; male extending wing; male raising crest; either mate tipping head upward, but not usually both together. Contact ranges from 1 to 20 seconds.

#### *Parenting*

Both mates may incubate for approximately 50-minute periods during day; female may spend

Nuttall's Woodpecker is only rarely known to nest in artificial structures. This successful nesting was documented by a nest-box monitor. (A) A recent hatchling with an unhatched egg. (B) A nestling male within 1 or 2 days of fledging. *Lee Pauser; May 12, 2008 (A) and May 25, 2008 (B), San Jose, CA.*

longer periods away. Brooding and feeding duties shared about equally between male and female.

### *Dispersal*

Adults eventually drive young from natal territory, occasionally giving spread-wing display and agonistic call to discourage young. Little is known about dispersal of juveniles, with seasonal ranges varying widely depending on habitat quality.

## NONBREEDING BEHAVIOR

### *Feeding*

***Preferred Foods:*** Diet is primarily insects, also other arthropods and up to 20 percent vegetable matter. Beetle larvae, especially long-horned beetles and click beetles, constitute about one-third of diet, with true bugs and caterpillars constituting another third; wide range of other arthropods. Plant foods include elderberries, seeds of various shrubs and large forbs, and rarely flower buds and apples. Also reported eating almonds and visiting Red-breasted Sapsucker sap wells.

***Foraging Behaviors:*** Feeds mostly by probing and gleaning from bark surface; will tap to pry open flakes and cracks, rarely bores beneath bark or into cambium. Often moves rapidly along trunk and branches, peering side to side into small crevices. Will glean from leaf surfaces and clumps of small twigs; may engage in a sort of hover-gleaning behavior, using wings for balance among small foliage clusters; also will gather fruits while hanging upside down.

Males may prefer larger trunks and branches than females and may typically forage closer to ground. Males more frequently tap for deeper prey, whereas females more often glean from surface and bark flakes.

### *Territory Defense and Sociality*

Wide array of territorial displays includes pointing bill toward aggressor; swinging, turning, or bobbing head; flicking or spreading wings; tail spreading, including flashing underside of tail at aggressor; supplanting aggressor from drumming perch; and direct attack in flight, either preceded by or in absence of other displays.

During extended encounters, wing spreading typically includes zig-zagging "hop-flight" toward opponent, with head and bill turned away. Frequently performs mothlike, wing-fluttering aerial display when flying over or displacing antagonist.

Male will raise rear of crest during encounters with other males or when present during female–female encounters. Males of adjacent territories will engage in progressively intense series of actions, closing ground between them with each display and building up to an attack or chase; such altercations often begin with alternate drumming near territory boundary and may include any of the above displays accompanied by vocalizations.

### *Interactions with Other Species*

Conflicts with other woodpeckers are most likely over feeding areas. Will respond to drumming or calls of closely related woodpeckers such as Ladder-backed, Downy, and Hairy, possibly to help maintain exclusive territories. Hairy Woodpecker typically prevails in conflicts with Nuttall's, but Downy and Nuttall's may displace each other equally. Known to interact with Red-breasted Sapsucker and Acorn Woodpecker at their sap trees.

Conflicts with cavity-nesting passerines typically occur when other species attempt to nest in tree with active Nuttall's nest. Ash-throated Flycatcher is dominant over Nuttall's nesting in same tree,

**OPPOSITE:** An adult male Nuttall's Woodpecker feeds an earwig to a nestling male. *Tom Grey; May 2008, Santa Clara Co., CA.*

but Nuttall's is dominant over Oak Titmouse and House Wren. Other conflicts observed with Western Scrub-Jay.

#### MISCELLANEOUS BEHAVIORS

Climbing behavior of Nuttall's Woodpecker differs from that of other woodpeckers: Nuttall's rarely moves more than about 1 ft. (0.3 m) without circling or working on another side of a trunk or branch. Readily hops to nearby limbs or trunks, even into tangles of small twigs (with or without use of tail). Tends to creep rather than hitch, with bill thrust ahead or at slight angle.

### CONSERVATION

**HABITAT THREATS:** Clearing of oaks for agricultural and urban development likely has affected regional populations; Nuttall's may have ranged into OR before westward settlement. Sudden oak death syndrome in cen. CA oak woodlands may create short-term increase in suitable nesting substrate but long-term decline in foraging habitat as trees fall. Fire suppression has led to increased fuel loads, which have in turn driven unnaturally large fires in oak woodlands, decimating large swaths of suitable Nuttall's habitat. Clearing and fragmentation of vegetation in riparian floodplains degrade nesting habitat.

**POPULATION CHANGES:** Early Breeding Bird Atlas studies estimated 2,500–5,000 pairs in Monterey Co. (1993) and up to 2,100 pairs in Orange Co. (1997). Sparse breeder in Marin Co.; probably extirpated from San Francisco. Despite decline in quality of riparian floodplains, Monterey Co. populations appear stable over last 100 years.

**CONSERVATION STATUS AND MANAGEMENT:** Yellow on 2007 WatchList. California Partners in Flight Focal Species in oak woodland; listed as critically dependent on oaks or oak woodland in Sierra Nevada. All listings are generally due to restricted range and potential threats to oak woodland habitats. Has responded positively to habitat restoration efforts that retain broad matrix of age classes in s.-cen. CA.

#### REFERENCES

American Ornithologists' Union 1947, 1976; Bailey 1914; Ballard et al. 2004; Block 1991; Browning and Cross 1994; Carpenter 1919; Coston 1999; Graham 1940; Heindel and Heindel 2005; Jenkins 1979; Kemper 2002; Lowther 2000; Miller and Bock 1972; Morlan 2002; Palmer 1928; Short 1965; Unitt 1986; Withgott 2002; Zack 2002.

## DOWNY WOODPECKER
### *Picoides pubescens*

L: 6.75–7 in. (17.1–17.8 cm)
WS: 11–12 in. (27.9–30.5 cm)
WT: 0.7–1.2 oz. (21–33 g)

**The little Downy Woodpecker** adorns backyard gardens and parks throughout North America. And this is one tiny woodpecker. Not only is it the smallest on the continent, but the Downy also ranks as one of the ten smallest woodpeckers in the world. Despite the bird's Lilliputian stature, birders from coast to coast well know the Downy's descending whinny heard from within their local woodlands.

More common in the East than in the West, the Downy is the second most abundant woodpecker in North America (behind the Northern Flicker). Benefiting from its ubiquity, the Downy Woodpecker has been the subject of numerous behavioral studies, one of the most notable describing sexual niche partitioning.

Downies generally reside year-round throughout their range, although regular seasonal movements occur among some populations breeding in the far north and at high elevations. These birds tend to move southward or downslope as the frost level lowers in elevation, often overwintering in urban woodlots and frequently visiting well-stocked feeding stations.

### DISTRIBUTION

**PERMANENT RANGE:** Resident through most wooded regions of N. Am., from boreal tree line south to Atlantic and Gulf Coasts and west to s. CA. Distribution very patchy, especially in West, based on suitable habitat.

**SEASONAL MOVEMENTS:** Generally sedentary, but regularly moves into habitats and locations not occupied when breeding, including increased presence at feeding stations from late summer through early spring.

Long-distance postbreeding dispersal (300 mi. [500 km] or more) may be mistaken for migration. Field data confirm such movements, but no known records definitively show return to point of origin that defines migration. Females may show a greater tendency than males for long-distance dispersal.

### HABITAT

Strongly favors riparian and other fairly open deciduous woodlands with low canopies, especially

**OPPOSITE:** An adult male Downy Woodpecker, Pacific Northwest rainforest subspecies (*P. p. gairdnerii*). *Scott Carpenter; Feb. 2006, Washington Co., OR.*

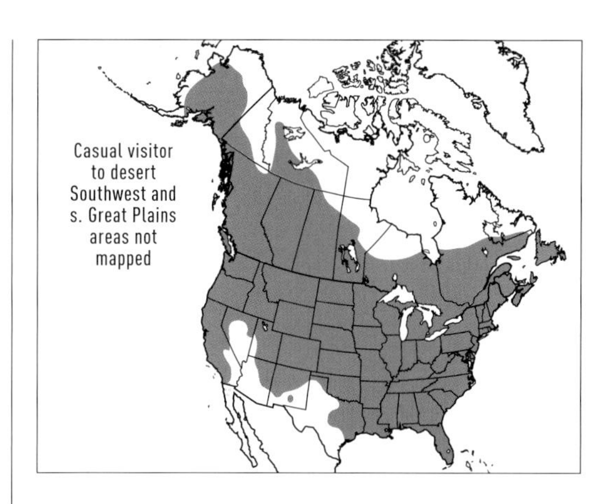


moist bottomlands and swampy aspen groves. Easily adapts to developed areas, including citrus groves, parks, and neighborhoods. Often occupies smallest size and age class of shrubs and trees among all woodpeckers in its home territory. Nearly always prefers deciduous over coniferous woodlands, but found in conifers near riparian areas, as well as burns in mixed-conifer forest with or without returning saplings and shrubs. Favors cottonwoods on plains and aspens in mountains.

### DETECTION

Often vocally conspicuous, calling frequently while feeding. Small size may hinder visual detection until bird moves into view.

**VOCAL SOUNDS AND BEHAVIOR:** Adults of both sexes have relatively simple repertoire of calls, with many subtle variations; generally call note, rattle, and churring. Intensity increases when associated with courtship and territorial displays.

Typical call note is sharp *pik* or *peet*; most often given singly or in uneven series; pitch increases with intensity. Sometimes given in rapid series during courtship flights.

Most diagnostic vocalization is classic descending whinny, sometimes named rattle call; series of notes often preceded by single or double call note, speeding up slightly and descending in pitch toward end. Little if any seasonal variation is noted in rate, though male may whinny more frequently during nest excavation. Considered important in forming pair bond, announcing location, and showing territorial dominance.

Other sputtering and churring calls are given during territorial or predator encounters, as well as sparrowlike *chirp* or *brip* as greeting between mates while foraging or at nest exchange.

**NONVOCAL SOUNDS:** Drum is a standard, steady, even series of beats, with broad uniformity across geographic range; occasionally performed on large,

An adult male Downy Woodpecker, nominate subspecies. *Alan Murphy; Aug. 2011, Jefferson Co., KY.*

resonant bark flakes, snapped-off treetops, or branch stubs. Both sexes drum year-round, more intensively and more frequently from late winter through spring. Females may drum more often than males earlier in breeding season.

Dueting between mates is performed during nest-site selection and pair-bond maintenance. Both sexes may not be able to differentiate their own drums from those of neighboring pairs, though vocalizations may be more individually recognizable.

Methodical tapping during nest excavation, typically slower than drum, may aid in attracting mate to site. Tapping is also heard during persistent foraging.

## VISUAL IDENTIFICATION

**ADULT:** Tiny black-and-white "pied" woodpecker, with striped head, whitish back and belly, and mottled wings. Short bill and puffy, dusky-colored nasal tufts. Variable red nape patch in male. Amount of white in wings varies geographically, greatest in eastern birds showing two bold wing bars. White outer tail feathers may show minimal black spotting on outer edges.

In flight, black wings show strong white flecking. Broad vertical white stripe through center of back is completely bordered by solid black. Much lighter underneath than above, flashing solid whitish belly and mostly white underwing.

**JUVENILE:** After about 2 weeks, nestlings appear nearly as feathered as adults. Juvenile female is distinguished from adult by short tail and shorter bill, with some red possible in forehead. Juvenile male shows red forehead flecked with black onto crown.

**OPPOSITE:** An adult female Downy Woodpecker, *P. p. medianus* subspecies, which occurs from the north-central Great Plains east to the Mid-Atlantic Coast. *Robert Royse; Oct. 2006, Scioto Co., OH.*

**INDIVIDUAL VARIATION:** Multiple reports of birds with malformed or damaged bills adapting and surviving. Albinistic, leucistic, and melanistic plumages are occasionally observed, very rarely including birds with solid black back. White parts of plumage occasionally acquire unusual coloration from contact with environmental stains, including tree resin and berry juices. Males rarely show yellowish nape or forehead.

**PLUMAGES AND MOLTS:** Preformative and prebasic molts generally occur June–Oct.

**DISTINCTIVE CHARACTERISTICS:** Smallest woodpecker in N. Am. Nasal tufts are disproportionately prominent. Downy and Hairy are the only two North American woodpeckers with solid white back. Retains relatively high levels of body fat year-round, allowing species to remain active in very cold temperatures.

**SIMILAR SPECIES:** Looks almost exactly like a small Hairy Woodpecker. Three key characteristics best separate the two species:

1. *Sound*—Call note is "thinner" and more plaintive than Hairy's loud squeak; sweet, descending whinny contrasts with Hairy's more sustained, scratchier, and rarely descending rattle. When nesting in close proximity, Downy's drum is generally notably shorter, slower, and softer than Hairy's.

2. *Foraging habitat and behavior*—Downy favors riparian strips and generally smaller trees, branches, and shrubs than Hairy; often forages among weeds and shrubs smaller than its body width; more often in ornamental plantings and developed areas than Hairy. Hairy generally favors larger trees in mature stands; rarely forages on substrate smaller in diameter than its own body width.

3. *Bill shape and size*—Downy's bill is more conical, and generally shorter than depth of its head (lores to nape).

Other distinguishing factors are more subtle and variable. Downy's outer tail feathers usually show black bars; Hairy's are usually all white. Downy's white mustache does not extend to white back as in Hairy; Downy lacks recurved extension of black malar stripe onto sides of upper chest; male Hairy

A fledgling male Downy Woodpecker, likely of the *P. p. medianus* subspecies. *Gene McGarry; June 2012, Woodstock, NY.*

A fledgling male Downy Woodpecker of the Pacific Northwest rainforest subspecies (*P. p. gairdnerii*). *Ollie Oliver; July 2013, King Co., WA.*

may exhibit split red nape patch, whereas Downy's is always solid.

Downy call note is also reminiscent of American Three-toed Woodpecker's call.

**GEOGRAPHIC VARIATION:** Size varies clinally across continent, with larger birds in north, in cooler areas, and at higher elevations. Exceptions correlate to variability in these three factors; for example, populations along upper Mississippi R. are smaller than those in s. Appalachians.

## SUBSPECIES

Six to 10 subspecies are recognized depending on source. AOU currently lists seven.

*P. p. pubescens*. Generally se. U.S. Smallest race; drab gray underparts. Includes "*minimus*" as synonym and brownish-breasted "*meridionalis*" of s. Atlantic and Gulf regions.

*P. p. medianus*. N.-cen. Great Plains to Mid-Atlantic Coast. Medium-sized, with large white spots in wing coverts. Typically smaller than *nelsoni* and whiter underneath than *pubescens*. Includes "*microleucus*" ("Newfoundland" Downy Woodpecker) of NL and Anticosti I., and possibly obsolete "*verus*," said to be intermediate in size and range between *nelsoni* and "*meridionalis*."

*P. p. nelsoni*. Interior AK south through n. BC, east to w. ON; possible records from CO. Large, white underneath, with minimal black barring on outer tail feathers.

*P. p. glacialis*. Coastal AK, Kenai Peninsula east through Prince William Sound, southeast to Juneau. Plumage characters are intermediate among *nelsoni, leucurus,* and *gairdnerii*.

*P. p. leucurus*. Interior Kenai Peninsula east and south to Bella Bella, BC, east to AB Rockies; e. Cascades and e. Sierra Nevada, east to cen. and s. Rockies. Reduced white in upperparts and reduced black in outer tail feathers, white underneath; Rocky Mts. birds may lack any markings in undertail coverts.

Includes "*parvirostris*" and "*homorus*," the latter of which includes "*oreoecus*" ("Batchelder's" Downy Woodpecker) of interior Rockies.

*P. p. gairdnerii.* Coastal rainforest, w. BC south to nw. CA. Almost solid black wing coverts and inner secondaries, occasionally with only small white spots. Birds of Vancouver I. and coastal OR through CA show white spotted tertials and wing coverts; includes "*fumidus.*"

*P. p. turati.* WA and OR Cascades to interior and n. coastal CA, occasionally to Rogue R. valley. Smaller and paler than *gairdnerii.*

**HYBRIDIZATION:** Several hybrids intermediate between Downy and Nuttall's in both size and plumage have been reported in cen. and s. CA. One possible Ladder-backed x Downy individual reported from upper TX coast, but the two species rarely come into contact.

Characters common to hybrid individuals may include aberrant back pattern (white back in Nuttall's, barred in Downy), aberrant position of red on head (Nuttall's with red nape, adult Downy with red crown), or aberrant markings below (Downy with blackish specks in flanks, Nuttall's pure white below).

## BEHAVIOR

### BREEDING BIOLOGY

#### *Nest Site*

Female most often selects nest site, but selection requires approval from mate. Preliminary selection may be followed by drumming, after which mate comes to inspect and drums in approval. May excavate several "starts" before selecting final cavity site.

Typically excavates cavity on underside of dead branch or leaning limb of dead tree, in slightly open-canopy forest; may select live trees with advanced heart rot. Preferred tree species vary widely by region; some favorites include pine, maple, oak, aspen, and poplar. Rarely nests in walls of human dwellings or in excavated fenceposts. Reported nesting in artificial polystyrene snags placed in cutover forest.

#### *Courtship*

Monogamous, although some helping behavior reported (see Parenting and Young Birds). Pair bond is renewed in late fall, with increased drumming and paired foraging. By early winter, female disperses, rebonding again in late winter. In habitats with ample food supply, pairs may remain together through winter. Courtship drumming begins as early as Jan. in southern states.

Wide variety of courtship displays. Most impressive, known as butterfly flight, involves mates chasing each other through open woodland (below canopy) with wings held high and flapped with slow, weak beats, sometimes in large loops through territory. Female provokes copulation by flying to male and perching crosswise on limb near nest cavity. Copulation sequence is typical of that of many woodpeckers. Contact lasts 10–16 seconds.

#### *Parenting and Young Birds*

Female begins laying 1–10 days after cavity is completed. Incubation typically begins after last egg is laid, occasionally sooner. Rarely double-brooded, and only in southern regions. Hatching is asynchronous, with high mortality among late-hatching chicks. Male may make more frequent and longer brooding and feeding visits than female.

Nonparental "helper" females are rarely observed incubating and feeding nestlings of mated pair. Helpers are typically not juveniles of prior year but may be more closely related to male of mated pair than to female.

Fledglings may first fly to nearest tree, perching motionless for over an hour and renewing begging behavior as adults return with food; within 2–3 days, juveniles beg more aggressively. Adults shepherd young until at least 3 weeks old, feeding and leading them to food sources. Each adult may temporarily "adopt" one or more young for short period and later trade fledglings with mate.

#### *Dispersal*

Young begin cavity roosting at 3 weeks, occasionally in natal cavity. Juveniles remain in natal territory for at least a few weeks, during which parents will chase away young fledged from neighboring nests.

Individuals usually breed in first full summer (second year), and may establish a territory within 1.6 mi. (2.6 km) of their natal site.

### NONBREEDING BEHAVIOR

#### *Feeding*

***Preferred Foods:*** Generally omnivorous, but favors arthropods, especially insects. Will eat fruits, seeds, and sap in season. Averaged over year, consumes about 80 percent animal food, 20 percent plant food. Among animal foods, favors beetle larvae and ants, followed closely by caterpillars, with lesser amounts of other insects, spiders, and snails. Specialized insects may include corn earworm, tent caterpillar, cecropia moth, and scale insects; adept at extracting insect larvae from goldenrod and poplar galls.

Sap consumption is generally limited to late winter and early spring. Favored plant foods are mostly nut mast and fruits of poison ivy, mulberry and re-

An adult male Downy Woodpecker of the intermountain West subspecies (*P. p. leucurus*). Note the small wood-boring beetle larva this individual has extracted from a very small-diameter burned ponderosa pine tree. *Kristine Falco; Oct. 2009, Deschutes Co., OR.*

lated berry producers, with lesser amounts of grains and bast.

***Foraging Behaviors:*** Feeds mostly at bark surface or just below; rarely bores into cambium. Forages actively, gleaning and probing among shrubs or large weed stems, and boring small excavations into tree bark. Sharp, pointed bill is well adapted for piercing insect tunnels and picking tiny insect eggs from plant surfaces.

Rarely seen flycatching or ground gleaning; observed collecting tent caterpillars off ground, even when prey are more abundant in trees. Eats small fruits whole; takes suet and peanut mixtures at feeders; occasionally drinks sugar water from feeders. May cache small berries in bark crevices.

Notable and well-studied sexual variation in foraging substrates and feeding styles (sexual niche partitioning). Depending on location, one individual may excavate more than the other, while mate gleans and probes from bark or ground-gleans. Niche may also be separated into concentric circles, with one bird feeding in a smaller circle and the other outside that circle; into semicircles, one bird feeding more in one direction from center of territory, the other in the opposite direction; or by canopy height, one feeding higher in the canopy, the other closer to the ground.

Downy sap wells are smaller, more circular, and shallower than those of sapsuckers. Downies do not tend their wells like sapsuckers do, and likely use existing sapsucker wells and tree injuries far more often than creating their own wells.

### *Territory Defense and Sociality*

Outside breeding season, territories range from 5 to 30 ac. (2–12 ha), depending on habitat quality. Territories may be established by a pair or by individuals. Mates may segregate microhabitats during winter foraging, with male high in canopy and female lower; however, individuals are generally tolerant of opposite sex feeding in close presence. During breeding season, male may escort mate away from intruder before engaging in defense behavior.

Both sexes defend territories, but interactions with intruders are usually same-sex. Defending individual may fan tail, raise crest, and point bill up or forward, waving it back and forth. More rapid version of courtship butterfly flight may be used to chase an intruder. Downy may flick wings out and up slightly, or extend fully in more aggressive posture. Final attempt to drive away intruder may result in midair, bill-grappling struggle.

Close contact at feeders may cause birds to erect crown feathers, tip up bill, and bob head, sometimes hopping back and forth or circling, always in facing posture.

Downy occasionally reacts to small passerines by dodging around trunk and raising wings; believed to be practice behavior.

### *Interactions with Other Species*

Red-bellied, Red-headed, and Hairy Woodpeckers are known to usurp and enlarge Downy nests. Yellow-bellied Sapsucker will also usurp, but is small enough that it does not need to enlarge cavity. Downy has nested within 50 ft. (15 m) of Red-cockaded Woodpecker without conflict, but Red-cockaded may dominate in conflicts away from nest site.

European Starlings will harass Downy parents tending a nest but are too large to fit inside without modification; will usurp cavity if entrance has been enlarged.

Downy competes for cavities with flying squirrels and numerous passerines, including nuthatches, chickadees, titmice, House Sparrows, and House Wrens. Downy is known to physically evict birds attempting to enter its cavity when occupied.

Downy regularly joins passerine flocks in winter, especially groups dominated by titmice and chicka-

An adult female Downy Woodpecker, nominate subspecies, in full spread-wing display. *Cissy Beasley; Oct. 2011, Nueces Co., TX.*

dees; favors flocks of three or more birds. Will also join mobbing groups of songbirds but typically does not engage in flushing the predator.

Red-headed Woodpeckers dominate winter foraging territories, forcing Downy to feed lower in canopy. Downy is known to forage behind Pileated Woodpecker, gleaning small food items from Pileated's excavations (commensalism).

Yellow-rumped Warbler and Pinyon Jay both observed chasing Downy from feeding areas. Downy is known to rob food stores of White-breasted Nuthatch and at least attempt the same from Lewis's Woodpecker. Hairy Woodpecker may rarely feed a begging Downy fledgling.

### MISCELLANEOUS BEHAVIORS

Downy flies with more frequent wingbeats and longer-duration bounds than most woodpeckers. Observed drinking from many different sources, including bases of epiphytes, dripping icicles, and bird baths. Will take dust, snow, or water baths, and occasionally sunbathes.

Individuals excavate new roost cavities, typically shallower than nest cavities, in fall. Reported excavating roost cavities in building walls, and occasionally roosts in nest boxes. Winter cavity entrances tend to face away from prevailing wind, rather than randomly as during nesting.

## CONSERVATION

**HABITAT THREATS:** Degradation of riparian habitats for agricultural development eliminates nesting and roosting trees, as well as foraging substrate, potentially precipitating local population declines. However, clearing of coniferous forests and subsequent planting of hardwoods and ornamental trees, often associated with residential development, may result in local population increases.

Large-scale forest clearing for dense, even-aged monoculture counters Downy's preference for open stands and eliminates nesting habitat in short term. Downy may be one of the few tree gleaners that responds positively to forest management that combines habitat conservation and logging.

**POPULATION CHANGES:** Notable local population increases correlate with bark beetle outbreaks and fires, especially in West and South.

**CONSERVATION STATUS AND MANAGEMENT:** Listed as moderate conservation priority in AL, AZ, MT, and WA, and as nongame indicator species in IL. Otherwise protected only by broad general protections for nongame birds and other native wildlife.

Management guidelines for optimal habitat include conservation of forested habitat, ranging from virgin bottomlands to sparsely stocked stands along ridges, with retention of snags and live trees large enough for nesting and maintenance of relatively open canopy.

## REFERENCES

American Ornithologists' Union 1947, 1976; Bailey and Niedrach 1938; Batchelder 1889; Brewster 1897; Browning 1995; Burchsted 1987; Confer and Paicos 1985; Contreras, A., pers. comm.; Corben 1999; Cronenweth, S., pers. comm.; Dodenhoff 2002; Fisher 1902; Forsyth 1988; Gabrielson 1924; Greenway 1978; Grinnell 1923, 1928; Harrison 2002; Hicks 1939; Hubbard 1963; Jackson and Ouellet 2002a; Lincoln 1917; Linsdale 1936; Mills et al. 2000; Ostry et al. 1982; Petit et al. 1985; Rett 1918; Schroeder 1983; Smith 1996; Soule 1899; Westnedge 1891.

# HAIRY WOODPECKER

## *Picoides villosus*

L: 8.5–10.5 in. (21.6–26.7 cm)
WS: 15–17.5 in. (38.1–44.5 cm)
WT: 1.5–2.8 oz. (42–80 g)

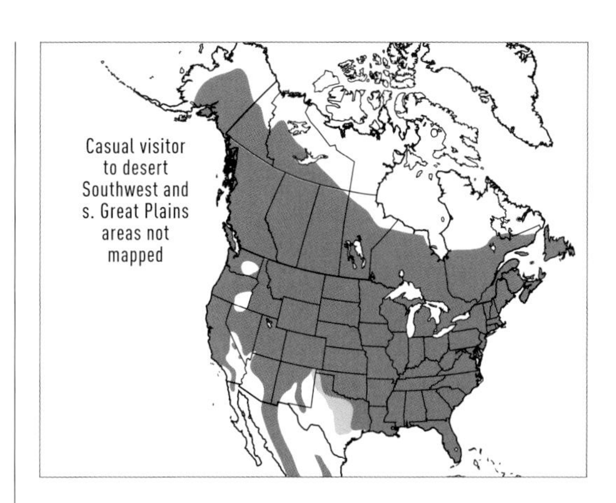


**Ubiquitous, adaptable, and opportunistic,** the Hairy Woodpecker epitomizes the North American woodpecker. Coniferous forest landscapes from the Rockies westward frequently support more Hairies than any other woodpecker species. East of the Rockies, the Hairy is less common than it is in the West, but this quintessential woodpecker is still widespread and locally abundant, from the northern treeline south to the Bahamas. Mountainous regions from Mexico to Panama host still more Hairy Woodpecker populations.

The Hairy's sedentary nature and occupation of such an immense range have led to some incredibly complex subspecific divisions in the last 100 or so years. As few as 11 and as many as 21 subspecies have been described by various authors, with up to 12 forms inhabiting North America alone.

Hairy and Downy Woodpeckers have nearly identical plumage and distribution, but a relatively wide genetic distance separates the two species. Where they overlap locally, Hairies and Downies exploit complementary feeding niches. Their similarity in plumage can be described as convergent evolution.

### DISTRIBUTION

Most widespread *Picoides* on the continent. Permanent resident throughout its range, with some northern and high-elevation withdrawal in winter and occasional seasonal wandering.

**PERMANENT RANGE:** W. AK, eastward along northern boreal tree line to NL. South into western mountains and eastward to Atlantic and Gulf Coasts. Rare at northern tree limit, but abundance of these populations is difficult to document because of remoteness of habitat. Restricted to islands of treed habitat in plains and Great Basin. Absent from treeless deserts and agricultural areas, as well as s. FL and FL Keys, though present on northern Bahaman islands.

**SEASONAL MOVEMENTS:** Northernmost populations may wander sporadically and unpredictably as post-breeding dispersal, or may withdraw southward in winter. High-elevation breeders of western mountains may move downslope in winter. Some inland populations may move coastward outside breeding season. Anecdotal observations of fall flights or large wintering populations may be misinterpreted as migratory behavior, but no confirmation of return-to-origin movements implying true migration.

**OPPOSITE:** An adult female Hairy Woodpecker, nominate subspecies. *Robert Royse; May 2005, Chippewa Co., MI.*

### HABITAT

Prefers mature coniferous and mixed-forest habitats, from mountains to bottomland. Rangewide abundance may be correlated with forest stand age, patch size, and disturbance level. In northern hardwood forests, may be up to twice as abundant in mature, undisturbed forest as in disturbed and successional habitats.

East of Rockies, pine and cedar are locally preferred, but species is widespread among birch-oak savannas, mature riparian habitats, mixed hardwoods, and southeastern swamp forests. Often the most common woodpecker in burned forests. In West, occurs from sea level to 11,500 ft. (3,500 m), with small populations in desert deciduous and riparian woodlands.

### DETECTION

Because of ubiquity and vocal nature, one of easiest woodpeckers to detect and identify throughout its range. Calls frequently throughout year, most often in mornings and early in breeding season, least often during excavation, incubation, and late summer. Vocalizations are fairly distinctive; foraging sounds are relatively loud.

**VOCAL SOUNDS AND BEHAVIOR:** Most vocalizations are fairly simple and fall into three categories: call note, rattle, and contact call. Both sexes give all calls.

Call note is simple sharp *peek* or *queek*, described in many ways and quite variable, but generally a single syllable and usually repeated. This is most common call, given in wide variety of circumstances while flying or perched.

An adult male Hairy Woodpecker of the intermountain West subspecies (*P. v. orius*) feeds a nestling male. *Steve Brad; July 2010, Mono Co., CA.*

Rattle, like *jee-jee-jee-jer-jer* or *chee-chee-cher-cher-cherk*, is rapidly repeated, variable-length series of notes, all on same pitch or slightly descending at end, usually at steady cadence. Rattle is usually preceded by, and slightly lower pitched than, opening single or double call note. Given all year but with increased frequency during early nesting season.

Contact call resembles *wicka-wicka* or *ricker-ricker* or some variation. Given all year between individuals upon or following close approach, and in breeding season as territorial call. At nest exchange, mates may also give softer and truncated version in single syllables, like *ruck-ruck-ruck* or *tuk-tuk-tuk*.

Adults also observed giving a quavering whistle when in distress. Juveniles can often be distinguished from adults into early fall by their harsher call.

**NONVOCAL SOUNDS:** Drumming is typical of other *Picoides* and often impossible to separate from that of other species in local habitats. Roll maintains steady cadence with little variation at start and finish. Drums throughout year but far more often during breeding season. Male and female drums are inseparable, and individually quite variable. Given in territorial, courtship, or family-group communications.

Mates perform mutual tapping as a few slow deliberate blows, often during nest-site selection. Aggressive feeder and easily heard chipping or flaking bark and boring into cambium.

## VISUAL IDENTIFICATION

One of the most variable woodpeckers on continent, though variation can be subtle. Description here applies to nominate *P. v. villosus*. See Subspecies for distinguishing characteristics of 11 additional forms.

**ADULT:** Typical pied *Picoides* plumage. Bill about as long as head is deep. Male shows bright red patch or band across center of nape, sometimes broken into two smaller patches by black stripe. Black malar stripe recurves onto sides of chest. White back is bordered by black, with two scalloped white wing bars in wing coverts. Solid black tail is framed by white outer tail feathers.

In flight, appears checkered black and white above; white diamond-shaped patch on back contrasts with solid black rump and central tail. Underparts white or off-white.

**JUVENILE:** Similar to adult, but black parts duller and white parts sometimes tinted light buff. White eyebrow does not reach front of eye, and forehead may be speckled with white; eye-ring absent or faint until at least 6 weeks old. Flanks, sides of upper chest, and white back are often finely barred black.

Male generally shows some red in forecrown and upper forehead, rarely orange or yellow, broken with black streaks, but no red nape. Female with much reduced red in forehead, sometimes only a few red-tipped feathers.

**INDIVIDUAL VARIATION:** Males average larger than females in bill, wing, and leg length; females may have longer tail than males, possibly related to foraging differences. Adult females and immatures of both sexes may show more prominent streaking in flanks than adult males.

White underparts may be stained from soot of burned bark, pitch, or other environmental sources. Pigment-related plumage variations are occasionally reported. Very rarely, individuals may be gynandromorphic, with red nape patch only on one side.

**PLUMAGES AND MOLTS:** Prebasic molt timing is similar to that of other woodpeckers. Juvenile males may begin acquiring red nape by late summer, achieving first basic plumage within about 4 months after fledging. Juveniles may also show obvious black bars in center of back before preformative molt is complete.

**DISTINCTIVE CHARACTERISTICS:** Hairy, Downy, and Red-headed are the only North American woodpeckers with solid white chest and belly, Hairy only rarely with faint streaking. Hairy and Downy are the only two with solid white stripe down entire back.

**SIMILAR SPECIES:** Classic identification challenge is comparison with Downy Woodpecker; see Downy Woodpecker, Similar Species, page 152, for detailed discussion.

Hairy Woodpecker is very similar to Rocky Mountain form of American Three-toed Woodpecker (*P. d. dorsalis*), which shows nearly solid white back and boldest facial striping of the American Three-toed Woodpecker subspecies. Hairy's facial pattern is superficially similar to that of some other *Picoides* species, but other distinguishing characteristics allow for easy separation.

Drumroll is often impossible to separate by

An adult female Hairy Woodpecker delivers a large wood-boring beetle larva to nestlings. This individual exemplifies the difficulty in assigning subspecies in the field. Its parents are most likely of the *P. v. monticola* subspecies, but because of location, either of them may have been an intergrade with *P. v. orius* or *P. v. harrisi*. *Dick Tipton; June 2009, Deschutes Co., OR.*

ear from that of Northern Flicker or other *Picoides* species sharing Hairy's local range and habitat. Vocalizations may resemble those of Arizona Woodpecker in style, but latter's voice is scratchier. Hairy's rattle may sometimes be confused with that of White-headed Woodpecker; latter's is more liquid in quality and often preceded by multiple doublet or triplet call notes.

**GEOGRAPHIC VARIATION:** Plumage and size vary clinally across Hairy's range. Generally decreases in size from north to south and from higher to lower elevations. Western birds are generally darker, males with broader red nape patch; populations in high-humidity regions show relatively dark breast and minimal spotting on upperwing. Island races of Northwest and Northeast show some streaking in flanks, barring on back, and black spotting on outer tail feathers. Underparts range widely, from white to yellowish to cinnamon.

## SUBSPECIES

**NORTH AMERICA:** Historically, 11 to 21 subspecies recognized, depending on source. AOU currently recognizes 12 subspecies in N. Am. alone (described below), with up to 9 in Neotropics. Broad introgression among adjacent populations makes field identification to subspecies problematic if not impossible.

*P. v. villosus*. East of Rocky Mts. to Atlantic Coast, northeast to NS and south to cen. TX. Intermediate in plumage between *septentrionalis* and *audubonii*.

*P. v. septentrionalis*. Cen. AK south and east through boreal forest to cen. ON; largest, whitest subspecies.

*P. v. harrisi*. Coastal Northwest, from extreme sw. YT to nw. CA. Underparts, head stripes, and sometimes back are washed dark grayish brown; usually some streaking on sides and flanks.

*P. v. sitkensis*. Coastal parts of se. AK and cen. BC. Dark like *harrisi* but buffier (less gray) below, with more white spotting and barring above and black parts paler than *harrisi*; may also have shorter bill than *harrisi*. Likely very difficult or impossible to separate these two subspecies in the field.

*P. v. picoideus*. Endemic to Haida Gwaii (Queen Charlotte Is.), BC. Darker below than any mainland population; back patch and outer tail feathers are strongly barred with black, and flanks have heavy black streaks. Similar in size to *harrisi*; upper mandible averages slightly shorter than that of Vancouver I. populations of that subspecies.

*P. v. hyloscopus*. Coast Ranges, n. CA south to mountains of n. Baja CA, east to s. Sierra Nevada.

A fledgling male Hairy Woodpecker, just out of the nest, that superficially appears to represent the Pacific Northwest rainforest subspecies (*P. v. harrisi*)—despite the location of its natal territory on the east slope of the Cascade Mountains. Without a positive ID of both parents, the location alone precludes a positive subspecies identification of the juvenile. *Dick Tipton; July 2008, Deschutes Co., OR.*

A fledgling male Hairy Woodpecker, nominate subspecies. *Gene McGarry; June 2012, Woodstock, NY.*

Paler, buffier, and smaller than *harrisi*; smaller at higher latitudes. N. Baja CA populations, once named "*scrippsae*," are slightly smaller still.

*P. v. orius*. Inland mountains and plateaus, s. BC south to AZ. Drab or creamy white below; faint wing markings. Larger and buffier than *hyloscopus*; smaller and darker than *monticola*.

*P. v. monticola*. Rocky Mts., BC south to NM, east to western plains, and west to e. slope of Cascades. Large; bright white below with minimal white in wings. Sometimes considered a variant of *septentrionalis*; originally proposed as "*montanus*."

*P. v. icastus*. Se. AZ and sw. NM, south through Sierra Madre Occidental to n. Jalisco. Plumage like *leucothorectis* but notably smaller.

*P. v. leucothorectis*. Se. CA east to w. TX and north into NV and UT. Much less white above and much smaller than *monticola*. Sometimes considered part of *orius*.

*P. v. audubonii*. E. TX east and north to se. VA, south to Gulf Coast and FL. Smallest eastern continental population, and representing end of cline. Darker and smaller than *villosus*, light buffy below, minimally white above.

*P. v. terraenovae*. Endemic to NL forests. Distinctive; very little white above, narrow back patch, variable black markings on outer tail feathers, sides and flanks with fine black streaks, especially in females and juveniles.

Females from NS (*villosus*) and Sapelo I., GA (*audubonii*), have exceptionally white face with white supercilium joining white malar stripe in front of eyes, similar to Bahaman races.

**NEOTROPICS:**

*P. v. intermedius*: San Luis Potosi, s. Tamaulipas
*P. v. enissomenus*: s. Jalisco south to cen. Guerrero
*P. v. hylobatus*: central Mexican highlands

*P. v. jardinii*: cen. Veracruz to cen. Oaxaca
*P. v. sanctorum*: cen. Chiapas south to s. Guatemala
*P. v. fumeus*: s.-cen. Honduras and n.-cen. Nicaragua
*P. v. extimus*: n.-cen. Costa Rica to w. Panama
*P. v. piger*: n. Bahamas
*P. v. maynardi*: s. Bahamas

**HYBRIDIZATION:** Two museum specimens debated as possible hybrids; thought to be Hairy x Ladder-backed, the most likely of which was collected from Coahuila, Mex., at 7,000 ft. (2,134 m)—above the typical elevation for Ladder-backed but still within range of local Hairies. A third bird observed in the field in OR in 2010 showed intermediate plumage traits between Hairy and White-headed Woodpeckers.

## BEHAVIOR

### BREEDING BIOLOGY

#### *Nest Site*

Typically selects snag or live tree with fungal heart rot, often favoring larger-diameter trees and apsens in western regions. In areas with low snag density and high competition for available cavities, may move into atypical habitats. Nests as high as 11,000 ft. (3,350 m) in s. Rockies.

#### *Courtship*

Monogamous, at least long term, possibly for life. Some pairs remain bonded year-round, others separate in fall and renew bond in Feb.

In typical courtship display, male repeatedly vocalizes, raising red nape feathers into a crest and fanning tail to flash white outer feathers. Other display behaviors include bill pointing with extended neck, bobbing dance, and wing flicking; also looping chase flights and solo mothlike flights.

Pair begins copulating well before egg laying begins. No favored distance from nest for copulation; male typically shuffles up trunk and out onto branch where female waits. Female may also perform flutter-flight toward male while calling, then land on a branch as invitation.

#### *Parenting*

May begin laying 1–14 days after cavity completion; full-time incubation begins after last egg is laid, but male roosts in nest cavity before that, possibly resulting in asynchronous hatching. Second broods are not confirmed, but Hairy will renest following a nest loss.

**OPPOSITE:** An adult male Hairy Woodpecker of the southernmost subspecies (*P. v. extimus*), from Costa Rica and Panama. *Stephen Shunk; Mar. 2015, Bosque de Paz, Costa Rica.*

An adult female Hairy Woodpecker, northern Bahamas subspecies (*P. v. piger*). Note the solid white lores and creamy buff wash to the throat and upper chest. *Gerlinde Taurer; Jan. 2013, Great Abaco, Bahamas.*

Division of parental duties varies widely among individual pairs. Nest exchange is preceded by soft churring calls from approaching bird, or tapping or soft drumming from mate inside nest.

#### *Dispersal*

First flight is often less than 33 ft. (10 m) from nest, juvenile landing on vertical trunk, then resting for up to 3 hours in landing spot. During first week, juveniles explore trunk surfaces and eat opportunistically, but they depend almost completely on parents for food. Parents may split up fledglings for feeding duties before dispersal at 3 or more weeks.

An adult male Hairy Woodpecker (most likely *P. v. monticola* subspecies) with a large wood-boring beetle larva, possibly of the Cerambycidae family. *Scott Carpenter; June 2012, Deschutes Co., OR.*

After young disperse, range of parental pair can be very large and may overlap with that of others, but often observed singly outside breeding season.

## NONBREEDING BEHAVIOR

### Feeding

***Preferred Foods:*** Consumes wide range of arthropods, mostly insects, and various fruits and seeds. Year-round, diet averages more than 75 percent insects, a third of which are long-horned and metallic wood-boring beetle larvae. Plant foods include various unidentified fruit, seeds, and other plant materials (possibly mast or cambium).

Local populations are known to increase during bark beetle outbreaks; this is one of three woodpecker species documented reducing local beetle and moth larvae outbreaks by 50–98 percent.

***Foraging Behaviors:*** Scales bark from trees by pecking or prying to remove flakes. Excavates tunnels into cambium in search of larval galleries. Also gleans from bark surface or crevices. Known to tap rapidly along branch or trunk in search of resonant prey tunnels beneath surface.

Some estimates of 75 percent gleaning and scaling, 25 percent excavating. Males excavate more often than females; females peck at bark or glean more often. Flycatching is rare and mostly in males, limited to warm, calm days. Will feed at sapsucker wells and excavate into sugar cane for sap. Known to hang from branches to excavate gall parasites and rarely to excavate into large, hollow stalks of various forbs.

Fall through spring, occasionally bores into large live pine cones to extract seeds, juveniles more than adults. Rarely takes sunflower seeds from feeders, extracting mast by wedging seed into bark crevice and hammering into shell; feeds regularly at suet and peanut feeders.

Observed actively using wings to catch dropped food items while feeding on vertical trunk; known as direct wing catching. Some evidence of ability to cache and refind food.

### *Territory Defense and Sociality*

Local nesting densities vary widely, from 6 to 15 pairs per 25 ac. (10 ha); in ne. OR reported as 1.7–4.7 birds per 247 ac. (100 ha) in grand fir forest. Highest densities are nearly always in large contiguous forest stands, lowest in areas cleared of large stands for urbanization or agriculture.

Territorial behavior is always same-sex opposition, typically peaking twice annually, at start of breeding season and just before pair initiates excavation.

Variety of displays includes bill waving, with extended neck and wings, and exaggerated threat display, with head tipped way back and wings held over back at steep angle. Drumming is a frequent component of territorial announcement by both sexes.

### *Interactions with Other Species*

Reports of conflicts with other woodpecker species over nesting sites or winter feeding territories; occasionally nests in same or adjacent tree with Red-naped Sapsucker.

Known to feed opportunistically among other woodpecker species, and especially with Black-backed Woodpecker in burned forest. Observed supplanting Black-backed while roaming and feeding through forest stand, occasionally after Black-backed begins an excavation, Hairy then withdrawing targeted prey. When foraging with Black-backed, Hairy may spend more time on branches than Black-backed, possibly because of Hairy's greater agility. Like Downy, known to follow Pileated Woodpecker, gleaning remnant foods left behind from large excavations.

May join mixed-species flocks in cold winters. May engage in bizarre displays in presence of small passerines, explained as possible practice for evasion from true predators.

European Starling and various squirrels are known to usurp nest and roost cavities.

## MISCELLANEOUS BEHAVIORS

Drinks water collected on tree surfaces, at bases of epiphytes, from bird baths, and in small streams. Observed bathing in snow and shallow-water pools. Scratches head both directly (under unlowered wing) and indirectly (over lowered wing).

## CONSERVATION

**HABITAT THREATS:** Snag removal and fragmentation of woodlands and forests are detrimental to local populations. Fire suppression in western forests leads to large-scale stand-replacement fires that provide high-quality short-term habitat that decreases in productivity for nesting and foraging in long term.

**POPULATION CHANGES:** Opportunistic nature of species correlates with local population increases during bark beetle outbreaks and immediately following fire episodes. Remains in burned forest for several years following fire.

**CONSERVATION STATUS AND MANAGEMENT:** Listed as moderate conservation priority in AK, MT, and eight states east of Rockies, including in s. New England. Indicator Species in IL and on Carson National Forest, NM. Closely associated with many other sympatric species that are listed at higher levels of concern in many areas, therefore increasing need for monitoring.

Forest fragmentation, loss of old-growth, and competition with European Starling may be significant factors in local population declines. Similar forestry practices for all woodpeckers would benefit Hairy, including delayed salvage of burned forest and retention of large snags, although Hairy may depend less on snags for nesting than other woodpeckers. Retention of large forest patches is essential for Hairy habitat, whether conifers in West or hardwoods in East.

Through sheer abundance, Hairy plays a major ecological role in local forests, but it is not formally listed as threatened and therefore not monitored well; even a minor decline in populations can have a significant impact on dependent species, especially secondary cavity nesters. Has responded positively to habitat restoration efforts that retain a broad matrix of age classes in s.-cen. CA.

## REFERENCES

Allen 1911; American Ornithologists' Union 1947, 1976; Anthony 1896, 1897; Burtch 1923; Chambers 1979; Conner 1975, 1977; Conner and Adkisson 1977; Daniel 1901; Dunham 1963; George 1972; Greenway 1978; Hailman 1959; Hempel 1922; Irons 2010; Jackson and Ouellet 2002b; Jenkins 1906; Jensen 1923; Kilham 1960, 1966b, 1968, 1969, 1974a, 1974b, 1979b; Kirkpatrick et al. 2006; Kisiel 1972; LeBaron 2004; Miller et al. 1999; Morrison and With 1987; Munro 1943; Nielsen-Pincus 2005; Parks 1944; Ramp 1965; Reynolds and Lima 1994; Selander 1965; Shelley 1933, 1938; Short 1969a; Stallcup 1969; Sutton and Edwards 1941; Swarth 1911; Villard and Beninger 1993; Weikel and Hayes 1999; Wood 1905a.

# ARIZONA WOODPECKER

***Picoides arizonae***

L: 7–8 in. (17.8–20.3 cm)
WS: 14–15 in. (35.6–38.1 cm)
WT: 1.2–1.8 oz. (34–51 g)

**At the northern end of the Sierra Madre** Occidental, a series of "sky islands" rises out of the parched Sonoran Desert. The oak-dominated encinal gives way to Madrean pine-oak woodland, and a sharp *keek* rings out over the morning chorus of the Painted Redstart, Hepatic Tanager, and Scott's Oriole. Just beneath the shallow canopy, the sharp call note is appended with the distinctive tattoo of a woodpecker drumming from the snapped-off limb of a large Emory oak. A second *keek* emanates from the creek bed below, and the chocolate brown Arizona Woodpecker exits his nest cavity in a sycamore snag, finished with his nighttime incubation duty, and ready for another day in paradise.

Occurring in North America only from southwestern New Mexico's Animas and Peloncillo Mountains west to the scattered ranges of southeastern Arizona, the Arizona Woodpecker is the most range-restricted of our continent's Picinae. This species is sometimes frustratingly difficult to see, but its loud scratchy rattle and sharp call note, both reminiscent of a Hairy Woodpecker, are easily recognized in oak-covered mountain slopes across its home region.

Likely because of its restricted range, the Arizona Woodpecker escaped description in the early works of Audubon, Wilson, Coues, and others, being first described as a North American resident in 1875 by Henry Henshaw, who considered it a subspecies of Strickland's Woodpecker of Mexico. In 1886 Edward Hargitt described the northern form as distinct from Strickland's Woodpecker, and he named it *Picus arizonae*, the Arizona Woodpecker. Today, the Arizona is one of the most sought-after by birders—and least studied by ornithologists—in North America.

## DISTRIBUTION

**PERMANENT RANGE:** Permanent resident throughout its range. Found only in se. AZ and extreme sw. NM; western limit at Quinlan Mts.; northern limit in Santa Catalina, Galiuro, and Pinaleno Mts.; east to Animas and Peloncillo Mts. of NM.

Occurs south into Mex. throughout Sierra Madre Occidental and across central volcanic belt of w. Jalisco and n. Colima. Disjunct populations occur in s. Jalisco and Michoacan.

**OPPOSITE:** An adult female Arizona Woodpecker. *Doug Backlund; Feb. 2013, Santa Cruz Co., AZ.*

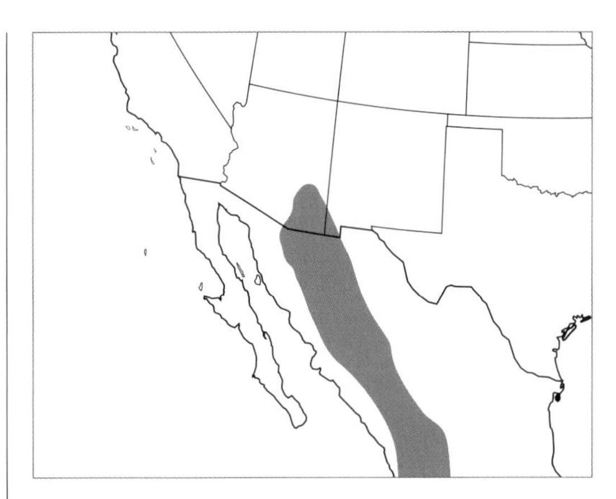

**SEASONAL MOVEMENTS:** Nonmigratory; some downslope movement into encinal—habitat dominated by oaks—and canyon mouths, possibly in response to seasonal or episodic food shortage. Individuals rarely move into adjacent lowlands.

## HABITAT

Mid-elevation Madrean pine-oak woodlands and mountain riparian dominated by oak, but also containing pine, juniper, and sycamore. Especially associated with sycamore-walnut riparian areas, and live oaks adjacent to riparian woodland; occasionally mesquite. Found primarily from 4,500 to 7,500 ft. (1,370–2,285 m) and locally to just below 3,900 ft. (1,190 m).

Favors oaks in pine-oak woodland in north, transitioning to pine woodland in southern part of range (Mex.). Occurs at elevations just below Hairy Woodpecker and just above those of Ladder-backed Woodpecker, maintaining narrow overlap zones with both species.

## DETECTION

**VOCAL SOUNDS AND BEHAVIOR:** Gives three primary calls: call note, rattle, and churring. Most frequently heard is call note, a sharp *keek*, slightly more grating than call of Hairy Woodpecker; often given in pairs or run together as *peedeek*. Given year-round by both sexes.

Rattle is a series of notes given in fairly rapid succession, with notes sharper than single call note and trailing at end, as *keek, keek, keek, kek, kek, kuk*. A shorter version often begins with call note but is burrier than typical rattle and contains only a few notes, like *keet, tchur, tchur, jurr*.

Churring is a plaintive, variable call, like *chuwit* or *shureet* followed by *tchuur, tchuur, tcherr*.

Also gives a chatty *chk, jrk, jurk* or *chruk, churk* at nest site, especially when feeding nestlings.

Calls immediately before nest exchange or in response to mate's vocalizations or drumming; also during pair formation and bonding and during

An adult male Arizona Woodpecker. *David Hollie; Aug. 2009, Santa Cruz Co., AZ.*

flight display. Most vocal during territory establishment and into nesting period.

**NONVOCAL SOUNDS:** Drum is typically short but firm, increasing slightly in volume toward end. Both sexes drum, usually in morning during nesting season; male drums more frequently and with longer rolls than female; incubating bird may drum upon departure from cavity. Drumming is sometimes interspersed with or superseded by rattle or *keek* call.

Ritual tapping is performed infrequently, and only by male after display flight, to attract female; ranges from 2 to 11 taps.

## VISUAL IDENTIFICATION

**ADULT:** Relatively small, dusty brown pied woodpecker, well camouflaged in sunny oak environment. Somewhat typical head pattern of North American *Picoides,* with dark auricular patch nearly completely encircled by prominent white. Solid brown crown extends dorsally to solid brown back, broken only by broad rear-crown patch in male, which tapers at each end to barely meet auriculars.

Back, rump, and central tail are solid brown, becoming progressively darker toward tail, with tail feathers darkest of all. White outer tail feathers are strongly marked with up to five dark brown or blackish bars. Spotted upper chest transitions to barring through belly and flanks.

In flight, solid dark brown above except for fine white flecking through flight feathers. Dark and heavily marked below.

**OPPOSITE:** This fledgling male Arizona Woodpecker was observed leaving the nest just moments before this image was captured. Note the wear in the tail feathers due to tight quarters inside the nest cavity. *Stephen Pollard; Jul. 2014, Cochise Co., AZ.*

**JUVENILE:** Duller overall than adult; underparts are grayish with smaller but more elongated spots than adult's. Male shows more extensive and more fully red feathers in crown than female, usually extending forward of eyes; red-tipped feathers in female are confined to central crown and usually do not reach forward of eyes. Most, but not all, juveniles can be sexed by crown pattern.

**INDIVIDUAL VARIATION:** Less than one-quarter of individuals in AZ show some white barring on rump, or even less frequently on back (similar to Strickland's Woodpecker of e. Mex.). Some individuals show leucistic blond patch in center of back. Adult female rarely has one or two red crown feathers.

**PLUMAGES AND MOLTS:** Preformative molt occurs June–Oct.; juvenile males may retain some red crown feathers into Sept. Subsequent prebasic molts occur mid-July–early Sept., occasionally into Oct.

**DISTINCTIVE CHARACTERISTICS:** The only North American woodpecker with an obviously solid brown back. Amount of white in head is exceeded only by White-headed and Red-cockaded Woodpeckers.

**SIMILAR SPECIES:** Unlikely to be mistaken for other woodpeckers in its range. Adult vocalizations most closely resemble those of Hairy Woodpecker in pattern and use, with characteristics of other *Picoides* among the variations; call is generally hoarser or scratchier in tone than Hairy's. Also similar to Hairy Woodpecker in behavior.

**GEOGRAPHIC VARIATION:** Size generally decreases from north to south, with southernmost populations (Mex.) averaging smallest. May also become darker to south as it approaches range of Strickland's Woodpecker, but this varies greatly.

### SUBSPECIES

Three subspecies recognized, with only *P. a. arizonae* occurring north of Mex. This is the largest subspecies, with greater sexual dimorphism in bill size than in other forms. *P. a. fraterculus* and *"websteri,"* the latter not recognized by all authorities, occur in Mex.

**HYBRIDIZATION:** Limited hybridization with Strickland's Woodpecker in Tzitzio, Michoacan, Mex., at southeastern limit of Arizona Woodpecker's range. However, about one-quarter of individuals in this region are believed to be Arizona Woodpeckers with heavily barred backs. No hybridization documented with Arizona's second-closest relative, Hairy Woodpecker, despite some habitat overlap in mountain canyons.

## BEHAVIOR

### BREEDING BIOLOGY

#### *Nest Site*

Typically nests in dead branch of riparian walnut, live oak, sycamore, maple, or cottonwood. May also excavate in agave stalk or apple tree.

#### *Courtship*

Probably monogamous. Displays include drumming and tapping; male performs flight display, holding wings in place and gliding toward mate at nest. Very quiet during nesting, but conspicuously vocal during pair formation or reaffirmation in spring. Although paired, mates stay unusually distant from each other, remaining in contact by vocalizing and drumming.

#### *Parenting*

Nest exchange during incubation and brooding occurs immediately upon return of foraging adult, preceded by vocal signal. Feeding visits by parents are often rapid.

Typically 2–3 young hatched. May hatch asynchronously, resulting in young of different sizes and probable mortality of third chick. Midday temperatures up to 100° F (37.8° C) inside active, unshaded cavity; chicks may stretch out along opposite sides of cavity to dissipate heat.

#### *Young Birds and Dispersal*

Fledglings peck at bark, call, and preen; make short flights within a day after fledging. Parents call as they approach juveniles with food; later use same call to summon juveniles. Young begin foraging on their own after about 4 days but remain dependent longer. Parents may divide up fledglings for feeding duties, young tagging along with respective adult. Small family groups remain together until at least late July, possibly longer. Complete nest failure is rare, although some nests produce only one fledgling.

### NONBREEDING BEHAVIOR

#### *Feeding*

***Preferred Foods:*** Larval, pupal, and adult insects, especially long-horned and snout beetle larvae. Also *Formica* ants, fruits, and acorns.

***Foraging Behaviors:*** Northern (North American) populations forage mostly in oaks, but occasionally sycamore, pine, walnut, willow, Arizona cypress, and juniper; sometimes on agave stalks and blos-

An adult male of the Mexican endemic Strickland's Woodpecker (*Picoides stricklandi*), once considered the same species (conspecific) with Arizona Woodpecker. Note this bird's smaller stature, shorter bill, and strongly barred back. *Amy McAndrews; Sept. 2012, Perote, Veracruz, Mex.*

soms. Foraging time on pine increases southward into Mex., being nearly all pine at southern edge of range. Frequently probes and gleans from deep within bark crevices. Also flakes off bark, often scratching off flakes with both feet simultaneously.

Male, with longer bill, excavates into cambium more often than female. Generally works from near base of a tree to about two-thirds of tree height, then flies down to next tree, in rapid succession, frequently spiraling up trunk. May cover up to 70 trees per hour regardless of tree size or species. Occasionally jumps instead of flying between trees.

### *Territory Defense and Sociality*

Encounters are usually same-sex, while mate of defender observes. Intruding females may persistently invade a territory, to be met by resident female; both display by waving their upward-pointing bill,

An adult female Arizona Woodpecker, at home in the encinal woodlands of the northern Sierra Madre Occidental of southeastern Arizona. *Dick Hartshorne; June 2011, Santa Cruz Co., AZ.*

along with partial wing flapping, swaying, and fanned tail. Intruder may also bob up and down while calling repeatedly.

Actively defends nest area from intruders. Unknown if larger foraging area is defended. May be observed in family groups of up to five birds. Adult males feed together during breeding season.

### *Interactions with Other Species*

Known to attack Acorn and Hairy Woodpeckers in nest vicinity, especially shortly after young hatch. Display may include raised crown and horizontal extension of wings, but often is silent. Northern Flicker is known to usurp Arizona Woodpecker nest.

May forage in same habitat as Ladder-backed Woodpecker at lower elevations; Hairy is known to move downslope into Arizona Woodpecker habitat.

Occasionally observed in flocks of mixed passerines outside breeding season. Observed in fall feeding among Acorn Woodpeckers and Mexican Jays.

## CONSERVATION

**HABITAT THREATS:** Groundwater depletion and riparian grazing adversely affect sycamore habitats in Southwest, resulting in potential long-term impact to Arizona Woodpecker habitat. Many decades of fire suppression have produced volatile forest conditions, placing habitat in jeopardy of large-scale stand-replacement fire.

Habitat fragmentation in Mex. resulting from logging and widespread rural development may have adversely affected the contact zone between Arizona and Strickland's Woodpeckers, countermanding efforts to clarify systematic relationship between the two species.

**POPULATION CHANGES:** Extremely small sample sizes limit scientific utility of standardized surveys, especially for trend analysis, but surveys are still useful for local monitoring purposes. Secretive nesting habits cause difficulty in assessing population numbers. Species is not common anywhere in its North American range, but populations are believed to be stable, in part because of low commercial value of preferred habitat. If Ivory-billed Woodpecker is extinct, Arizona Woodpecker is rarest member of its family in N. Am.

Typically found in densities of 1 pair per 25 ac. (10 ha) in oak-juniper-pine woodland in Chiricahua Mts. (half the density of Acorn Woodpecker and Northern Flicker) and 1 pair per 100 ac. (40 ha) in pure oak woodland.

**CONSERVATION STATUS AND MANAGEMENT:** Considered "moderate ecological specialist" by Arizona Partners in Flight, ranking highest priority for conservation among the state's breeding woodpeckers; priority for biodiversity conservation in NM. Yellow on 2007 WatchList.

Firefighting in typical habitat is difficult because of rugged terrain and should not be counted on as an indirect solution to protecting habitat.

## REFERENCES

American Ornithologists' Union 1947, 1976, 2000; Davis 1965; Fowler 1903; Johnson et al. 1999; Lane 1977; Latta 1999; Ligon 1968b.

# RED-COCKADED WOODPECKER

***Picoides borealis***

L: 8.5 in. (21.6 cm)
WS: 14 in. (35.6 cm)
WT: 1.4–1.9 oz. (40–45 g)

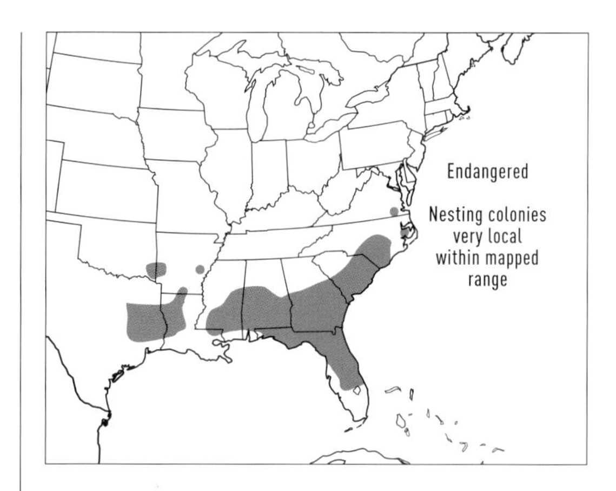


**In the mid-1800s,** John James Audubon considered the Red-cockaded Woodpecker abundant from New Jersey south and west to Tennessee and Texas. The species evolved as an extreme pine specialist, and it once thrived in the expansive pine forests across the southeastern corner of the continent. Over more than 200 years, hundreds of thousands of acres of pine trees in the Southeast were leveled to support expanding human civilization. Rapidly dwindling populations of Red-cockaded Woodpeckers persisted among isolated islands of habitat—mostly on public lands.

In 1970, the U.S. Fish and Wildlife Service formally listed the Red-cockaded Woodpecker as a federally endangered species. Since then, it has become one of the best-studied woodpeckers on Earth. At first, we learned how to maintain the small extant populations, and slowly these populations grew. As we learned more about the species' ecology, we began to manage the forests so that its habitat could grow. We learned about the critical role that fire plays in these habitats, and we learned much about the species' complex social system.

The Red-cockaded Woodpecker cannot endure without intensive management of its remaining populations and the restoration of healthy southeastern pine forests. Recognizing the species' tenuous existence and its defining role in forest ecology, we may hope to pave the course of recovery for this spirited little woodpecker.

## DISTRIBUTION

**PERMANENT RANGE:** Nonmigratory species, with occasional wandering because of limited carrying capacity of isolated habitats and pioneering when offspring need to disperse; generally resident throughout its range. Individual populations are widely scattered, and most occur on public lands, including military bases. All known populations are intensively managed. Found in all southeastern states, NC through s. FL, and AL to e. TX piney woods. Small, somewhat tenuous populations in se. VA, e. OK, and AR.

Rare records of females, and very rarely males, dispersing far beyond natal territories, likely in search of suitable habitat; one female dispersed 210 mi. (338 km), AR to LA, then her granddaughter returned the same distance to the grandmother's natal site. Another female was observed in IL in 2000, approx. 404 mi. (650 km) from the nearest known populations. Long-distance dispersal—generally less than 62 mi. (100 km)—may be more common than once believed, and may be an adaptation that enhances genetic diversity.

**OPPOSITE:** An adult Red-cockaded Woodpecker, of indeterminate sex. *Greg Lasley; June 2008, Jasper Co., TX.*

## HABITAT

Ecologically linked to broad stands of fire-maintained, mature, open pine forest. Four elements must exist for habitat to support sustainable populations: (1) extensive contiguous forest, to facilitate effective dispersal of offspring once they reach breeding age, as well as sustainable succession of age structure among trees; (2) regular fire regime—a 1- to 5-year ground-clearing regime is critical for controlling encroachment of hardwoods and maintaining open understory; (3) mature trees, old enough to have developed red-heart fungus, which makes the interior of trees easier to excavate and provides a sap-free chamber for nest or roost cavities; one AR study in Ouachita Mts. showed mean age of trees with heartwood decay at 110 years; and (4) live pine, facilitating long-term use of each cavity (important since excavation requires high energy expenditure) and so birds can maintain a free resin flow around cavity entrance, which helps prevent predation by tree-climbing rat snakes; live trees are also more fire-resistant.

Overwhelmingly favors pine trees, feeding 10 percent of time or less among hardwoods, although this depends on region and season. Favors larger pines, likely because of higher surface area and deeper bark furrows providing increased prey base. Preferred pine species vary locally, but longleaf pine is widespread preference wherever it exists. At edges of Red-cockaded's range, it will inhabit atypically blended habitats, such as slash pine mixed with bald cypress and grassy wetlands. Extirpated

Restored and meticulously managed slash pine flatwoods at the Fred C. Babcock/Cecil M. Webb Wildlife Management Area in southern Florida. This stand is managed as habitat for the Red-cockaded Woodpecker. The "RCW" sign marks the boundary of a Red-cockaded "cluster" site, with white rings painted around trees that contain artificial nest cavity inserts. *Stephen Shunk; Apr. 2013, Charlotte Co., FL.*

populations from s. Appalachian Mts. occupied mixed woodlands with pitch and shortleaf pine, oak, and hickory, with nearly closed canopy.

Age class of trees is more important than size; for example, longleaf pines that are 100 ft. (30 m) tall in SC could be the same age as those in cen. FL that barely reach 40 ft. (12 m), yet both are suitable age for heart-rot fungus to develop.

## DETECTION

One of chattiest North American woodpeckers, almost constantly communicating while foraging. Most easily detected in early morning and late evening when group activity around cluster site is greatest. (See Territory Defense and Sociality for discussion of cavity clusters.) Once in birds' presence, observers can find birds on specific trees by watching for bark flakes showering to the ground from high in canopy.

**VOCAL SOUNDS AND BEHAVIOR:** Frequently heard call note sounds like *beewr* or *peew* or *wheew*, slightly hoarse, becoming clearer in tone at higher frequency (when more excited). Also gives a sharp *peet* note, similar to Downy Woodpecker's but softer; often extended into rattle that is less organized than Downy's, or less consistent in cadence. Also a *rhut* or *rhet* or *eret*, likened to low-pitched waterfowl flight call. Before and after feeding young, gives an excited *peep* reminiscent of Pygmy Nuthatch.

Many other sounds described may be variants on above calls. Will call in flight or when perched. Calls often accompany territorial or courtship behaviors, such as open wing display, flutter-flight display, or erratic flight performance. Most vocal on leaving roost in morning and on return to cavity-tree cluster in late afternoon; least vocal during laying and incubation.

**NONVOCAL SOUNDS:** Most conspicuous nonvocal sound is stripping of large flakes of bark from pine trees; size of flakes and open, "hollow" nature of foraging stands make flaking, tapping, and shuffling around tree easily heard. Rapid wingbeats are heard easily during displays and when bird leaves nest site.

Adults of both sexes drum, mostly breeding male in spring. Typically drum on trunk, sometimes on resonant branch, usually within cavity-tree cluster. Drum pattern is muffled and short.

Reported "tongue drumming," sounding similar to a rattlesnake's rattle, is produced by rapid vibration of tongue on tree surfaces, especially when excited; may be related to foraging.

**OPPOSITE:** An adult female Red-cockaded Woodpecker at a "natural" nest cavity—that is, one excavated by the woodpecker in a live pine tree. Note the heavy resin flow around the cavity entrance. *Paul Bannick; May 2007, Richmond Co., NC.*

## VISUAL IDENTIFICATION

**ADULT:** Birds often occupy upper parts of canopy, where they are difficult to see. Bold white face, unique within species' range, is best visual cue. Evenly spaced black-and-white bars through center of back typical of "zebra-backed" *Picoides*. Proportionately long tail is apparent when perched vertically on trunk. Head may appear all white because of bold white cheek patch.

Tiny red "cockade," almost never visible in field, adorns each side of male's head at rear border of crown and face patch; comprises 12–16 short red feathers often tucked under black crown feathers. Position of red resembles what would be the forward tips of a red nape patch. Female identical but lacks red cockades; separation of sexes is usually impossible in the field.

**JUVENILE:** Generally duller than adult, including cheek patch. Sexes may be separable in the field, as red forehead patch of juvenile male is more obvious than adult male's red cockade. Female shows white flecking on forehead.

**INDIVIDUAL VARIATION:** Leucism is very rare, possibly exhibited as salmon-colored flight feathers and coverts and tawny crown.

**PLUMAGES AND MOLTS:** First basic plumage is acquired late May–mid-Nov. Gray cheek and red spot or white flecks on forehead are last feathers to be replaced. Subsequent prebasic molts occur early July–early Oct.

A clutch of Red-cockaded Woodpecker nestlings less than 1 week old. These birds are part of the species' northernmost population, at The Nature Conservancy's Piney Grove Preserve in southeastern Virginia. Nestlings like these are temporarily removed from the nest for data collection and banding and then placed back in the nest cavity. *Brian Watts, courtesy of Center for Conservation Biology; June 2014, Sussex Co., VA.*

**DISTINCTIVE CHARACTERISTICS:** Tail often appears disproportionately long when bird is observed perched against tree trunk. Only White-headed Woodpecker shows a bolder white face. Adult Red-cockaded shows least amount of sexual dichromatism of all North American *Picoides*.

**SIMILAR SPECIES:** White face and highly gregarious behavior separate Red-cockaded from other species in its range. Red-bellied Woodpecker is also "zebra-backed" but noticeably longer, with bright red nape (and crown in male). Occurs within 100 mi. (160 km) of Ladder-backed Woodpecker in cen. TX, but plumage similarity may be due to convergent evolution; recent DNA studies indicate these two species may be from disparate lineages.

**GEOGRAPHIC VARIATION:** Size varies clinally, larger birds inland and north, decreasing in size coastward and south.

**SUBSPECIES:**

Two subspecies are recognized: *P. b. borealis*, from n. FL north and west through remainder of range; and "*hylonomus*," from n.-cen. to s. FL. Not all authorities recognize the latter, citing identical plumage characters and steady cline in physical measurements from north to south.

## BEHAVIOR

### BREEDING BIOLOGY

#### *Nest Site*

Red-cockaded is often mistakenly described as the only woodpecker to excavate in live trees. Many other woodpeckers occasionally excavate in live trees, and some do so frequently, but this species is the only one that does so almost exclusively; also known to excavate in dead pines or decayed hardwoods. Natural cavities most often are in pine trees 80–120 years old, range of 60–250 years. Readily adapts to nest-box inserts installed in younger trees (see Conservation).

Entrance tunnel angles slightly upward through sapwood 2.5–3.5 in. (6–9 cm) to decayed heartwood. Older cavity entrances are much wider than new ones and beveled at lower edge. Breeding male typically occupies newest cavity as roost site, which doubles as nest site. Cavities from prior years may be used for decades by individuals from surviving generations.

#### *Courtship*

Pairs may form any time a female arrives at a cluster occupied by unmated males, usually in early spring. Monogamous; only one record of polygyny. Usually mates for life, but occasionally switches mates between breeding seasons.

Adults perform flutter-flight display, with at least three birds usually present, one of which does not participate; birds often call during display. Also performs erratic "corkscrew" flight accompanied by rapidly repeated call, culminating in copulation and drumming. Pairs may copulate any time of year, but more often in late spring.

### *Parenting*

Cooperative breeder; groups include one breeding pair and zero to four helpers, usually male offspring of previous seasons; only about 5–30 percent of helpers are females. First breeding of males is usually delayed, since most males serve as helpers in natal group. Helpers primarily assist in feeding of young, and groups with helpers generally fledge more young.

Not all groups nest every year; up to, and rarely more than, 25 percent do not breed. Rarely a cluster may support multiple synchronous nests. Up to 5 percent of broods contain a runt egg, which may be correlated with communal nesting.

### *Young Birds*

Female nestlings are generally smaller than males, possibly explaining lower female survival rate and hence disparate sex ratio among groups. Nests rarely fledge more than two young. Heavy rain or drought may cause widespread nest failure by limiting available food supply.

Young are able to feed independently soon after fledging, but breeding male is responsible for most feedings; older group members may feed young for up to 5 months.

The pale patch of feathers erupting in the center of this nestling's crown will eventually become the red forehead patch of a juvenile male Red-cockaded Woodpecker. *Brian Watts, courtesy of Center for Conservation Biology; June 2014, Sussex Co., VA.*

Although much less vocal than adults, fledglings are capable of producing adult array of vocalizations; may continue to remain quieter than adults into fall.

### *Dispersal*

Five months after fledging, juvenile male may usurp older female's roost cavity. Young remain with family group into fall. Females typically disperse by winter; males stay longer and maintain strong fidelity to cluster site into next breeding season.

Dispersal distance ranges from 1.1–3.6 mi. (1.8–5.4 km); influencing factors include local population density, quality of foraging habitat, and cavity-tree availability.

Home ranges often exceed 200 ac. (80 ha) and may exceed 1,000 ac. (400 ha) in poor-quality habitat; difficult to generalize because of erratic daily feeding distances, extremely varied habitat quality, and variable proximity of neighboring clusters. Extended home range is used through winter, typically small sections each day; daily foraging sites may be up to 1.2 mi. (2 km) from cluster site.

## NONBREEDING BEHAVIOR

### *Feeding*

***Preferred Foods:*** Year-round diet averages 86 percent arthropods (more than half of which are ants), 14 percent plant materials. In some seasons, ants account for up to 70 percent of diet. Also other arthropods, including adult and larval insects and eggs.

Will take prey items based on their abundance; favors southern pine beetle (adults and larvae) over ants in areas of beetle infestation; consumes corn earworms when habitat abuts cornfields. Vegetable materials include pine seeds and wide variety of fruits, based on availability, especially berries. Laying females will cache bone fragments, likely as calcium source for nestlings (or to support egg development, if before laying). Young are most often fed insect larvae, wood cockroaches, and centipedes.

***Foraging Behaviors:*** Forages almost exclusively on bark surface of live trees. Extracts prey by a combination of scaling and probing, with some excavation into rotting wood. Scaling is most prominent; pries or pulls loose bark plates off pine trees with bill; female, and rarely male, may rip bark loose with feet.

Probing involves visual and sensory exploration of bark crevices, stump tops, snapped branch ends, and needle and cone clusters. Rarely visits suet feeders.

An adult male Red-cockaded Woodpecker feeds on a small pine branch by probing beneath the peeling bark for prey. *Robert Royse; Mar. 2007, Citrus Co., FL.*

Some evidence of sexual niche partitioning; males may prefer branches and upper trunk of pines, females favoring trunk below lowest branches.

### *Territory Defense and Sociality*

Mated pair and sometimes helpers will attack extragroup intruders; other times, nonfamily members occasionally feed and roost among unrelated group.

In aggressive territorial display, extends and raises wings, then closes them quickly as either aggressor jabs with bill. Fluttering aerial display, similar to courtship, begins with canopy-level chase and lasts 15–30 minutes. Often holds wings in steep dihedral, sometimes followed by spiraling chase around tree trunk.

Social system is highly developed, usually including two to five individuals, and only one female before breeding. Birds constantly move together, calling while feeding. Group size increases with addition of fledglings. Tightly clustered cavities may contribute to reduced competition among individual group members since group members all defend same territory; also aids in predator defense.

Males often stay with group as long as there are cavities in area or until they find a mate and establish a new territory, then excavating cavity at new site while roosting and feeding with natal group.

### *Interactions with Other Species*

Aggressive toward Hairy Woodpecker and Blue Jay, occasionally toward Downy Woodpecker when foraging closely, but both Downy and Hairy sometimes feed and nest a short distance from Red-cockaded cavity; known to roost in same tree as Hairy.

Up to 50 percent of cavities in a given cluster may be occupied by other cavity nesters (including flying squirrels) and many invertebrates, but highest-quality cavities are usually occupied by Red-cockadeds.

Red-bellied Woodpecker is able to enter and exit Red-cockaded cavity without enlarging entrance, and has been observed pulling adult Red-cockaded out of cavity; may even injure or kill Red-cockaded during conflict over cavity. Great Crested Flycatcher will nest in and defend abandoned Red-cockaded cavity.

In winter, often followed by flocks of small passerines, including Brown-headed Nuthatch, Pine Warbler, and Eastern Bluebird, which collect missed or newly exposed prey items.

## MISCELLANEOUS BEHAVIORS

Exits roost around sunrise, later if overcast, departing with group members to feed; may return during inclement weather or after predator encounter. Group returns just before sunset, individuals tapping at bark surface to maintain resin flow at cavities before roosting for night. May roost in cavity made by other species, natural tree cavity, snapped-off treetop, or under limb.

Rarely comes to ground. Bathes and drinks in shallow puddles on limbs; also takes water from leaves or pine needles. Head scratching does not include wing movement. Group members may preen together (but not each other).

Sunning occurs with or without prior bath, after which preening may resume. Stretching is common, accompanied by fanned tail feathers.

**OPPOSITE:** An adult Red-cockaded Woodpecker performs a courtship or territorial display with fully raised wings and a spread tail. *Alan Murphy; Feb. 2009, Montgomery Co., TX.*

The southern flying squirrel aggressively competes with the Red-cockaded Woodpecker for available cavities in the southeastern pine forests. *Mark Herse, courtesy of Avon Park Air Force Range; Sept. 2012, Polk Co. FL.*

## CONSERVATION

**HABITAT AND POPULATION HISTORY:** Prior to European settlement, an estimated 247 million ac. (100 million ha) of southeastern pine forest supported between 920,000 and 1.5 million breeding pairs. Considered "abundant" by Audubon in mid-1800s. Fire suppression and rangewide clearing of old-growth pines over several hundred years brought populations to an estimated low of 4,029 active territories in early 1990s.

First recorded breeding in MD in 1939, extirpated there as a breeder by 1958; also extirpated from KY (2001), MO (1946), NJ (one breeding record, before 1866), and TN (1994). Isolated records from IL (2000, 2001), OH (1872, 1975), and PA (before 1877). Peripheral VA populations declined from 23 active clusters in 1977 to 2 in 2000; up to 13 as of 2014. Many local declines and extirpations have continued into 21st century.

**CURRENT HABITAT THREATS:** Restriction of favorable habitat to managed "islands" of public land profoundly affects species' viability, inhibiting dispersal of young birds and limiting genetic diversity.

Cutting of large timber and subsequent repeated logging of younger stands limits sustainable regeneration of mature trees required for nesting. Loss of large trees also decreases bark-surface area required for foraging. Conversion of forested lands to agriculture or residential development summarily eliminates habitat. Fire suppression allows growth of dense ground cover and hardwoods favorable to other woodpeckers and to flying squirrels and other predators, as well as proliferation of unfavorable arthropod prey species.

**CURRENT POPULATION STATUS, DISTRIBUTION, AND TRENDS:** U.S. Fish and Wildlife Service (USFWS) 2003 recovery plan estimated 14,068 Red-cockaded Woodpeckers in 5,627 "known active clusters" in 11 states—less than 3 percent of estimated population prior to European settlement.

As of 2013, rangewide population was estimated at approximately 7,000 active clusters (occupied territories with at least a single male). On average, rangewide, 89 percent of active clusters consist of a primary breeding group (PBG)—a minimum of one adult male and one adult female in the same breeding area, or cluster. Recovery plan defines 39 designated recovery populations (DRPs) across the species' range; as of 2012, rangewide, DRPs included 5,853 active clusters.

Among the 39 DRPs, 13 are considered "primary core recovery populations," each with a population goal of 350 PBGs; 4 of the 13 have surpassed this recovery objective: Eglin Air Force Base, FL; Ft. Bragg, NC; Ft. Stewart, GA; and Francis Marion National Forest, SC.

One of the early conservation measures for the Red-cockaded Woodpecker was the installation of metal cavity-entrance restrictors, which prevent larger woodpeckers from enlarging the entrance and outcompeting the Red-cockaded for the nest site. *Jim Johnson; May 2006, Vernon Parish, LA.*

Largest aggregations of PBGs according to 2006 and 2013 figures include Apalachicola National Forest, FL (451 PBGs in 2006, 718 in 2013); NC sandhills (374 PBGs in 2006, 649 in 2013); and Francis Marion National Forest, SC (approximately 344 PBGs in 2006, 445 in 2013).

After 2005 translocation of birds from other regions, population at VA's Piney Grove Preserve grew from 2 breeding pairs in 2000 to 10 breeding pairs in 2012. Habitat restoration in native shortleaf pine forests of Ouachita National Forest of AR and OK yielded a population increase from 24 birds on 14 active territories in 1998 to 87 birds on 38 territories in 2005.

Growth rates are estimated in 5-year periods. For 2008–2012, combination of all recovery populations showed growth of 3.8 percent per year; goal is 5 percent. Some individual populations are growing at a rate of up to 15 percent per year. Overall, DRPs are growing at a rate of approximately 200 per year.

Elaborate snake traps are designed to capture tree-climbing rat snakes, one of the primary predators of the Red-cockaded Woodpecker. *Stephen Shunk; Apr. 2013, San Augustine Co., TX.*

Meticulous monitoring by wildlife management agencies makes data collected by standardized surveys such as Breeding Bird Surveys and Christmas Bird Counts somewhat superfluous, but these surveys are still useful for local monitoring purposes and for engaging local birding communities.

**CONSERVATION STATUS:** First listed as endangered by USFWS in Oct. 1970, with subsequent protections under the first Endangered Species Act (ESA) of 1973. Species of Highest Continental Concern by American Bird Conservancy; Threatened in MS; Species of Concern in FL; and Endangered in eight other southeastern states.

Controversy on TX national forests beginning in 1977 led to 1985 and 1988 court rulings that U.S. Forest Service had violated agency's own Red-cockaded Woodpecker protections and ESA of 1973.

In September 1989, Hurricane Hugo leveled thousands of Red-cockaded Woodpecker nest trees in the Mid-Atlantic states, leading to the invention of nest cavity "inserts." *Stephen Shunk, courtesy of Joe Neal; Ouachita National Forest, AR.*

To mount a Red-cockaded Woodpecker nest cavity insert, a hole is cut in the tree, the box is inserted, and the edges are covered with wood putty. The fresh resin flow around this nest box indicates that it is likely in use. *Stephen Shunk; Feb. 2009, Charlotte Co., FL.*

First USFWS recovery plan published 1979; final plan issued 2003. Annual reports issued by federal and state agencies since 1998. Most recent 5-year review issued was 2006, but because of recent increase in recovery populations, further reviews will be issued only pending the anticipated change in listing status. Based on current trends, downlisting from Endangered to Threatened is expected for 2020–2030.

**MANAGEMENT ACTIVITIES:** Use of metal plates as cavity-entrance restrictors began in late 1980s, preventing damage caused by larger woodpeckers enlarging entrances for their own use. Destruction caused by Hurricane Hugo in September 1989 led to invention of artificial cavity inserts and relocation of individual birds. Inserts are now used extensively in absence of mature trees. Translocation of young females into groups without females revitalizes isolated groups and enhances genetic diversity; young birds are also translocated when extremely small, isolated habitats reach carrying capacity. Prescribed fire is now used extensively in existing and potential habitats throughout species' range to clear understory and maintain suitable forest characteristics.

Safe Harbor program was initiated in NC in 1995 to engage private landowners in habitat conservation; now in place in all states with DRPs. Enrolled private lands generally support very small populations and are not included in federal management, monitoring, or reporting. As of 2013, 855 active clusters on about 2.5 million ac. (1.01 million ha) of private lands were enrolled in program rangewide. Program may eventually provide important corridors linking managed public lands.

Habitat goals for primary core recovery populations include contiguous forest patches of 80,000 ac. (32,375 ha). In eastern Gulf Coast region, general habitat prescription calls for minimum patch sizes of 125,000 ac. (50,000 ha).

## REFERENCES

American Bird Conservancy, undated; American Ornithologists' Union 1947, 1976; Bailey 1999; Boyne 2008; Conner, R.M., pers. comm.; Conner and Rudolph 1989; Conner et al. 2001, 2004b, 2006; Cordle 2006; Costa 2002, 2006; Costa et al. 2004; DeLotelle et al. 1987, 1992; Ertep and Lee 1994; Fournier 2005; Georgia Museum of Natural History 2000; Hooper and Lennartz 1982; Houston 2007; Jackson 1974, 1977a, 1977b, 1978, 1979, 1990, 1994; Jones 2004; Koenig 1980a; Lennartz and Harlow 1979; Ligon 1970; Ligon et al. 1986; Liles 2008; McDearman, W., pers. comm.; McFarlane 1992; Meanley 1943; Neal, J., pers. comm.; Nesbitt et al. 1978; Noecker and Corn 1997; Ouachita National Forest 2005; Phillips and Hall 2000; Repaski et al. 1991; Sherrill and Case 1980; Skorupa and McFarlane 1976; Snyder and Houston 2000; Springston 2004; Stone 1995a, 1995b; Sweet 2001; Tekin 2006; Thompson 1976; U.S. Army Environmental Center 2006, 2008; U.S. Fish and Wildlife Service 1999, 2001, 2003; Watts 2005, 2006; Williams 2002; Wilson 2006.

## WHITE-HEADED WOODPECKER

***Picoides albolarvatus***

L: 9–9.25 in. (22.9–23.5 cm)
WS: 16–17 (40.6–43.2 cm)
WT: 1.8–2.8 oz. (50–79 g)

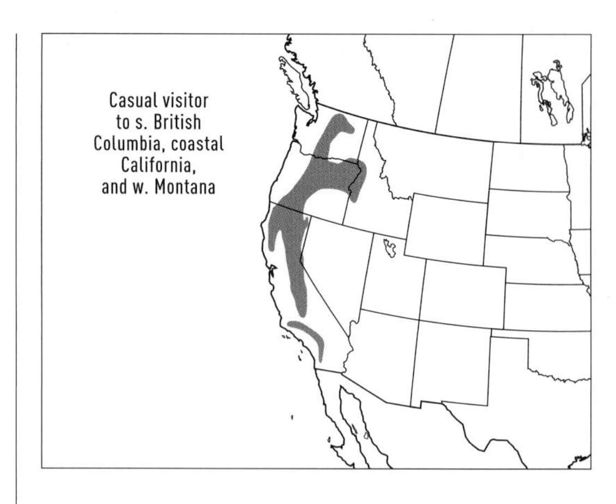


**Dressed for a black-tie dinner,** the striking White-headed Woodpecker adorns narrow stretches of pine-dominated forestland from British Columbia's southern Okanagan Valley to the peninsular ranges of southern California. When it is present in the right habitat, the White-headed is hard to miss, but its extreme specialization makes the "right" habitat sometimes hard to find.

In an interesting parallel with the Red-cockaded Woodpecker, the White-headed also requires large stands of mature pine trees and a fire-cleared understory. The feeding styles of the two species are also similar; they both frequently forage on the surfaces of large-diameter, live pine trees, and they are consummate gleaners and probers, easily twisting their heads upside down to reach for hidden prey items.

The White-headed also possesses the brightest white head among North American woodpeckers, with the possible adaptive significance of enhancing foraging by reflecting light deep into the dark holes and bark furrows of large pines.

In addition to probing for insects, the White-headed Woodpecker spends much of the year drilling the seeds out of large pine cones. This distinctive feeding style and the White-headed's penchant for probing require specialized pine-forest habitat. Unfortunately for the White-headed Woodpecker, large pine trees also bring a hefty bounty in the timber market, putting local woodpecker populations at risk, and making this beautiful woodpecker increasingly scarce, especially in the northern parts of its range.

### DISTRIBUTION

Permanent resident throughout range, though prone to short-distance wandering outside breeding season.

**PERMANENT RANGE:** Spotty distribution, based on strict habitat requirements (see below). Rare around northern fringes of range (endangered in BC), increasing southward, down east slope of WA and OR Cascades, east to Blue Mts. and w. ID. South through Warner Mts. and Sierra Nevada to Transverse and Peninsular Ranges of s. CA. Also, high and dry portions of nw. CA Coast Ranges, across to Siskiyou Mts.

**SEASONAL MOVEMENTS:** Generally sedentary, with strong fidelity to home territory; some wandering, especially in fall and winter, within and to edge of pine forests at lower elevations. Sporadic extra-range movements to coastal areas may be due to food shortages or unusual reproductive success, with local populations exceeding carrying capacity of local habitats. Dispersing juveniles may account for most movements.

### HABITAT

Coniferous, pine-dominated mountain habitats, with ponderosa pine occurring consistently through most of range; favors old-growth with moderately open canopy and light shrub cover maintained by regular ground-clearing fires. Shows strongest affinity for regions and stands with at least two pine species; in addition to ponderosa, favors sugar, Jeffrey, and Coulter pines. Wide range of other conifers may be present, as well as oak and aspen, but large-cone-producing pines are the critical link for this species.

In northern parts of range, may persist in areas with only ponderosa pines; healthiest populations are in areas with largest trees and snags for nesting. May select large, barkless, broken-top snags in some regions. May occur at higher elevations among western white pine, but generally absent from low-elevation gray pine in w. Sierra Nevada. May border on being locally abundant in forests managed with underburn and small tree thinning as long as large trees remain.

In some areas and outside breeding season, may move into chaparral and occasionally into developed areas with groves of non-native trees, especially if there are exotic, large-coned pines.

### DETECTION

**VOCAL SOUNDS AND BEHAVIOR:** Fairly vocal all year, with call note, rattle, and contact calls. Distinctive

**OPPOSITE:** An adult male White-headed Woodpecker brings food to nestlings. *Steve Brad; June 2010, Mono Co., CA.*

**ABOVE:** A nestling female White-headed Woodpecker. Note the faint amount of red in the crown compared with that of the fledgling male opposite. *Kristine Falco; June 2010, Deschutes Co., OR.* **OPPOSITE:** A fledgling male White-headed Woodpecker. *Alan Murphy; Aug. 2006, Deschutes Co., OR.*

double call note is sharp *br-deet* or *chick-it*, also given in triplets and by both sexes throughout year. Repetition pattern somewhat resembles double-note call of Nuttall's Woodpecker.

Often extends basic call notes into longer rattle during interactions with other individuals or other species, in flight or while perched; similar to, but more fluid and less burry than, rattle of Hairy Woodpecker. Delivers shorter version of rattle during territorial interactions, also starting with single or double call notes.

Often gives repeated squeak notes with rattle and in response to or accompanying drumming between mates. Mates give additional contact call during breeding season.

**NONVOCAL SOUNDS:** Typical drum is a steady series of beats with even cadence. Males drum more than females, especially after territorial encounters. Drumming is noted year-round but concentrated Mar.–June as territorial defense and mate contact, possibly varying geographically. Drum is difficult to separate in the field from that of Downy and Hairy Woodpeckers and Northern Flicker.

Tapping sounds associated with foraging are less noticeable than in other *Picoides* because of less aggressive feeding style. Males perform ritual tapping during nest-site selection; both adults tap from inside nest cavity while incubating and before nest exchange.

## VISUAL IDENTIFICATION

**ADULT:** Distinctive plumage, with bold white head and white wing patches contrasting against solid black elsewhere. Thin black eyeline extends from rear of eye to solid black nape; male shows narrow red band between rear crown and nape. White wing markings show as thin white line on central outer edge of folded wing.

In flight, solid black body is interrupted by flashy white wing patches. White head is visible even in poor light.

**JUVENILE:** Duller black with variable red patch on crown; white wing patches generally appear less solid in flight than in adults. Fledglings may show fine whitish barring on belly. Juvenile's bill is notably paler, broader, and shorter than adult's.

Young males may show pale orange-red patch on crown, varying in size; female crown is similar but red is paler and less extensive, often absent or limited to a few scarlet feathers on rear crown.

**INDIVIDUAL VARIATION:** Albinism is rare and may be limited to single feathers. Rare variant may show bright orange-yellow nape. Adult's white head often

Burned forests provide nesting substrate for woodpeckers for many years after a fire. This adult female White-headed Woodpecker is delivering a huge billful of ants to nestlings. *Stephen Shunk; July 2009, Deschutes Co., OR.*

stained with pine pitch by late summer. Males are slightly heavier than females.

**PLUMAGES AND MOLTS:** Preformative molt is completed by Sept. Spring and early summer adults are worn and faded; adult definitive prebasic molt July–Oct; central tail feathers may erupt as late as early Nov.

**DISTINCTIVE CHARACTERISTICS:** Strongly contrasting plumage compares only to Red-headed Woodpecker in North America.

**SIMILAR SPECIES:** Bold black and white contrast is unmistakable. Shares parts of range with other "black-backed" woodpeckers, including Black-backed Woodpecker and Williamson's Sapsucker, both light underneath and with darker head. The only bird in White-headed's range showing solid black above and below is much larger Pileated Woodpecker.

**GEOGRAPHIC VARIATION:** No variation noted within each of two subspecies populations.

## SUBSPECIES

Two subspecies are recognized. *P. a. albolarvatus* occurs everywhere except s. CA. *P. a. gravirostris* is resident from San Gabriel Mts. to southern limit of species' range in San Diego Co. Birds from Mt. Pinos region of s.-cen. CA are intermediate in measurements and may be intergrades.

*P. a. gravirostris* has slightly longer and deeper bill than *albolarvatus* and may be smaller overall; sexes show greater bill-length dimorphism southwest of Sierra Nevada. Larger bill of *gravirostris* is likely adapted to huge, spiked cones of Coulter pine.

**OPPOSITE:** An adult male White-headed Woodpecker excavates a nest cavity in a large ponderosa pine. *Paul Bannick; May 2005, Yakima Co., WA.*

## BEHAVIOR

### BREEDING BIOLOGY

White-headed Woodpecker often initiates nesting timeline approx. 2 weeks later than Hairy Woodpecker nesting in overlapping territories, which may help avoid competition for food resources as well as potential ambiguity between very similar drumming patterns.

#### *Nest Site*

Nest is generally low in a large-diameter, dead conifer, often snapped off, with partly decayed sapwood and often barkless at excavation site. Favors pine (ponderosa over others), but also aspen; species varies depending on availability. Occasionally excavates nest cavity in vertical branches of downed trees, in leaning trees, or in park bench or fencepost.

#### *Courtship*

Likely monogamous, remaining paired year-round, with some courtship-type activities in winter. Mates feed, drink, and forage together.

Elaborate display at nest; one bird, usually male, performs aerial fluttering, descending toward cavity entrance, giving squeaky call in flight and contact call on landing. Copulation typically occurs after nest is completed. Female flies to horizontal branch near nest tree, giving rattle call, with male at or inside nest cavity; male leaves cavity giving contact calls, and wing quivering ensues; after 3- to 10-second contact, male flies away and female flies to and enters nest cavity with rattle call. Mates may both fly off after copulation; flight often slow, with deep, labored wingbeats.

#### *Parenting*

Eggs are laid soon after cavity completion; possible intraspecific brood parasitism reported. Last egg in series often fails to hatch; parents toss eggshells just outside nest cavity. Mates are attentive during incubation, employing soft drumming from inside and outside nest cavity, and calling frequently. Nest exchange involves one mate calling during approach flight and then landing near cavity entrance.

Adults call and drum near nest cavity to coax nestlings out of nest. Parents split up fledglings for feeding duties. Calls of young are distinguishable from those of adults, being weaker and squeakier. Adults observed leading fledglings to suet-feeding stations and feeding young fresh pine seeds. Young may learn to feed at sap wells by observing adult.

#### *Dispersal*

Young typically associate with parents through fall, rarely into Dec.; will follow adults up to 4.7 mi. (7.6 km) from nest site for favorable foraging. Immature females may wander more than adults and young males.

Family groups observed dispersing up to 5 mi.

An adult male White-headed Woodpecker delivers food to nestlings. In a landscape that lacks standing snags, woodpeckers will typically excavate in whatever is available; this nest is located in a large snapped-off branch extending vertically from a large downed ponderosa pine. *Kristine Falco; May 2008, Jefferson Co., OR.*

The White-headed Woodpecker is a quintessential prober, exploring deep bark crevices and furrows for its arthropod prey. *Jim Burns; Oct. 2008, Riverside Co., CA.*

(8 km) from breeding territory immediately after fledging to exploit spruce budworm infestation; adults may travel up to 8 mi. (13 km) in fall in search of sugar pines.

Maintains permanent home range and strong nest-site fidelity. Typical home range size is 247–524 ac. (100–212 ha) on continuous old-growth sites and 741–845 ac. (300–342 ha) on fragmented sites.

## NONBREEDING BEHAVIOR

### Feeding

***Preferred Foods:*** Year-round, approximately half insects and half pine seeds, with wide seasonal fluctuations in each; in fall, arthropods may account for 75–85 percent of diet, remainder pine seeds; in OR, pine seeds may constitute up to 90 percent of diet from late summer through mid-winter.

In addition to ants, selects adults and larvae of bark beetles, scale insects, termites, and cicadas; feeds same to young. Preferred pine seeds include ponderosa, Jeffrey, sugar, and Coulter pines; also documented eating seeds of knobcone pine and white fir, and rarely seeds (and insects) from mullein stalks. Also consumes tree sap and readily attends suet and peanut feeders.

***Foraging Behaviors:*** Year-round, feeding technique is estimated to vary from 35 to 85 percent probing and gleaning, 5 to 30 percent extracting cone seeds, 24 to 40 percent bark excavation, 4 to 7 percent sap sucking; remainder flaking, flycatching, and ground gleaning.

Acrobatically twists head and neck while searching for prey among bark plates and furrows. Spends more than 50 percent of all foraging time on trunks of large, live pines. Remainder is divided among branches, needle clusters, and cones, rarely snags or downed logs.

In typical cone-feeding style, clings to branch above cone (*P. a. albolarvatus*) or to cone surface

(*P. a. gravirostris*) and chips into cone with lateral stabs to expose seeds; removes whole, or nearly whole, pine seeds; may eat seeds immediately or may fly to anvil site on same or nearby tree; wedges seed into crevice; breaks seed apart with bill and eats smaller pieces. Cone foraging generally peaks between early fall and early winter, but timing varies with cone maturity among different pine species. In years with low cone productivity, may increase consumption of wood-boring beetle larvae, among other arthropods.

In early spring, also bores small, round xylem wells in horizontal rings around small trees, much like sapsuckers, and eats sap from natural tree wounds.

Sexual niche partitioning is only noted sporadically but variability is notable. Seasonal variation may exist in favored tree species (one sex using ponderosa pine and the other favoring incense cedar), part of tree (one in upper canopy, other among lower trunk and branches), feeding style (one using foot scratching, the other bill striking), succession (one on fresh cones, the other on cones opened by mate), tree health (one on live trees, the other on snags), and food choice (one favoring insects, the other seeds). Both sexes are more often found on similar substrates in summer, disparate or complementary substrates in winter.

### *Territory Defense and Sociality*

Territorial behaviors are similar to those of other *Picoides*: rapidly flicks wings outward, flashes crown feathers (including red patch in male), swings head and bill back and forth, and rarely performs brief full-wing spread.

Males, and occasionally females, engage in persistent chase sequence, whereby two aggressors chase each other around bark surface of large tree, just a few feet from each other, shuffling in all directions, then suddenly freezing in place. Activity starts again spontaneously, with equal vigor, until next freeze posture. Repeating sequence may last up to 30 minutes. May perform other displays while shuffling, including head pointing or swinging and wing flicking, accompanied by series of call notes. Also performs mothlike flight display during same-sex encounters, usually early in breeding season. Physical contact is unusual throughout range, males rarely locking feet together in flight and falling nearly to ground before separating.

Some defense of entire home range occurs all year, with smaller territory defended during breeding season. Males may be socially dominant over females. Mates may feed in close proximity, but unpaired adults are normally much farther apart.

Intra- and interspecific aggression are minimal among southern populations.

### *Interactions with Other Species*

Frequently interacts with other cavity nesters, especially bluebirds, nuthatches, and swallows around nest site; may abandon excavation activities after such aggression. Observed chasing chipmunk around rocks near nest snag that contained newly hatched young; will also share snags with other cavity nesters and observe from cavity entrance while neighbors attack small-mammal intruder.

European Starling and flying squirrels known to usurp nests at various stages in breeding cycle; ground squirrels, tree squirrels, and chipmunks known to forage on eggs and nestlings, and rarely adults. White-headed Woodpecker avoids vicinity of Acorn Woodpecker colonies.

Will defend fresh cone crop from Pygmy Nuthatch and Red Crossbill, but Hairy Woodpecker is biggest competitor for live pine cones. Many Hairy–White-headed interactions are observed, neither dominating consistently. Otherwise, feeding niches are complementary; Hairy more often excavates into cambium, White-headed on and just below bark surface.

Known to bathe and drink closely with other species, including Hairy Woodpecker and Evening Grosbeak.

## MISCELLANEOUS BEHAVIORS

Exits cavity later in morning than other species in its habitat. Roosts mostly in cavities, but also under sloughing bark or in large trunk cracks or crevices. Individual birds may use up to 13 different roost sites. After first snowfall and through winter, all roosting is usually in same cavity. Roosts mostly in cavities in large-diameter, heavily decayed snags, but occasionally exposed in live ponderosa pine. Mates may use same roost tree. Often calls when approaching roost; lands high on roost tree and shuffles down to cavity entrance. Typically does not excavate new roost cavities.

Observed bathing and drinking from variety of sources, including puddles of rain or melting snow and leaky sprinkler heads. May drink more than other *Picoides* because of high proportion of plant matter in diet. Observed sunbathing atop snags.

## CONSERVATION

**HABITAT THREATS:** Large-diameter pines are declining throughout most of range because of logging, snag removal, maintenance of even-aged plantations, and uncharacteristic stand-replacement fire (occurring as consequence of long-term fire suppression). Past forestry practices have severely fragmented northern populations. North of CA, ponderosa pines tend to grow at relatively low elevations, in flat terrain, and closer to human habi-

Like many woodpeckers, the White-headed will forage for ants on the ground when it is busy feeding hungry nestlings. *Bill Hunter; June 2007, Deschutes Co., OR.*

tation, hence are more susceptible to forestry activities. Snags often pose human safety risk and are removed during timber operations.

Fire suppression is detrimental to habitat in numerous ways (See Introduction: Woodpeckers and Fire, p. 31).

Pine beetle outbreaks or other pathogens may threaten populations in areas with single-pine-species dependence.

**POPULATION CHANGES:** Abundance decreases with latitude north of CA. Remains fairly common through most of CA range. Surveys reveal densities as high as 5 pairs per 100 ac. (40 ha) in s. Sierra Nevada. Populations from cen. OR and north maintain highest densities in old-growth ponderosa pine, those to south in mature pine mixed with other conifers.

Dramatic fluctuations in BC populations are typical for many species at limit of their ranges; maximum populations are recorded in mild winters. Human settlement and recent cone crop failures likely precipitated BC decline.

**CONSERVATION STATUS AND MANAGEMENT:** Populations outside CA listed as conservation priorities at various levels, with greatest concern in northernmost habitats, especially BC, s. Okanagan region, where endangered. Listed as critically dependent on late successional/old-growth forest in Sierra Nevada.

Management prescriptions in ID and OR have focused on ponderosa pine forest, and include modifying timber harvesting, supporting regular fire regime, and encouraging snag retention and population monitoring.

In BC, snags over 16 in. (40 cm) in diameter are restricted from firewood cutting. Goal is to increase breeding pairs to 35 by 2050. This will require management and restoration of 34,600 ac. (14,000 ha) of habitat.

Estimated to require 100–550 large-diameter snags and 100–120 large trees per 247 ac. (100 ha), more than most other woodpeckers. Food rather than nest sites may be primary limiting resource.

## REFERENCES

Alexander and Burns 2006; American Ornithologists' Union 1916, 1923, 1947, 1976; Blood 1997; Bryan and Sarell 2006; Buchanan et al. 2003; Cannings 1995; Dixon 1995a, 2012; Duncan 1933; Frenzel, R., pers. comm., Frenzel 2004; Garrett et al. 1996; Gebauer 2004; Grinnell 1902; Heindel and Heindel 2005; Heindel 2005; Kozma 2010; Krannitz 2006; Lewis and Rodrick 2003; Ligon 1973; Mellen-McLean et al. 2013; Milne and Hejl 1989; Morrison and With 1987; Robinson 1957; Rogers 1978; Shunk 2004b; Stone 1993; Visher 1910.

# AMERICAN THREE-TOED WOODPECKER

***Picoides dorsalis***

L: 8–9 in. (20.3–22.9 cm)
WS: 14–16 in. (35.6–40.6 cm)
WT: 1.7–2.3 oz. (48–65 g)

**Unpredictable, enigmatic, and discreet,** the American Three-toed Woodpecker (hereafter referred to in this account as "Three-toed," for ease of readability) is for many birders the most sought-after woodpecker on the continent. Rarely is a Three-toed found far from dead or dying timber, whether insect- or fire-killed, but even in optimal habitat this quiet little bark beetle specialist can easily go undetected.

The Three-toed's affinity for boreal and montane spruce forests tucks it away in some remote and challenging habitats. In the Cascade Mountains, one must frequently enter roadless areas on foot or horseback to find it—in Oregon alone, more than 1 million acres (405,000 ha) of the Three-toed's range lie in federally protected wilderness areas. Canada's boreal forest carpets 1.3 billion acres (526 million ha) of continuous Three-toed habitat, much of it so remote it would take lifetimes to explore.

A little homework and a lot of patience can go a long way toward adding this bird to one's life-list. As with most woodpeckers, the best time to find the Three-toed is at the peak of the breeding season, when adults are most vocal and nestlings can be heard loudly calling from their cavity entrance. Burned forests reach their peak of productivity for this woodpecker in the first two years after a fire, so timing is everything. Finally, learning its habitat preferences—locally favored tree species and corresponding elevations—will greatly increase the odds of success.

## DISTRIBUTION

Found across boreal forest, from w. AK to NL, and south throughout Rocky Mts. and Cascade Mts. of OR and WA; rare in n. Warner Mts., possibly into n. CA. Generally a permanent resident throughout its range, but prone to unpredictable and opportunistic wandering.

**PERMANENT RANGE:** South of tree line throughout most of AK, east across forested interior Canada to NL; south into n. New England and Adirondacks; rare breeder in Upper Midwest.

Most of BC except southern coast and islands, south through OR and WA Cascades; throughout Rocky Mts. to Mogollon Rim. Local in Snake Range, NV, and Black Hills, SD.

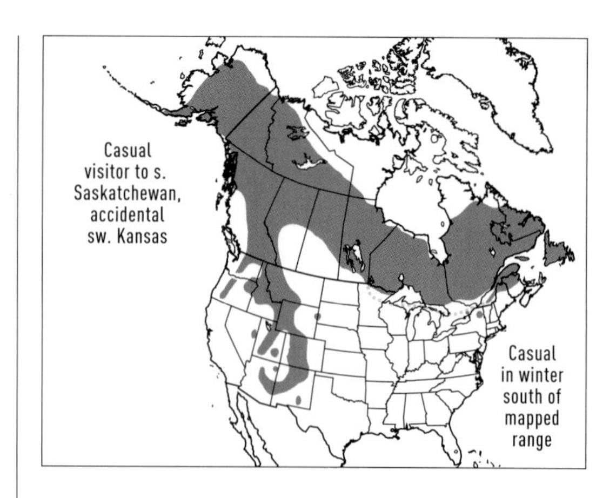


**SEASONAL MOVEMENTS:** Movements outside breeding season vary widely across species' range; small rangewide population size may limit ability to accurately document dispersal.

Southwestern populations may move downslope in winter, with regular movements seen in Rockies and cen. Cascades. East of Rocky Mts., rare, erratic winter resident from s.-cen. SK south to IA; also n. New England south to Mid-Atlantic.

Eastern populations are known for local sporadic irruptions, often associated with large-scale burns, insect outbreaks, or disease. Decline in insect food sources in normal range may also catalyze movements. Three-toed irruptions are less frequent, later in season, and occur over shorter distances than those of Black-backed Woodpecker, likely because of differences in foraging behavior.

## HABITAT

High-latitude and subalpine coniferous forests, often associated with spruce and lodgepole pine. In northeastern regions, also found in mature hardwood forests, and frequently associated with boreal-forest swamps. Regularly inhabits burned, insect-killed, and flooded forests, which offer abundant food supplies and nesting substrates. Elevations range from 4,250 to 11,000 ft. (1,300–3,350 m) in West and from 1,200 to 4,100 ft. (360–1,250 m) in East.

In far western forests, found mostly in lodgepole pine–Engelmann spruce subalpine habitats; rarely drops below this zone into Douglas-fir or ponderosa pine mixed-conifer forest, more frequently into large swaths of stand-replacement burned forest. Known to frequent burned forests especially high-intensity stands in first 3 years after fire, with highest level of nesting activity in unsalvaged stands.

**OPPOSITE:** An adult male American Three-toed Woodpecker excavates a fresh cavity. This is the nominate subspecies of the central and southern Rocky Mountains. *Doug Backlund; June 2006, Black Hills, SD.*

An adult female American Three-toed Woodpecker, western boreal forest subspecies (*P. d. fasciatus*). *Doug Backlund; June 2010, Brooks Range, AK.*

In northern boreal forest, found in black spruce bogs as well as diverse stands with other conifers or quaking aspen. Rarely inhabits riparian willow thickets. Individuals found out of range are often associated with forest disturbance, such as recent fires, windthrow episodes, or beetle outbreaks.

## DETECTION

Possibly the most difficult of all North American woodpeckers to locate, for multiple reasons: relative scarcity throughout range, unpredictable seasonal wanderings, difficulty of accessing habitats, vocally quiet nature in all seasons, and near silence outside breeding season.

**VOCAL SOUNDS AND BEHAVIOR:** One of least vocal woodpeckers in N. Am. Calls are generally similar to those of other *Picoides*, and all calls are probably given by both sexes. Some references cite up to 13 distinct vocalizations, but these can be summarized as three basic calls: call note, rattle, and twitter.

Call note is like that of Downy Woodpecker but less sharp; described as squeaky *pik* or *peet* or *teet*. Given singly or in variably spaced series, year-round, and accompanying wide array of behaviors. Also gives a sharp *kweek* call.

Rattle is a fluid series of evenly cadenced notes, usually preceded by one or two call notes. Given during territorial displays and when interacting with other species, generally early in breeding season, often in flight. Young birds give a short rattle. Second version of rattle given as a unique series of hurried, squeaky high-pitched notes.

Twitter is a low-pitched, plaintive courtship call given between mates and sometimes in mate defense; often accompanies head swinging and is sometimes given in flight.

**NONVOCAL SOUNDS:** Drumming repertoire is diverse. Drum is generally shorter in duration than that of most woodpeckers and is performed more often. One of few species fairly easily identified by its drum.

Both sexes probably give all types of drums. Standard drum is a brief roll that speeds up in cadence but drops in pitch at end; similar to that of

Black-backed Woodpecker, but generally shorter in duration and less resonant. It is often performed at slow cadence, especially when given between mates, and rarely remains on same pitch and frequency; may rarely begin with three or four slow taps disconnected from main roll. Occasionally performs a stuttered drum similar to that of sapsuckers, but even more varied in pitch and more erratic in phrase patterns.

Tapping sounds are most easily heard when individuals excavate roost or nest cavities. Feeding sounds are typical of aggressive bark flaking rather than heavy drilling.

## VISUAL IDENTIFICATION

Fairly typical "pied" woodpecker, black and white throughout. Well camouflaged against charred tree trunks and in filtered light. Characteristics below describe intermediate *P. d. fasciatus* subspecies.

**ADULT:** Head is blackish with two white facial stripes; thin white eye stripe often widens and curves downward around rear of ear patch. Long nasal tufts may give short-billed appearance. Female has black crown speckled with white; male has white speckling fore and aft of nickel-sized yellow crown patch, which is often difficult or impossible to see from below.

Dull black wings contrast with light center of back, which sometimes appears more mottled than barred, especially when worn. Narrow barring almost through lower belly.

In flight, blackish above, strongly mottled gray underneath.

**JUVENILE:** Duller than freshly plumaged adult. Crown is dull black with small yellow patch; buffy underneath, with brownish-spotted flanks. Juvenile female may show a few yellow-tipped feathers in center of crown, rarely no yellow, usually not extending behind eyes.

**INDIVIDUAL VARIATION:** Males average about 10 percent heavier than females. Other variation is mostly attributable to subspecies and introgression.

**PLUMAGES AND MOLTS:** Preformative and prebasic molts occur June–Oct. Unlike in most woodpeckers, juvenal primaries 1–6 are replaced while in nest; primaries 7–10 are replaced before they fully develop. Also, juvenal rectrices may not molt sequentially.

**DISTINCTIVE CHARACTERISTICS:** Three-toed and Black-backed Woodpeckers are the only two *Picoides* in N. Am. that show yellow rather than red in male plumage. Also, they are the only two North American woodpeckers with three toes per foot (missing the hallux).

**SIMILAR SPECIES:** Both Black-backed and Hairy Woodpeckers could be mistaken for Three-toed under poor viewing or hearing conditions (bad light, long distance, loud wind, etc.). Eastern *P. d. bacatus* most closely resembles Black-backed; Rocky Mt. *P. d. dorsalis* most closely resembles Hairy.

**GEOGRAPHIC VARIATION:** Notable variation among geographically distinct populations, with introgression at contact zones.

### SUBSPECIES

Three subspecies generally recognized, differing in size and extent of white on back and face.

*P. d. dorsalis*: Generally s. Rocky Mts.; w. MT south through nw. NM. Largest subspecies; long, heavy bill; faint white spotting on forehead; back either solid white or with minimal narrow black bars; little or no black on outer tail feathers; boldest white facial markings of all subspecies.

*P. d. fasciatus*: Northwestern boreal forest, from AK south through Canadian Rockies to OR Cascades and Blue Mts., n. ID, and nw. MT. Intermediate in size; heavily spotted on forehead; mostly white back with narrow black bars; outer tail feathers nearly pure white. Includes formerly recognized Alaskan "*alascensis*."

*P. d. bacatus*: From n. MB eastward. Smallest subspecies; nearly solid or solid black forehead; white eye stripe faint or absent; narrow white, gray, or brown barring on back; outer tail feathers sometimes with black bars. Includes formerly recognized and larger "*labradorius*" from Labrador.

## BEHAVIOR

### BREEDING BIOLOGY

Knowledge of breeding biology is limited for American Three-toed Woodpecker; far more data are available for Eurasian Three-toed Woodpecker, with which Three-toed was formerly considered one species. Lack of data from N. Am. may result in part from difficulty in accessing habitat, especially early in nesting season, when snow impedes access to higher elevation habitats.

#### *Nest Site*

Often excavates cavity through bark; favors coniferous and deciduous snags over live trees; may also prefer trees with heart rot. Favors wide variety of tree species, but mostly conifers. As with Black-backed Woodpecker, bottom edge of cavity entrance may be strongly beveled inside and out. Cavity entrance is often on underside of leaning tree; orientation may be random or preferentially east through south.

### *Courtship*

Seasonally monogamous, but possibly not long term.

### *Parenting*

Adults may consistently land below nest before entering cavity to feed. Nest success rate is relatively low; may be lowest in areas subject to active logging during nesting season.

Parents may divide brood for postfledging feeding rituals. Regular vocal contact maintained between parents and young.

### *Dispersal*

Young may return to breed in their first spring within 820 ft. (250 m) of natal site.

## NONBREEDING BEHAVIOR

### *Feeding*

***Preferred Foods:*** Year-round diet is estimated to be up to 94 percent animal foods: 85 percent beetle larvae; 8 percent ants; also moth pupae and other arthropods. Wide variety of beetles consumed, depending on region and tree species; some regional favorites include larvae of bark, engraver, long-horned, and metallic wood-boring beetles. Minimal vegetable material includes cambium and sap.

Almost inextricably linked with burned or insect-killed forest, and well known for its ecological role in stemming insect outbreaks. Forests experiencing local bark beetle epidemics host high concentrations of Three-toeds, especially in winter; individual birds may consume thousands of larvae per day.

May very rarely visit suet feeders.

Adult male American Three-toed Woodpecker of the northeastern subspecies, *P. d. bacatus*, tolerates a light spring snow as it works the burned conifers of an Ontario forest. *Paul Jones; Mar 2014, Timmins, ON.*

An adult female American Three-toed Woodpecker, western boreal forest subspecies (*P. d. fasciatus*), leaves a nestling male begging as she departs for another food-gathering foray. *Stephen Shunk; June 2013, Anchorage, AK.*

***Foraging Behaviors:*** Flakes and pecks to remove bark plates or chunks, exposing etched or tunneled galleries of bark beetle larvae. Feeds primarily on conifer trunks, infrequently on branches, rarely on downed wood or ground, and almost exclusively on dead or dying trees. Species of favored trees is probably less important than tree morbidity.

Bores through unburned patches of bark on lightly to moderately burned trees (same substrate favored by bark beetle larvae). Will also peck obsessively at same area of a single dead trunk for long periods, pausing quietly for frequent breaks, continuing until tree is stripped completely bare of bark before moving to next tree.

Bark removal is achieved with a combination of lateral and direct blows; does not typically excavate deeper than cambium. Probably gleans occasionally, but not reported flycatching. Known to drill xylem wells in rings around trees; also observed feeding at Yellow-bellied Sapsucker wells in fall.

Sexual niche partitioning occurs in some regions. Males may favor larger-diameter stems and trunks than females. Females may feed more frequently at edges of burned and unburned forest; males more common inside burn.

### *Territory Defense and Sociality*

Very little, if any, intraspecific territoriality is exhibited during breeding season. Males establish territory, which they may reuse for multiple years; sexes may partition their territory internally. Territorial displays are similar to those of other North American *Picoides*. Engages in bill pointing, head swinging, and wing and tail spreading, often with raised crest and accompanying vocalizations. Also performs mothlike flight display.

Concentrations of individuals may be found feeding together in stands with high food productivity, especially in recently burned forest. Mates remain paired during breeding season but are generally solitary in winter.

Winter territories may range from 74 to 740 ac. (30–300 ha).

### *Interactions with Other Species*

Highly territorial toward Black-backed Woodpecker, generally near Black-backed nest site. Encounters are usually same-sex; Black-backed is generally dominant. Reports of occasional aggression with Yellow-bellied Sapsucker and Hairy Woodpecker. Conflicts include chasing, wing flapping,

An adult female American Three-toed Woodpecker, western boreal forest subspecies (*P. d. fasciatus*), carries bark beetle larvae to nearby nestlings. *Dick Tipton; June 2009, Deschutes Co., OR.*

and chattering, rarely culminating in midair clasping of feet and tumbling to ground.

Likely some conflict with other species when sharing a roost tree that contains multiple cavities. In some cases, Three-toed will forage near sapsuckers and other woodpeckers, especially in areas of abundant food supply.

Tree Swallows may attempt to usurp active nests. Boreal Chickadees are known to forage among bark chunks dropped by foraging Three-toeds.

### MISCELLANEOUS BEHAVIORS

After feeding episodes, adults may fly to branch, drum slightly, then preen. Will also fly to branch and preen when flushed from cavity.

Typically roosts in cavities in dead trees, rotating cavities from every few days to a few weeks; may favor unlogged stands for roosting. Known to roost in abandoned Pileated Woodpecker cavities.

## CONSERVATION

**HABITAT THREATS:** With a smaller population than most North American woodpeckers, Three-toed may be more sensitive to timber harvesting, post-fire salvage logging, and forest fragmentation, all of which reduce availability of foraging and nesting substrates. In forests with short-term harvesting rotations, this species may be excluded permanently in regions where it depends on old-growth.

Forestry activities frequently remove snags and unhealthy timber on which this species depends. In some parts of range, especially Rockies and Cascades, species frequents high-intensity burned stands, especially in first 3 years after fire. Suppression of such fires limits available habitat; it also allows unnatural buildup of fuels, which can catalyze stand-replacement events far larger than those that occurred under presettlement regimes. These scenarios provide large swaths of optimal habitat over short term, often with consequent explosion in woodpecker population, which cannot be sustained over long term.

**POPULATION CHANGES:** Greatest concentrations of this species occur in North American boreal forest, most of which is not covered by standardized

**FIGURE 12.** American Three-toed Woodpecker Defense Posture. An adult female observed in Oregon mimicking fungal conchs to escape detection by a Common Raven. *Denny 2001.*

The American Three-toed Woodpecker finds optimal habitat in beetle-infested pine-spruce forests like this stand in the Pacific Northwest. *Stephen Shunk; July 2013, Deschutes Co., OR.*

surveys. Local populations are consistently more abundant in recently burned and diseased forest than in unafflicted forest; highest densities usually occur within first 3 years following event. Opportunistic use of disturbed habitats and irruptive tendencies suggest evolutionary tie to exploitation of constantly changing food supply following episodic natural disturbance.

**CONSERVATION STATUS AND MANAGEMENT:** Targeted for conservation priority throughout range south of Canada. Sensitive in NB; Threatened in NH. Removed from U.S. Forest Service Sensitive Species list in 2011 for Rocky Mountain region.

Difficulty in assessing population numbers and unpredictable distribution make monitoring of response to conservation measures problematic.

Studies in OR recommend a variety of management practices that retain dead, dying, and decaying trees; ideal prescription for OR would protect 530 ac. (214 ha) for each pair of woodpeckers in old-growth mixed-conifer or lodgepole pine forests above 4,500 ft. (1,385 m).

Recommendations for Rocky Mts. populations prior to 2011 (see above) included maintaining age-class diversity, enhancing snag density, eliminating or limiting salvage logging, leaving burned forest intact 3–5 years before salvage, interpreting insect infestations at forestwide scale rather than individual stand scale, and allowing natural fires to burn in suitable habitat.

## REFERENCES

American Ornithologists' Union 2003; Anderson and McGee 2001; Bangs 1900; Bevier 1990; Bock and Bock 1974; Bryens 1929; Clement 1943; Denny 2001; Dixon 2012; Fleming 1902; Goggans et al. 1988; Gregory 1923; Jensen 1923; Kibbe and Boise 1984; Leonard 2001; New Brunswick Natural Resources 2015; Rogers and Jaramillo 2002; Rottenborn and Morlan 2000; Shainin 1939; Shaw 1925; Short 1974; Thiel 1978; Trochet et al. 1988; USFS 2011; West and Speirs 1959; Wiggins 2004; Wood 1913, 1921; Wood 1959; Yunick 1985; Zink et al. 2002.

# BLACK-BACKED WOODPECKER

## *Picoides arcticus*

L: 8.1–9.8 in. (20.5–24.8 cm)
WS: 16–17 in. (40.6–43.2 cm)
WT: 2.2–3.1 oz. (61–88 g)

**One of the handsomest members** of the Picidae worldwide, the Black-backed Woodpecker sports a glossy black tuxedo, blending perfectly against the charred trunks of burned conifers across its range. On the extreme end of the specialization spectrum, the Black-backed may be the best adapted woodpecker in the world for extracting wood-boring beetle larvae from the trunks of infested trees. This species possesses a suite of anatomical adaptations that facilitate the delivery of maximum force when striking a tree.

Once recognized as a "conservator of the forest" with its insatiable hunger for wood-destroying beetle larvae, the Black-backed Woodpecker receives little credit today for its role in managing insect outbreaks. Recent research may once again elevate the profile of this species by emphasizing its ecological link to burned forests. Planning for controversial postfire salvage operations has not typically included provisions for the Black-backed Woodpecker. However, monitoring of local populations may be critical for the long-term conservation of the forest and this unassuming forest denizen.

Ecology aside, the Black-backed Woodpecker's engaging behavior and its stunning wardrobe make it simply a fun bird to watch in North American forests.

## DISTRIBUTION

Permanent resident in northern and western coniferous forests. As with American Three-toed Woodpecker, this species' prevalence across expansive boreal forest makes it difficult to study and to accurately assess distribution and abundance.

**PERMANENT RANGE:** Cen. AK and mid-McKenzie R., NT, south across Canada to s. NL; decreasing in abundance south into n. Great Lakes and n. New England. Isolated populations in Adirondacks, Black Hills, and northern lower peninsula of MI.

Nw. BC, east of Coast Ranges southward through n. Rockies to Grand Tetons. East slope of Cascades in WA and OR, south from Siskiyous into CA Cascades, Warner Mts., and Sierra Nevada. Rare to sw. OR Coast Range.

**SEASONAL MOVEMENTS:** Not truly migratory, although some populations move south of breeding territories in fall into northern Great Plains and Mid-Atlantic. Highly variable and irregular in annual numbers and movements. Known for erratic seasonal irruptions, notably winters of 1860–1861, 1923–1924, 1956–1957, and 1974–1975, with reports as far south as IL, NJ, and PA. Up to 470 reports collected locally during major irruption events.

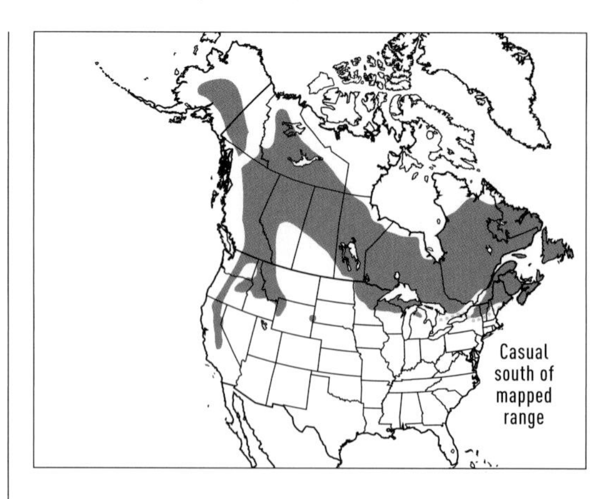


**HABITAT:** Coniferous, subalpine, and mid-elevation mixed-conifer forests throughout range, with variety of locally favored tree species. Occurs at wider range of elevations, and is more dependent on burned forest, than American Three-toed Woodpecker. Typically favors heavily disturbed stands following bug kill, burn, or flooding.

Prefers relatively recent burns to unburned mature forest, with abundance increasing in first 4 years after fire, then declining rapidly as adult beetles emerge and disperse; sometimes rare in live forest surrounding burn. Unburned habitats in Midwest may include black spruce–tamarack bogs and cedar swamps. In Northeast, frequents boreal forest with conifers and aspen. Prefers relatively dry coniferous forests in Pacific states and ID.

During unpredictable winter irruptions, may be found in completely different habitats, including suburban areas affected by Dutch elm disease.

## DETECTION

Typically an active and vocal forager, often heard before seen. Also found by watching for bark flakes falling in woods. Can be easy to find if present in burned stands, where sound travels better than in live forest.

**VOCAL SOUNDS AND BEHAVIOR:** Unique vocalizations are easily detectable and separable in the field from those of other woodpeckers. Three basic call types: sharp call, screechy rattle, and scratch call.

Sharp call, like *jeek* or *jik,* resembles squeak of

**OPPOSITE:** An adult male Black-backed Woodpecker pauses between feeding bouts. *Steve Brad; June 2012, Mono Co., CA.*

An adult female Black-backed Woodpecker with a wood-boring beetle larva. *Paul Bannick; June 2007, Nevada Co., CA.*

rubber-soled shoe on a wooden gym floor. Given singly or in loose series by both sexes throughout year. This is general alarm note or contact call, given in faster series during courtship, and often combined with scratch calls when agitated or particularly active.

Rattle is like Downy Woodpecker's descending whinny in structure, but faster, much burrier, and more metallic, resembling screeching-stop sound of a speeding cartoon character: *jee-jhee-jhee-jhee-jhee-jee-jur*, sometimes purring briefly at end of series. Also given by both sexes, year-round, often during head-swinging display.

**OPPOSITE:** The Black-backed Woodpecker thrives in burned coniferous forests, such as this stand in the Oregon Cascades, as long as the burned trees are left standing. *Stephen Shunk; June 2009, Jefferson Co., OR.*

Scratch call comprises multiple (usually three or four) harsh grating notes, often preceded by sharp call and descending slightly in pitch on final note; resembles a scratchy chuckling sound, *shi-ji-jett* or *jee-ji-jitt*. Sometimes called snarl call. Typically given in breeding season to address mate, defend territory, or when feeding young.

Unique gurgling call given by both adult and fledgling may be signal to feed or be fed.

**NONVOCAL SOUNDS:** Drumming is fairly distinctive and forceful; tends to descend in pitch and increase in tempo toward end of roll, like a rapidly accelerating Ping-Pong ball. Like American Three-toed, Black-backed typically drums more often than most woodpeckers; pattern is less variable than in other *Picoides* and faster and longer than somewhat similar American Three-toed drum. Individually variable.

Both sexes drum for courtship and territorial communication, most frequently Apr.–June but also in late winter. Female's roll averages slightly longer than male's. Some individuals may have favorite drumming post, but there is no consistent substrate type across range.

Tapping is performed occasionally, with single rap on trunk before roosting or multiple taps after sharp call. Male observed smacking its mandibles together when agitated, making a clicking sound. Adults often drum in association with scratch call, when flushed from nest, or when nest is closely approached.

## VISUAL IDENTIFICATION

**ADULT:** Almost entirely black dorsal surface, with contrasting single white facial stripe, faint spotting in remiges, and white outer rectrices. Very well camouflaged against charred tree trunks, especially when perched on shaded side of tree. Thin white "nick" often extends behind top-rear of eye. Male shows circular lemon to golden yellow crown patch, but this is often difficult to see unless bird tips its head sideways. Bluish gloss may be apparent in black upperparts. White flecks on primaries may blend with strongly barred sides and flanks.

In flight, almost entirely black above, with faintly spotted flight feathers; underparts mottled gray and white.

**JUVENILE:** Similar to adult but duller, especially on crown. Underparts buff with brownish barring on sides and flanks. Both sexes may show yellow crown patch, male's bolder yellow and more forward than female's; female may lack yellow entirely.

Nestling male (left) and female (right) Black-backed Woodpeckers. Note the extent of yellow in the crowns. *A, Kristine Falco; July 2009, Deschutes Co., OR; B, Paul Bannick; June 2007, Nevada Co., CA.*

**INDIVIDUAL VARIATION:** Some adults show a few white tips to rump feathers. Males are slightly larger than females, most notably in 5–10 percent longer bill. White outer tail feathers and belly often become sooty brown from foraging in burned forest. Melanistic birds rarely reported, and markings are difficult to distinguish from charcoal stains.

**PLUMAGES AND MOLTS:** Preformative molt is completed by Oct. Later prebasic molts occur June–Oct.

**DISTINCTIVE CHARACTERISTICS:** Darkest head of all North American woodpeckers. Along with American Three-toed Woodpecker, only "three-toed" woodpecker in N. Am. (missing hallux), and one of two *Picoides* in N. Am. with no red in male plumage.

**SIMILAR SPECIES:** Closely related to American Three-toed Woodpecker; until 1976, the only two North American species in *Picoides* genus. May be confused with American Three-toed in e. N. Am., where American Three-toed subspecies *P. d. bacatus* exhibits nearly solid black back with thin and sometimes faint white barring. *P. d. bacatus* also shows reduced to absent white eye stripe, approaching that of Black-backed, but typically has thinner white mustache than Black-backed. Throughout range, Black-backed is generally larger with longer bill.

Vocal repertoire is diverse and conspicuous compared with subtle, plaintive calls of American Three-toed. Black-backed is also more likely to be found at lower elevations, and in more diverse forest types than American Three-toed.

Other species with black or all-dark backs that share this species' range include Williamson's Sapsucker and White-headed and Pileated Woodpeckers, all of which have some white visible above compared with dorsally all-dark Black-backed. Also, all three give distinctive vocalizations easily separable from those of Black-backed. Lewis's Woodpecker may appear all black in poor light, but its behavior and vocalizations could not be mistaken for those of Black-backed.

**GEOGRAPHIC VARIATION:** Additional subspecies was proposed in 1900 but not generally accepted; named *"tenuirostris,"* "Slender-billed" Black-backed Woodpecker, based on small sample size from s.-cen. OR averaging 1.5 percent (male) and 0.8 percent (female) longer wings, darker nasal tufts (less blending with white mustache), and 21 percent narrower bill (at widest point) than northern and eastern birds.

Some bill-size variation may be associated with local foraging substrates—for example, variable bark thickness of different conifer species—but geographic variation in this character probably remains too slight to warrant subspecies designation.

## BEHAVIOR

### BREEDING BIOLOGY

#### *Nest Site*

Nests in wide variety of tree species, live and dead (burned or unburned). Also known to nest in utility poles. Rarely reuses cavity from prior year.

May prefer dense, unlogged forest stands and less decayed, relatively small-diameter trees compared with other woodpeckers in its range, but larger diameter than average snag conditions. In burned forests, nest trees are surrounded by high snag densities averaging 129 snags per ac. (319 per ha). May also favor trees with intact tops and those infected with heart rot. May prefer barkless portions of trees (or will remove bark around cavity entrance), though CA populations may leave bark intact more often than elsewhere in range.

Cavity entrance is more oval than that of most woodpeckers and is beveled on lower edge, inside and out.

#### *Courtship*

Seasonally monogamous. Some pairs remain together year-round, others observed switching mates shortly after fledging period and remaining with newly paired mate through next breeding season.

#### *Parenting*

In some regions, female may defer all feeding duties to male as fledging date approaches (when nestlings may outweigh female). Other observations show that male feeds less often but provides more food per visit.

Female may renest in new cavity if first clutch is lost. Nest success rate is high in unlogged burned forest of mixed burn severity. Up to 5 percent of clutches may contain a runt egg, an occurrence that is more typically correlated with communal breeding.

Young are fed insect prey collected within about 0.5 mi. (0.8 km) of nest; foods include adult and larval insects, including pine beetle larvae during outbreaks.

Parents are known to communicate to nestlings when other woodpecker species approach nest site, causing nestlings to stop calling. Nestling begging sounds are occasionally audible from 650 ft. (200 m) or more.

#### *Young Birds*

Young usually fledge later than other woodpeckers in range. Adults may split fledglings to simplify feeding duties, young occasionally switching from one adult to the other. Fledglings are known to mimic adult movements and behavior as they tag along.

#### *Dispersal*

No data on dispersal of juveniles, but distances likely depend on food conditions in local region; young may remain closer to natal site when fledged among a recent burn or insect outbreak.

### NONBREEDING BEHAVIOR

#### *Feeding*

***Preferred Foods:*** About 75–95 percent of diet is larvae of long-horned (Cerambycidae) and metallic (Buprestidae) wood-boring beetles; stomach contents of 51 birds showed 15–20 larvae per bird, with high of 34 larvae in one stomach. Also consumes various other beetle larvae, as well as other arthropod prey, depending on region and tree species diversity. Vegetable foods typically make up less than 15 percent of diet; include wild fruits, plant seeds, mast, and cambium.

Large-scale salvage logging of burned timber may pose a serious conservation threat to local Black-backed Woodpecker populations. *Stephen Shunk, courtesy of Institute for Bird Populations; July 2011, Lassen Co., CA.*

***Foraging Behaviors:*** Despite primary diet of wood-boring beetle larvae, feeding style varies based on most readily available prey; typically some combination of: 80–85 percent pecking and boring, 20–70 percent bark scaling, 1–30 percent trunk-surface gleaning or probing, and 0–5 percent ground feeding. Like American Three-toed Woodpecker, may spend considerable time on each trunk before moving to next tree.

Will forage on heavily burned trunks for some prey species, or on unburned patches of bark on moderately burned trees for other species. Often feeds among small-diameter trees and snags, occasionally on downed logs. Local population numbers diminish within 4–5 years after fire, depending on burn severity and incidence of secondary beetle outbreaks. In mixed-conifer forest, may choose Douglas-fir for foraging 6 or more years postfire, since this species hosts beetle larvae longer after a fire than other conifers.

Some sexual differences in foraging habits, variable by region. Males may generally feed lower on trees and on less severely burned trees than females. Possibly owing to smaller bill size, females may favor smaller-diameter trees than males.

### *Territory Defense and Sociality*

Nest density averages 1 nest per 16.5–25 ac. (6.67–10.12 ha). Mean densities reported from ne. OR of 1.1–3.5 birds per 247 ac. (100 ha), highest in mature Douglas-fir–ponderosa pine forest. Nonbreeding home ranges of 500–1,000 ac. (200–400 ha) for males 6–8 years after fire. Size of winter home range increases with decrease in local food availability.

Typically defends territories from other Black-backeds in breeding season. Performs wide range of displays: points bill at aggressor; raises bill high in air; lowers bill and shows yellow crown; raises crest, often in conjunction with other displays; hunches back in submission; and swings head side to side. Wing-spreading display may accompany any of the above, along with a combination of calls.

Like other *Picoides*, performs fluttering, mothlike flight display; may also spread tail during flight or other displays.

Both sexes are known to engage in physical attack, usually not involving contact but occasionally culminating in wing batting or rarely locking feet in flight and tumbling downward with opponent.

### *Interactions with Other Species*

Conflicts occur in nest-site vicinity, especially with other cavity nesters, which occasionally usurp Black-backed nests. Known to interact frequently with Hairy Woodpecker, with Hairy most often the aggressor. Black-backed has larger winter home range and generally prefers less severely burned trees than Hairy Woodpecker. May segregate foraging habitat by excavating for wood-boring prey more than Hairy, with Hairy favoring bark beetles.

Where ranges and habitats overlap, Black-backed dominates American Three-toed Woodpecker, often forcing it out of its preferred habitat.

### MISCELLANEOUS BEHAVIORS

Described as lacking grace when climbing. May frequently alternate among preening, drumming, and scratching, even while feeding young. Known to roost in trunk scars, branch forks, mistletoe clumps, "witches' brooms," or pressed up against trunk.

## CONSERVATION

**HABITAT THREATS:** Forest management practices over last 100 years have adversely affected Black-backed habitat throughout species' range. Fire suppression, thinning, and especially postfire salvage logging are all detrimental to this species. Fire suppression inhibits natural patchwork of burned and unburned forest, resulting in large-scale stand-replacement fires. These episodes may facilitate rapid short-term population growth that cannot be sustained in long term, without more frequent, intermediate-scale fires. Black-backed abundance in forests that were commercially thinned and then burned is lower than in burned forests that were not thinned prior to fire.

Postfire salvage logging frequently involves removal of large trees, whether burned or not, to optimize value of timber operation; this may minimize potential of burns to provide long-term sustainable habitat.

**POPULATION CHANGES:** Extremely erratic nature of local populations makes data from standardized surveys equally erratic, with numbers booming and busting over periods of 3–5 years following episodic forest disturbance.

Short-term population increases are documented in burned forests, but there are no data on where these birds come from or where they go after postfire habitat suitability has peaked for optimal nesting and foraging. Nearly all studies show higher densities in recently burned or beetle-infested forest than in adjacent unburned or uninfested forest; numbers generally decrease 4–5 years after burn. Reported as 20 times more abundant in burned versus unburned forests of ne. WA.

**CONSERVATION STATUS AND MANAGEMENT:** Targeted for conservation priority in 13 states; Critically Sensitive Species in OR; candidate for Threatened listing in WA and CA.

Stand-replacement fires should be allowed at a scale and frequency proportionate to those of presettlement fire regimes; salvage logging should be postponed until 5 years after a fire whenever possible. Road closures and prohibition of snag removal for firewood would benefit local populations.

Postfire salvage recommendations include retaining snags greater than 9 in. (23 cm) diameter in clumps at greater than 42–50 snags per ac. (104–123 per ha). In OR and WA, Woodpecker Management Areas—generally below 4,500 ft. (1,372 m)—are recommended to be exempt from variety of timber operations.

### REFERENCES

American Association for the Advancement of Science 1957; Apfelbaum and Haney 1981; Bagg 1919; Bangs 1900; Blackford 1955; Bock and Lynch 1970; Brewster 1884; Capen 1926; Chandler 1997; Covert 2003; Dixon and Saab 2000; Dudley 2005; England 1940; Forristal, C., pers. comm.; Forristal et al. 2004, 2005; Goggans et al. 1988; Griscom 1924; Hall 2008; Hasbrouck 1890; Helmuth 1937; Herbert et al. 1926; Herrick 1884; Howitt 1927; Jung 1927; Kilham 1966a; Koenig 1980a; Lewis et al. 2003; Marshall 1992a; Mayfield 1958; Nappi et al. 2003; Nichols 1904; Nielson-Pincus 2005; Oberholser 1918; Olson 2002; Robbins 1900; Robinson 1926; Saab et al. 2004, 2009; Sawyer 1916; Schorger 1940; Shainin 1939; Short 1974; Siegel 2011; Siegel and DeSante 1999; Stevenson and Anderson 1994; Tucker 1926; Van Tyne 1926; Villard and Benninger 1993; West and Speirs 1959; Wood 1959; Wright 1905, 1919; Yunick 1985.

# NORTHERN FLICKER
## *Colaptes auratus*

L: 10–14 in. (25.4–35.6 cm)
WS: 19–21 in. (48.3–53.3 cm)
WT: 3.7–5.9 oz. (105–167 g)

**The Northern Flicker** is more widespread, more conspicuous, and more abundant than any other North American woodpecker. Ironically, it is the least woodpecker-like of them all. Flickers feed almost entirely on the ground, and they eat more ants than any other North American bird.

A flicker's bill is decurved slightly, making it excellent for probing but inferior for excavating, and unlike in the tree-climbing woodpecker species, a flicker's feet are nearly always in the zygodactyl, or perching, formation. Both the Northern and Gilded Flickers are our only woodpeckers with brightly colored underwings and undertail, which are excellent plumage characters for defending a territory or attracting a mate.

From its intricate plumage to its engaging behavior, the Northern Flicker has fascinated humans for centuries. Native Americans adorned their baskets and ceremonial garb with its bright wing and tail feathers; it is the only woodpecker to boast "state bird" status (in Alabama); and more than 150 folk names have been assigned to this ubiquitous ant-eater.

The Northern Flicker's ubiquity has afforded unparalleled opportunities for field study. It was a "Yellow-shafted" Northern Flicker shot in Sabine Parish, Louisiana, on Christmas Day 1905 that made the record books as the first ever banded-bird recovery in North America. And to date, more than 70,000 Northern Flickers have been banded in North America. This species has also been the subject of intensive research on the sharing of parental traits across taxonomic lines, with "Yellow-shafted" and "Red-shafted" Northern Flickers interbreeding freely from southeastern Alaska to the Texas panhandle. These two forms of the Northern Flicker were once considered separate species, with "Yellow-shafted" birds in the North and East, and "Red-shafted" birds in the West. Shared plumage traits among birds migrating southward from the broad intergrade zone may be very difficult to detect in the field, and many birds may be impossible to assign to one form or the other.

## DISTRIBUTION

Widely distributed, from northern edge of boreal forest to Atlantic and Pacific Coasts and south into Mex. and Cen. Am.

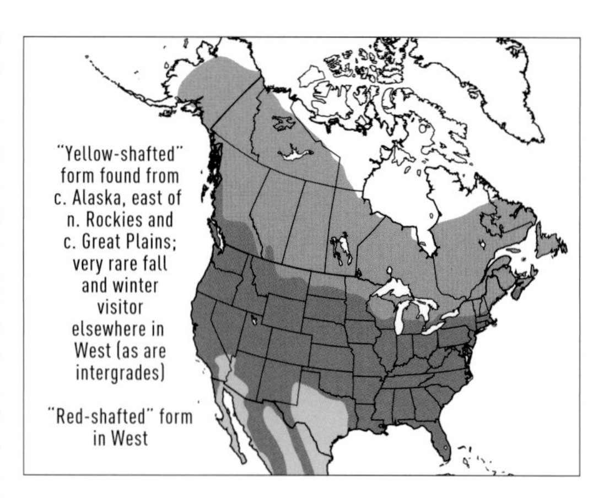


**BREEDING RANGE**

***"Yellow-shafted" Northern Flicker:*** Wooded parts of w. AK, eastward across boreal treeline to NL; e. Great Plains south to e. TX and eastward to Gulf and Atlantic Coasts, including W. Indies.

***"Red-shafted" Northern Flicker:*** Se. AK through s. interior BC, east to AB Rockies and south through western Great Plains, w. TX, and into Mexico. Disjunct populations in n. Cen. Am.

**SEASONAL MOVEMENTS:** Widespread and locally abundant migrant throughout breeding range. In fall, AK and n. Canadian populations, and some midcontinent migrants, generally move south through s. CA and TX into nw. Mex. From Atlantic Provinces west to Prairie Provinces, generally funnels southward. Populations from northern parts of intergrade zone mostly mix with Great Plains breeders in southward movements.

Mostly nocturnal migrant, though there are some records of large-scale morning movements in fall. Weather likely influences length of migration as well as departure dates. Small groups of males may precede peak movements. Usually moves in loose flocks of a few to 100-plus birds; rarely up to 1,000 birds recorded per hour at individual monitoring sites.

**WINTER RANGE:** "Yellow-shafted" subspecies generally winter from s. Canada (plains and east) south through breeding range to n. Mex.; "Red-shafted" subspecies generally from coastal and s. BC and s. AB south through remainder of breeding range, to w. TX. Southerly breeding birds are more or less sedentary. Intergrades are common and can be expected anywhere south of contact zone.

**OPPOSITE:** An adult male "Yellow-shafted" Northern Flicker, showing all the typical plumage traits of this subspecies group. (See Subspecies, p. 219.) *Alan Murphy; Nov. 2009, Polk Co., FL.*

An adult female "Yellow-shafted" Northern Flicker doing what flickers do best: foraging on the ground. *Marie Read; Apr. 2007, Tompkins Co., NY.*

## HABITAT

Most widespread woodpecker in N. Am., occupying broad range of habitats, including riparian woodlands, swamps, suburbs, subalpine forest, and oak woodland. Often found at ecotones, especially forest edges at open woodlands and grasslands. Commonly found in burned and cutover forests with remnant snags.

Nearly as abundant as Hairy Woodpecker in unburned western forests, at least equally so at lower elevations, and more common in arid pine and juniper woodlands.

Seasonal habitats overlap substantially, but populations that depend on winter-hibernating ant species usually withdraw from these areas in winter. Winter populations can border on abundant in urban and suburban environments.

Able to withstand extreme winter temperatures on single nights, but has difficulty when temperature remains below 0° F (–18° C) for extended periods; some local population declines in Midwest are attributed in part to severe winters.

## DETECTION

Occupation of open habitats, extensive breeding range, widespread migration, and easily heard vocalizations make this the most conspicuous woodpecker in N. Am.

**VOCAL SOUNDS AND BEHAVIOR:** Generally quite vocal, with diverse repertoire. Primary calls include single note, "jungle call," churring, and wicka call, also a subtle flight call.

Single note is a variable and familiar *kee'er* or *kee'ew,* descending on second syllable, audible at some distance, and given year-round by both sexes. Considered a call of general self-announcement.

Jungle call is a sometimes long, rapid series of notes like *kee, kee, kee, kee . . . ,* generally all notes on same pitch and cadence until dropping slightly on last one or two. Given more often by males than females, usually from a high perch, and primarily early in breeding season as courtship or territorial proclamation; less often during excavation, rarely during incubation, and sometimes heard through

After drumming on the electrical insulator atop the pole, a male "Red-shafted" Northern Flicker lets out its signature "jungle call" as its ultimate pair-bonding and territorial declaration. *Stephen Shunk; Apr. 2009, Deschutes Co., OR.*

onset of fall migration, rarely in winter. Series is often shorter (fewer notes) after young fledge.

Churring occurs, often in combination with wicka call, just about any time two individuals come in contact, whether a mated pair or two unrelated birds at a winter feeding station.

Wicka call includes repeated *weeka, weeka, weeka* . . . or *ricka, ricka, ricka* . . . , first note slurred upward, second falling quickly. Complex and highly variable; given by two or more adults in close proximity, especially during courtship bobbing behavior and other elaborate dances and territorial interactions.

Additional vocal call is described as soft w*hurdle* or *whirr* (with rolled "r") given in flight and rarely heard; believed to be a response to territorial breach by another individual, but also heard when approaching artificial feeding stations.

**NONVOCAL SOUNDS:** Highly variable drumroll is a rapid series of blows at steady cadence and is performed on wide variety of substrates, including artificial surfaces. Northern Flicker is less prone to select fixed drumming and calling sites than other woodpeckers. Males drum more often than females. Performed in conjunction with jungle call during territorial proclamation.

Rolls are usually shorter in duration than those of similar pattern given by most *Picoides*, but this

An adult female "Red-shafted" Northern Flicker approaches a small pond. *Stephen Shunk; July 2009, Deschutes Co., OR.*

is difficult for human ear to distinguish. Unknown whether drumming patterns of "Yellow-shafted" and "Red-shafted" forms can be distinguished from each other. Instead, drumming patterns vary within the two forms, and often by habitat.

Ritualized tapping is rarely observed; steady bursts of single taps, usually three to five, much slower than drumming; performed during nest exchange. Since flickers most often feed on ground, other tapping sounds are usually associated with nest excavation.

## VISUAL IDENTIFICATION

**ADULT:** Third-largest North American woodpecker (behind Ivory-billed and Pileated). Brownish and "zebra-backed" above, light peach ("Red-shafted") or light golden ("Yellow-shafted") background to flanks and sides, ashy in center of belly, on flanks and sides, with prominent black breast shield; numerous and variable black spots throughout underbody. Brightly colored underwing and tail linings and white rump are conspicuous in flight. See Figure 13, page 219, for variation between the two forms.

Head and throat generally show two-toned background with variable malar and nape colors; male's bold malar stripe (red in "Red-shafted," black in "Yellow-shafted") forms elongated teardrop. Spotted belly contrasts against striped back; belly spots transition to black bars or smudges on flanks and undertail coverts.

In flight, appears lightly mottled brownish. Underwing linings are brightest at coverts, progressively fading outward.

**JUVENILE:** Resembles adult but smaller and duller overall. Black bars on upperparts are broader than in adult, head and neck grayer; crown and nape showing variable amount of red, mostly in males; black breast shield smaller and more oval-shaped; larger spots on belly.

Malar stripes of juvenile males paler than those of adult males.

**INDIVIDUAL VARIATION:** Only males show boldly colored malar stripe; otherwise no plumage difference between sexes. Males average slightly larger than females, but this is usually indistinguishable

Two nestling male "Yellow-shafted" Northern Flickers—near fledgling age—wait for the next food delivery. *Marie Read; July 2014, Tompkins Co., NY.*

| | YELLOW-SHAFTED | RED-SHAFTED |
|---|---|---|
| Crown | gray | rich brown |
| Nape | bold red "V" or "V" on gray | brownish gray |
| Auriculars | light tan, gray at rear | gray |
| Malar | black in male, light tan in female | red in male; brownish in female |
| Throat | light tan | gray |
| Flanks & Sides | light golden | light peach |
| Wing & tail shaft/linings | lemon yellow | salmon red |
| Light parts of back | light yellowish brown | dark reddish brown |

**FIGURE 13.** Identifying "Yellow-shafted" and "Red-shafted" Northern Flickers. The plumage characters described above serve as a general guideline for field identification. Outside the breeding season, south of the contact zone, and especially in the West, all characters should be diagnosed to confirm identification; birds in flight are typically impossible to identify as anything other than "Northern Flicker." *Stephen Shunk 2015.*

in the field. Size and concentration of belly spots highly variable, with some individuals showing heart-shaped spots. Highly variable plumage is seen among intergrades; see "Yellow-shafted" x "Red-shafted" Intergrades, page 220.

More than other woodpeckers, individuals are reported with deformed bill, usually with one mandible overgrown, sometimes excessively so; possibly a consequence of non-excavatory feeding style, contrary to other woodpeckers that constantly "trim" their bill while feeding.

**PLUMAGES AND MOLTS:** Preformative molt begins at fledging, early June–late July, depending on nesting date; completed in Sept.–Oct. Second prebasic molt occurs during second summer.

**DISTINCTIVE CHARACTERISTICS:** Front of flicker esophagus stretches to form huge crop, up to 2.4 in. (6 cm) long, which aids in delivering food to nestlings. General plumage of Northern and Gilded Flickers is unique among North American woodpeckers.

**SIMILAR SPECIES:** Very similar in plumage to Gilded Flicker, though the two species typically occupy different habitats (see Gilded Flicker, Similar Species, p. 229, and Hybridization, below, for discussions). Most important plumage difference is bright yellow wing and tail linings of Gilded Flicker versus salmon red in "Red-shafted" Northern Flicker, with which Gilded overlaps ("Red-shafted" x "Yellow-shafted" intergrades are rare in Gilded Flicker range). Gilded Flicker crown is more cinnamon-colored and breast shield is more oval, versus crescent or semicircle in Northern Flicker, and Gilded Flicker is generally smaller overall.

**GEOGRAPHIC VARIATION:** Noticeable difference between northern and eastern "Yellow-shafted" group and western and southern "Red-shafted" group, but subtle plumage variation within each subspecies group.

**SUBSPECIES:**

Four subspecies groups: two from w. Oaxaca, Mex., north to species limits; one from s. Chiapas, Mex., south to n. Nicaragua; and one in Caribbean; most with multiple subspecies (see Figure 14, p. 221).

**"Yellow-shafted" (*auratus*) Group**
*C. a. luteus*: Largest subspecies, though decreasing in size north to south; w. AK east to NL, northern Great Plains to ne. U.S.; includes "*borealis*."

*C. a. auratus*: e. TX, east and north to Mid-Atlantic, south to Gulf Coast and n. FL Keys.

**"Red-shafted" (*cafer*) Group**
*C. a. cafer*: Darkest subspecies; Pacific Northwest, including se. AK, south to n. CA, west of Cascades.

*C. a. collaris*: Pale western subspecies; west of Rockies through Great Basin to cen. CA, south to w. Mex.; includes "*canescens*," "*chihuahuae*," "*martirensis*," and "*sedentarius*."

*C. a. nanus*: Smallest subspecies; Chisos Mts., w. TX, south through e. Mex.

*C. a. mexicanus*: Grades clinally from southern limits of *collaris* and *nanus* to Isthmus of Tehuantepec, Mex.

Includes now extinct *C. a. rufipileus*, formerly restricted to Guadelupe Is., Baja Cal., Mex.

***Mexicanoides* Group**—generally resembles *cafer* group

An adult male intergrade Northern Flicker, showing primarily "Red-shafted" plumage traits, with the tan in the chin and throat from its "Yellow-shafted" lineage. *Brent McGregor; Dec. 2012, Deschutes Co., OR.*

*C. a. mexicanoides*: "Guatemalan" Flicker, Chipas, Mex., to s. Guatemala, Honduras, and n.-cen. Nicaragua; includes "*pinicolus*."

***Chrysocaulosus* Group**—generally resembles *auratus* group

*C. a. chrysocaulosus*: Cuba
*C. a. gundlachi*: Grand Cayman

**"Yellow-shafted" (*auratus* group) x "Red-shafted" (*cafer* group) Intergrades**

A broad and well-studied zone of introgression occurs between these two subspecies groups (see Figure 15, p. 221). Zone extends from se. AK to TX panhandle and adjacent parts of OK and NM; overlap is broadest in n. Great Plains, narrowest along AB–BC border. Zone is at least 4,000 years old. Position and width of intergrade zone are especially correlated with ecological transition from generally wetter climate in e. N. Am. to generally drier climate in w. N. Am.

Field identification of migrating and wintering birds is especially problematic in w. N. Am. Variation across intergrade zone is smoothly clinal, so plumage characters of intergrades show widely varied combinations and variations of parental traits. Commonly crossed traits include red or absent nape patch, black or red malar stripe, buffy or gray throat and crown, and a spectrum of wing and tail colors. Winter identification within or south of contact zones is safest as simply "Northern Flicker."

This adult male intergrade Northern Flicker could be considered "mostly Yellow-shafted," with its red mustache and brownish crown the only traits borrowed from its "Red-shafted" lineage. *Stephen Shunk; Mar. 2009, Sonoma Co., CA.*

### HYBRIDIZATION

***"Red-shafted" Northern Flicker* x *Gilded Flicker***

Scattered, isolated hybrid populations occur in s.-cen. and sw. AZ and se. CA where small remnant riparian habitats connect mountains to deserts. Hybrids generally more closely resemble Gilded Flicker. See Gilded Flicker, Hybridization, page 230, for further discussion.

## BEHAVIOR

### BREEDING BIOLOGY

*Nest Site*

Wide diversity of nesting habitats and substrates. Throughout range, generally prefers open woodland with grassy understory. Excavates 50–90 percent of nest cavities in dead or diseased tree trunks and large branches. May generally nest lower than other woodpecker species. Cavity entrance larger than all except Pileated's and Ivory-billed's.

Frequently reuses old cavities, occasionally finishing cavities started in previous year; often excavates in knotholes formed by broken limbs. Only woodpecker species to readily nest in artificial boxes and will excavate in buildings. Rarely nests in old Belted Kingfisher or Bank Swallow burrows.

*Courtship*

Generally monogamous, but up to 5 percent of a given population may practice polyandry. Pair bond is maintained through breeding season, but no evidence beyond, although courtship can be drawn out in some areas; drumming is heard as early as Jan. in southern parts of breeding range.

Courtship displays are similar to territorial behaviors; dancing involves head swaying or head bobbing accompanied by muted but emphatic wicka calls. Males are known to perform aerial display of jerky, ascending spirals, followed by same in descent, landing near female and dancing. Same-sex interactions usually intensify in presence of third bird of opposite sex.

Copulation occurs hourly during egg laying and as female returns to nest site. Female squats on horizontal branch, male mounts and spreads wings over her; act lasts a few seconds, after which male quickly leaves nest area and female enters nest cavity.

*Parenting*

Clutch size varies consistently between both subspecies and intergrading pairs; clutches are usually larger in warmer nests, which also tend to be in large-diameter live trees with south-facing entrance. Clutch size increases by 1 egg per 10 degrees latitude, and inland birds lay larger clutches than coastal birds. Large clutch size (average 7 eggs) compared with that of other woodpeckers is correlated with large cavity entrance, allowing multiple nestlings to beg from opening at once. Brood parasitism as high as 17 percent, and may be highest where cavities are scarce.

May experience locally scattered laying dates or abandonment because of competition with European Starling or other species. Eggshells may be left in nest; fecal material is usually carried 330 ft. (100 m) or more from nest site.

Nestlings may fledge early if nest is disturbed. Young may feed independently and adults may deny them food earlier compared with other woodpecker species.

*Dispersal*

Adults and juveniles typically leave breeding area shortly after fledging. Second-year birds often return to breed near natal site after their first spring migration; site fidelity is extremely high in older adults. Although nestling survival rate is high, annual survival among adults and juveniles reaches only 40 percent annually.

### NONBREEDING BEHAVIOR

*Feeding*

***Preferred Foods:*** Eats more ants than any other North American bird; favored genera vary regionally, but generally prefers smaller and less aggressive species over large, aggressive mound-building ants of genus *Formica*. Forages among grasslands with low vegetation, high percentage of bare ground, and high density of anthills or burrows. Also eats wide variety of other arthropods, including crustaceans and molluscs.

Common plant foods include wide variety of berries and large seeds. Prefers fruits in late fall and winter; "Red-shafted" is more reliant on animal foods in winter than "Yellow-shafted." Eats most readily available foods (e.g., sumac) over those more nutritious but harder to access (e.g., corn), possibly resulting in mortality in severe winters. Observed feeding at sapsucker, along with peanut and suet feeders.

***Foraging Behaviors:*** Along with Gilded Flicker, the continent's only woodpecker that feeds primarily on ground. Collects ants by tilling through or probing into soil with long bill and longest tongue of all North American woodpeckers; tongue surface is like extrasticky flypaper. Will forage for extended periods at a single ant mound or colony.

Regularly observed hopping on ground over wide area snatching up ants or probing for beetle larvae. Occasionally forages in trees and among downed wood, especially in cutover forest patches.

This juvenile Northern Flicker could be considered "mostly Red-shafted," but with a faint red nape and brown mixed with gray in the head, both acquired from its parent(s) with "Yellow-shafted" lineage. *Stephen Shunk; Aug. 2013, Jefferson Co., OR.*

In breeding season, adults may forage communally with neighboring pairs, or singly, while mate attends nest; in other seasons, small flocks of up to 12 birds may feed together at a productive food source such as an orchard or berry patch.

### *Territory Defense and Sociality*

Mean densities reported from ne. OR ranged from 4.2 to 7.4 birds per 247 ac. (100 ha), highest in mature Douglas-fir–ponderosa pine forest.

Both sexes are highly defensive of mate and breeding territory. Most common same-sex territorial interaction is head-bobbing dance facing aggressor, resembling noncontact fencing duel; most aggressive and most common early in breeding season, and accompanied by mutual wicka calls. "Yellow-shafted" forms erect red nape patch, and in most aggressive displays, both birds flick wings and spread tail to flash colors at opponent. Home ranges are highly variable based on season and habitat.

Individual bouts of aggression may last only 5–10 seconds or up to 30 minutes, with repeated chasing between bobbing stints; some bouts alternated with breaks may last all day. Each element of display varies in intensity. Typically does not defend foraging

**OPPOSITE:** An adult male "Red-shafted" Northern Flicker feeds a female fledgling. *Stephen Shunk; July 2009, Deschutes Co., OR.*

areas from other flickers, except in close proximity to nest tree.

### *Interactions with Other Species*

Known to share nest tree with Williamson's Sapsucker and Red-headed Woodpecker, but latter is also known to depredate or usurp flicker nest. Will physically attack starlings over nest site, and frequently falls victim to starling usurpation. Also competes with American Kestrel for cavities; observed usurping active kestrel nest, but also may nest nearby or in same snag with adequate cavity availability. Observed nesting in tree supporting Osprey nest.

Flickers observed foraging on ground together with jays and heeding jay alarm system in presence of avian predators. Also observed defending artificial feeding station against jays.

Despite its anatomical deficiencies for strong excavation, may be responsible for creating nest sites for more cavity-roosting or cavity-nesting animal species than any other woodpecker.

### MISCELLANEOUS BEHAVIORS

Spends up to one-quarter of time on ground, especially in winter. Often hops, and may run short distances between hops. Flight is generally less undulating (more direct) than that of most woodpeckers.

Roosts in cavities less often than other woodpeckers; will sleep on vertical surface with head tucked under scapulars. Also roosts on or inside buildings and under bridges and eaves. May select a different roost about every 10 nights on average.

Drinks from variety of sources, even breaking ice to reach water; also seen breaking off and eating chunks of snow. Scratches head without lowering wing, and stretches wing without extending leg.

## CONSERVATION

**HABITAT THREATS:** Habitat destruction and urban development are primary human activities threatening flickers. Snag removal limits nest-site availability; construction activities destroy ant colonies; widespread development favors starlings, which may be most significant factor in declining populations.

Stand-replacement fire may open up forested habitats for optimal foraging, but salvage logging and high incidence of windthrow in burns limit snag availability.

**POPULATION CHANGES:** Species may be abundant in a local microhabitat, but that habitat may be rare in a given region. Cavity availability greatly limits population density.

**CONSERVATION STATUS AND MANAGEMENT:** Listed as priority for conservation in eight states; Focal Species in AZ.

Steadily declining population trend could have profound effects on local ecosystems where flickers serve as primary cavity excavators. No formal studies document reasons for population declines, but many sources attribute correlation to nest-site competition with European Starling. Habitat conservation activities addressing nest cavity availability could help maintain populations. Installation of flicker nest boxes has also proven to be effective.

May thrive in thinned or burned forest with open canopy and downed wood, as long as snags remain. Snag management should carefully select for all successional stages of snags, rather than solely those of a single decay-class.

### REFERENCES

American Ornithologists' Union 1973, 1995; Baker 1975; Baker and Hendrickson 1937; Baldwin 1910; Bales 1989; Blake and Blake 1902; Bock 1971; Bower and Ingold 2004; Brackbill 1942, 1955, 1957; Brewster 1893; Brodkorb 1928, 1942; Browning and English 1967; Burns 1900, 1901, 1910, 1916; Chapman 1891; Cruz and Johnston 1979; Culbertson 1936; Dales 1956; Dennis 1969; Diamond 1971; DuMont 1933, 1935; Duncan 1961; Dykstra et al. 1997; Elchuk and Wiebe 2002, 2003; England 1941; Errington 1936; Erskine and McLaren 1976; Ewins 1994; Fisher 1905; Fisher 1910; Fisher and Wiebe 2006a, 2006b, 2006c; Fletcher and Moore 1992; Floyd 1937; Ganier 1926; Gignoux 1921b; Giles 1958; Gilmore 1930; Grater 1939; Grudzien and Moore 1986; Gullion 1939; Hauser 1957; Hooper 1936; Howitt 1925; Huey 1932; Jackson 1970b; Johnson 1934; Jones 1992; Jones and Bock 2003; Kilham 1959c, 1973; Koenig 1986; Kohler 1911; Labisky and Mann 1971; Landin 1978; Law 1916; Leister 1919; Little 1920; Martin and Ogle 1998; McGregor 2010; Merriam 1903; Meyer 1981; Moore 1987, 1995; Moore and Buchanan 1985; Moore and Koenig 1986; Moore and Price 1993; Moore et al. 1991; Nielson-Pincus 2005; Noble 1936; Orcutt 1884; Peterson and Gauthier 1985; Potter 1930; Rathbun 1911; Richardson 1910; Rinker 1941; Rosene 1936; Rothschild 2005; Royall and Bray 1980; Rumble and Gobeille 1994; Ryser 1963; Salafsky et al. 2005; Schwab et al. 2006; Sedgwick and Knopf 1991; Shelly 1935c; Sherman 1910; Short 1965c, 1967, 1972; Spofford 1969; Stone 1937; Stoner 1922; Taverner 1906; Test 1939, 1940, 1942, 1945, 1969; van Rossem 1936a, 1936b, 1944; von Bloeker 1927, 1935, 1936; Wauer 1965; Wiebe 2000a, 2000b, 2001, 2002, 2004, 2008; Wiebe and Bortolotti 2001, 2002; Wiebe and Kempernaers 2009; Wiebe and Moore 2008; Williams 1900.

## GILDED FLICKER

***Colaptes chrysoides***

L: 10–12 in. (25.4–30.5 cm)
WS: 18 in. (45.7 cm)
WT: 3.2–4.6 oz. (92–129 g)

**At the top of the tallest saguaro,** the Gilded Flicker peers across the Sonoran sands, basking in the setting desert sun. Its unmistakable "jungle call" notwithstanding, this is a shy creature, ceding the local stage to its smaller, more boisterous relative, the Gila Woodpecker. But "handsome" is owned by *chrysoides*, the male with its golden wing linings, crimson mustache, and cinnamon forehead and lores.

When not surveying the landscape from above, the Gilded Flicker spends most of its time on the desert floor, like the lizard of the local avifauna, lapping up ants and termites with its long sticky tongue. The winter menu may include a few mistletoe berries or palm dates, and the pollen of the saguaro blossom is favorite spring fare.

Where ribbons of water trickle down from southern Arizona's sky islands, the Gilded Flicker runs head-long into its closest relative, the "Red-shafted" Northern Flicker. At a few scattered locations in south-central and southwestern Arizona and southeastern California, the two interbreed, challenging our definition of the word "species." Hybridization between the two flickers may once have been more widespread, but urban and agricultural development have left only small patches where riparian habitats meet the desert. From 1973 to 1995, these two forms (along with the "Yellow-shafted" Northern Flicker) were considered to be subspecies of what was then called the Common Flicker. Extremely limited hybridization and several other factors led to their ultimate separation.

### DISTRIBUTION

**PERMANENT RANGE:** Mostly s. and w. AZ, rare in extreme se. CA and s. NV. Largest concentrations occur from s.-cen. AZ into Mex., with small populations along lower Colorado R. valley. Scattered e. AZ populations along Gila and San Pedro Rs.

Very local and rare in nw. Mohave Co., AZ, and Joshua tree forests of e. San Bernardino Co., CA. Until 1999, very rare in extreme s. NV. Confirmed breeding that summer; now locally common breeder in foothills of McCullough Mts. and Joshua tree forests in and around Wee Thump, Joshua Tree Wilderness.

**SEASONAL MOVEMENTS:** Primarily sedentary, with little winter movement outside nesting range.

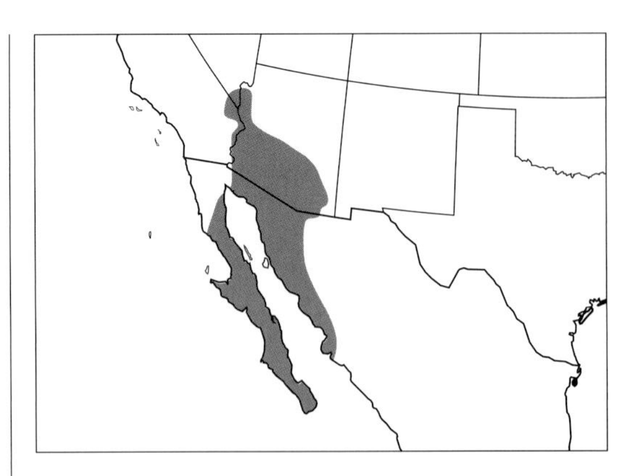

### HABITAT

Most common in lower Sonoran habitats. Greatest abundance is in high-density saguaro forests, therefore far more common in desert uplands than lowlands. Also found in riparian woodlands with cottonwoods and willows. Rare in mesquite habitat near Colorado R.

When riparian woodlands abut or approach saguaro habitat, selects riparian habitats more for foraging then nesting. Far less common than Gila Woodpecker in residential areas, despite presence of saguaros. Generally found from 200 to 3,200 ft. (61–975 m) elevation, rare in riparian habitats to 4,600 ft. (1,402 m).

### DETECTION

**VOCAL SOUNDS AND BEHAVIOR:** Gives typical flicker jungle call (see Northern Flicker, Detection, p. TK) in breeding season; also *kee-yer* or *keer* call note year-round. Courtship churring sounds may be more plaintive than those of Northern Flicker and are difficult to detect.

**NONVOCAL SOUNDS:** Drum is generally a short roll with even cadence. Easily separable from variably descending drum of Gila Woodpecker. Drum is easily heard in generally quiet desert habitats. Readily drums on artificial surfaces when available.

### VISUAL IDENTIFICATION

**ADULT:** Smaller than Northern Flicker but very similar in appearance. Brownish and "zebra-backed" above, buffy gray and spotted below, with slightly oval black breast shield. Head colors are sharply demarcated: brown above eye and gray below, lores and forehead cinnamon brown. Male shows broad red malar stripe, separated from base of lower mandible by cinnamon patch. Female lacks red malar stripe.

**OPPOSITE:** An adult male Gilded Flicker performs a wing and tail stretch. *Glenn Seplak; Dec. 2007, Pima Co., AZ*

This adult male Gilded Flicker is either melanistic or it is extremely soiled. *Jim Burns; Apr. 2006, Maricopa Co., AZ.*

In flight, mottled golden brown, with conspicuous lemon yellow wing and tail linings and white rump.

**JUVENILE:** Resembles adult, but smaller and duller overall.

**INDIVIDUAL VARIATION:** Only male shows malar stripe; otherwise no discernable plumage or size difference between sexes.

**PLUMAGES AND MOLTS:** Preformative molt begins at fledging, early June–late July, depending on nesting date; completed in Sept.–Oct. Second prebasic and subsequent molts occur during summer, earlier than in most woodpeckers.

**OPPOSITE:** An adult female Gilded Flicker cleans out a saguaro cavity in preparation for nesting. *Jim Burns; Mar. 2009, Maricopa Co., AZ.*

**DISTINCTIVE CHARACTERISTICS:** Like Northern Flicker, has extended crop for carrying food to young. May average narrower bill (measured across nasal openings) than other woodpeckers. Distinctive in its sharply intermediate plumage between that of "Red-shafted" and "Yellow-shafted" Northern Flickers.

**SIMILAR SPECIES:** Most easily confused with Northern Flicker, but the two rarely meet. Gilded is somewhat smaller and paler underneath, with much lighter back (narrow blackish bars). Most noticeable difference is seen in flight, with Gilded flashing bright yellow underwings and undertail and "Red-shafted" Northern showing bright salmon. Head coloration also differs, Gilded with cinnamon brown forehead fading to plainer brown on crown and narrowing down nape; "Red-shafted" is nearly all gray on head except for brown lores and crown. Gilded shows more oval breast shield, "Red-shafted" more crescent-shaped.

Gilded occurs primarily in lower Sonoran cactus forest, "Red-shafted" in pine-oak woodlands above desert floor.

Few confirmed records of pure "Yellow-shafted" Northern Flicker in AZ, but characters to watch for include larger size than Gilded; red nape in both sexes; black mustache in male; and more or less reversed head color, with gray crown and brown face, throat, and front of neck.

Superficially resembles Gila Woodpecker, but Gila is smaller and unmarked below, with black-and-white "zebra stripes" on back.

**GEOGRAPHIC VARIATION:** Subspecies vary in size, weight of barring, and tone of overall brownish coloration.

### SUBSPECIES

Up to four subspecies recognized, one in N. Am., *C. c. mearnsi*, or Mearns's Gilded Flicker. Occurs from North American range south to ne. Baja CA and w. Sonora, also off Sonora coast on Tiburón I. Largest and palest of the four subspecies; crown brighter cinnamon, black spots underneath smaller, black bars on back narrower.

*C. c. brunnescens:* cen. Baja CA; smallest subspecies.

*C. c. chrysoides:* resident in s. Baja CA.

*C. c. tenebrosus:* s. Sonora and n. Sinaloa, Mex.

When a saguaro cactus dies, it often leaves behind the remnants of a former Gilded Flicker nest cavity, known as a saguaro "boot." *Stephen Shunk, courtesy of Saguaro National Park; Aug. 2013, Pima Co., AZ.*

### HYBRIDIZATION:

**Gilded Flicker** x **"Red-shafted" Northern Flicker**

Small scattered hybrid populations in s.- cen. and sw. AZ and se. CA. Contact zones are where mountains connect to deserts by small remnant riparian habitats. Individuals generally resemble Gilded more than Northern; all hybrid populations are isolated from stable populations of either species. Small size and cinnamon forehead are important in diagnosing Gilded ancestry.

Hybridization was likely more extensive before riparian habitats were stripped for agricultural and other development. Emerging DNA evidence may corroborate species-level separation.

## BEHAVIOR

### BREEDING BIOLOGY

Species' limited range has restricted research on Gilded Flicker breeding biology, as has its former recognition as a subspecies of Northern Flicker.

#### *Nest Site*

Excavates about 90 percent of nest cavities in saguaro cacti, remainder in cottonwood and willow, rarely in tall mesquite snags. Cavity entrance is wider than tall (as in Wood Duck) and larger than that of Gila Woodpecker; also averages higher than Gila's, within 10 ft. (3 m) of top of saguaro. Large cavity size (compared with that of Gila Woodpecker) requires excavating into woody skeletal part of saguaro; relatively weak excavating ability of flicker may require these excavations to be near top of saguaro, where skeleton is relatively thin; these factors may often shorten life span of the cactus, whereas it continues to grow around lower Gila cavities, which are excavated only into softer tissue.

Nests earlier than Northern Flicker, possibly because of moderated temperatures in saguaro habitat earlier in season; will excavate in saguaro 3 or more months before nesting to allow drying time necessary for internal encasement of cavity.

**OPPOSITE:** An adult male Gilded Flicker atop blooming ocotillo. *Kristine Falco; May 2006, Pima Co., AZ.*

© Cindy Marple

**OPPOSITE:** A trio of Gilded Flicker nestlings in a saguaro cavity, two males (top and bottom) and a female in the middle. *Cindy Marple; May 2006, Pima Co., AZ.*

### *Courtship*

Courtship displays and other pairing behaviors are likely similar to those of Northern Flicker, but more research is needed.

### *Parenting*

Notably smaller clutches than Northern Flicker. Double-brooding reported in Colorado R. valley, with first brood fledging late May, second early July.

## NONBREEDING BEHAVIOR

### *Feeding*

***Preferred Foods:*** Ants and termites dominate diet year-round. Eats large quantities of saguaro pollen when blooming and mistletoe berries in winter.

***Foraging Behaviors:*** Probes ground for subsurface prey; also rotten, downed wood if available. Rarely gleans from bark of live trees.

### *Territory Defense and Sociality*

Territorial displays may be similar to those of Northern Flicker, but absence of migratory behavior may result in greatly reduced association with other Gilded Flickers outside breeding season.

### *Interactions with Other Species*

Known to nest in same cactus as Gila Woodpecker, Elf Owl, and Ash-throated Flycatcher. Nest cavities are used by Purple Martin, as well as many other desert cavity dwellers.

## CONSERVATION

**HABITAT THREATS:** Populations on lower Colorado R. declined substantially from loss of riparian habitat to agriculture. Not well adapted to urbanization; as development sprawls into saguaro habitats, range of Gilded Flicker is likely to continue shrinking. Threatened by desert wildfire, often fueled by non-native grasses, which destroys large saguaros. European Starling also noted as increasing threat because of competition for nest cavities.

**POPULATION CHANGES:** Standardized surveys show stable distribution primarily among saguaro habitats in s. AZ, especially in cen. Pinal, cen. La Paz, and extreme e. Pima Cos.

**CONSERVATION STATUS AND MANAGEMENT:** Endangered in CA. Red Listed on 2007 WatchList. Perhaps more than for any other North American woodpecker, additional monitoring and research are needed for a better understanding of this species' ecology.

Recommendations for habitat management include maintaining saguaros and reducing fire risk. Artificial nest boxes, used extensively for Northern Flickers, may help mitigate loss of habitat due to development.

## REFERENCES

American Ornithologists' Union 1912, 1973, 1995; Anthony 1895a, 1895b; Beffort 2008; Behle 1976; Brown 1904; Burns 2004; California Department of Fish and Game 2011; Corman, T.E., pers. comm.; Fletcher and Moore 1992; Floyd 2000; Gilman 1915; Inouye et al. 1981; Johnson 1969; Kerpez and Smith 1990a, 1990b; Koenig 1984; Korol and Hutto 1984; McAuliffe and Hendricks 1988; Monson 1942; Moore 1995; Phillips 1947; Short 1965b; Waters 2005; Zwartjes 1998.

## PILEATED WOODPECKER

***Dryocopus pileatus***

L: 16.5–19.5 in. (42.0–49.5 cm)
WS: 27–30 in. (68.6–76.2 cm)
WT: 8.5–12.0 oz. (240–341 g)

**King-o'-the-Woods,** Stump Breaker, Grand Pic—all are nicknames for the sixth-largest woodpecker on Earth. The Pileated Woodpecker is truly the reigning monarch of mature woods from the southern Yukon Territory to California's Kings Canyon, and from Nova Scotia to the Florida Keys. With territories spanning a mile (1.6 km) or more for a single pair, Pileated Woodpeckers are far more frequently heard than seen, but their obsessive quest for carpenter ants leaves conspicuous signs of their carpentry. Pileateds also excavate virtual caves in the trunks of large trees, leaving behind sheltered abodes for many cavity-nesting waterfowl and predatory mammals such as fisher and marten.

Always awe-inspiring, the fortunate glimpse of a Pileated Woodpecker generally comes in one of two ways: when the bird is flying above or high in the canopy with its distinct crowlike flight; or when it is at or near ground level, so focused on collecting ants that it can be closely approached. Even in its absence, its long gaping excavations and piles of large bark chips give away a Pileated's favorite feeding spot.

Regardless of how you pronounce the "i" in "pileated" (both long and short "i" are correct), this giant woodpecker deserves our attention for its critical dependence on mature forests and rotting wood. The ecological health of old-growth coniferous forests and bottomland hardwoods alike can be gauged by the presence of this monarch among the woodpeckers.

### DISTRIBUTION

**PERMANENT RANGE:** Throughout mature forests of Northwest, across boreal forest to s. Maritimes, and widespread in eastern woodlands.

Pacific Coast ranges from s. BC south through OR, decreasing into CA. Throughout Cascades, more common on west slopes in WA and OR. Interior ranges of s.-cen. OR, down west slope of Sierra Nevada.

Extreme se. YT and sw. NT to Maritimes and ne. U.S., including n. WI. Rocky Mts. from BC and AB southward through n. ID; rare in nw. WY and n. Great Plains.

Once extirpated from and possibly breeding again in se. NE; increasingly common south and east into MO, AR, and e. TX. Across valleys of upper Mississippi R. and tributaries, south to Gulf states and s. FL.

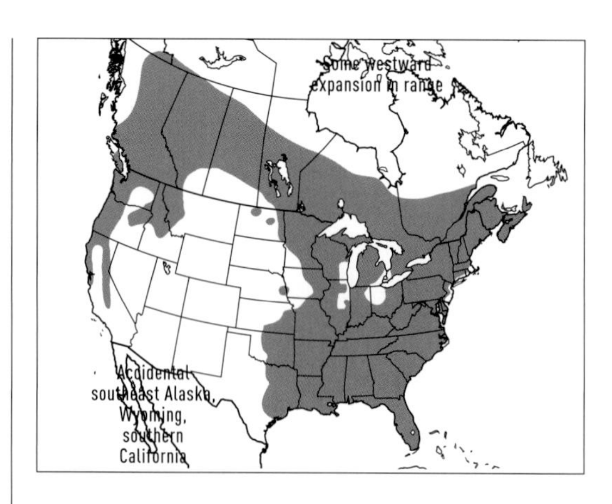


**SEASONAL MOVEMENTS:** Generally sedentary, but some eastern winter movements into s. ON and s. New England, including coastal islands, with some winter visitors possibly southward through Great Lakes. Seasonal movements are likely postbreeding juveniles wandering in search of optimal habitats for territory establishment. In some montane areas of West, may move to lower elevations in winter.

### HABITAT

Mature coniferous or deciduous forests, also some younger forests and riparian habitats with scattered, large snags. Typically found in stands with largest trees and high, closed canopy, or observed flying between such stands across cutover or developed patches. Population density is highest in broad forest expanses with sufficient downed wood and dead or dying trees to sustain carpenter ant populations. Frequently feeds in burned or cutover patches as long as downed wood remains.

In some eastern forests, selects mature oak-hickory stands with dense undergrowth; also dead pines and rotting logs. In Pacific forests, prefers stands more than 70 years old, favoring Douglas-fir, western larch, ponderosa pine, and grand fir as available; avoids most subalpine forests.

### DETECTION

Heard far more often than seen; loud resonant voice and slow, deep drum.

**VOCAL SOUNDS AND BEHAVIOR:** Best known for "jungle call," a loud, repeated series of *cuk, cuk, cuk* notes, similar to that of flickers but typically lower, more wavering in pitch, and tapering near end, often followed by slower irregular series of single call notes.

**OPPOSITE:** An adult female Pileated Woodpecker braves a nor'easter. *Marie Read; Dec. 2008, Tompkins Co., NY.*

Also utters similar call notes, erratic in pitch, volume, and succession, while feeding or in presence or on approach of another individual; series may continue for 30 seconds or more. Female's call is usually higher pitched than male's.

Also gives nasal *rut, rut, rut*—or *rutta, rutta*—greeting call during close interactions among pairs. Vocal repertoire is fairly simple compared with that of other woodpeckers, with variations of call note given in variety of circumstances.

Most vocal during courtship, territory defense, and in morning. Pairs communicate all year but are least vocal in winter; birds frequently call on way to evening roosts. Nestlings make loud rasping sounds, sometimes resembling a giant beehive.

**NONVOCAL SOUNDS:** Drum is relatively slow and deep, tapering in pitch and accelerating near end. Both sexes drum, males more than females. Rolls are typically short bursts; in breeding season, repeated once or more per minute for up to 3 hours. May end with a few hesitant single raps.

Adult will tap, double-tap, or perform short roll inside cavity on approach of mate for nest exchange or feeding; will also conduct short roll at prospective nest site. Disturbed birds also tap loudly one or more times. Pairs may tap away from nest site to maintain pair bond.

Feeding is often in soft wood and sounds are therefore muffled, but well-spaced deliberate blows are especially audible when excavating in standing trees.

## VISUAL IDENTIFICATION

**ADULT:** Distinctive in appearance throughout its range and generally huge, often seen flying above or just skimming top of canopy. Large head and bill are topped with long crimson red crest; red in male crown extends through forehead to upper mandible, female's forehead is speckled dark brown. Pied facial pattern with bold white mustache. Male also shows bar of red-tipped feathers at front of malar stripe. Buffy nasal tufts blend with white mustache. Bill is long with broad tip. All-dark central tail feathers are long, stout, and decurved, allowing spring effect for vertical climbing. Small white patch is sometimes visible in folded wing.

In flight, when viewed from below, most of underwing is white to grayish white with dark outer primaries and trailing edge. Thin dark line borders white along leading edge of wing. Body is dark brownish to black from upper breast through tail. Wings appear very broad, with neck, head, and long bill projecting well beyond leading edges.

Infrequently seen from above; entire dorsal surface appears mostly dark, with small amount of white at bases of outer flight feathers appearing as mottled white band in outer wing.

**JUVENILE:** Sex can be determined early in nestling cycle. Plumage is less glossy than adult's; feather vanes have fewer barbs, giving rougher appearance. Crest is shorter and red in both sexes, less extensive red in females.

**INDIVIDUAL VARIATION:** White-tipped barring on flanks is barely visible at close range and in fresh plumage, but this disappears with wear. Rarely melanistic or leucistic. Males are 10–15 percent heavier than females. Forehead color of females may get paler with age; blackish as fledglings, turning olive or even yellowish in older birds.

**PLUMAGES AND MOLTS:** Juvenal body plumage is molted gradually throughout late summer and early fall. Secondaries appear browner than those of adult. Adult prebasic molt occurs July–Nov. Fresh adult body plumage and dark flight feathers vary from dusky brown to blackish; overall, appears black, with darkest feathers in tail.

**DISTINCTIVE CHARACTERISTICS:** Except for Ivory-billed Woodpecker, which may be extinct, Pileated is only North American woodpecker with a crest, and except for White-headed Woodpecker, only one showing nearly all dark body when perched.

**SIMILAR SPECIES:** Unmistakable throughout most of range. In Southeast, much attention has been drawn to potentially problematic separation in the field from Ivory-billed Woodpecker. Sizes between the two do not overlap, but a fleeting glimpse of a very large black-and-white woodpecker in dense bottomland forest must be carefully evaluated. See Ivory-billed Woodpecker, Similar Species, page 248, for detailed comparison.

**GEOGRAPHIC VARIATION:** Western birds tend to have grayer throat and less barring than eastern and southern birds, and FL birds tend to be darker, but there is much overlap in all cases. Size generally decreases from north to south.

## SUBSPECIES

Up to four subspecies recognized: *D. p. pileatus* in se. U.S., except FL; *D. p. floridanus* throughout FL peninsula into FL Keys; *D. p. picinus* in w. N. Am.; and *D. p. abieticola* from Prairie Provinces east

**OPPOSITE:** An adult male Pileated Woodpecker approaches a nest cavity while a nestling male begs from the entrance. *Paul Bannick; June 2005, King Co., WA.*

Male and female Pileated Woodpecker nestlings wait at the nest cavity entrance for the next food delivery. *Paul Bannick; June 2009, King Co., WA.*

through e. Canada and n. U.S., north of s. Alleghany Mts.

Short (1982) recognized only *abieticola* and *pileatus* based on range and clinal size variation. Northern birds are 3–9 percent longer winged and longer tailed than FL birds and have 15–21 percent longer bill.

## BEHAVIOR

### BREEDING BIOLOGY

#### *Nest Site*

Variety of favored tree species, often regionally specific, including ponderosa pine, western hemlock, Douglas-fir, and wide range of broadleaf deciduous trees. May favor large conifer snags in West, live hardwoods with heart rot in East. Occasionally uses live trees, especially large aspens with heart rot or large branch-stubs of mature cottonwood, even if snags are abundant.

May use old nest cavities for roosting but rarely for nesting; will reuse same tree in successive years but with new cavity. Pair may start several excavations each spring, some of which may be completed and used for nesting in later years. Wide breadth of cavity entrance allows multiple young to beg from opening together.

#### *Courtship*

Monogamous, and presumed to mate for life. Increased courtship and territorial behaviors begin early Feb.–Mar. Behaviors include spreading wings, raising crest, head swinging, bark stripping, hopping, and bowing, with repeated single call notes, tapping, and drumming. Copulation is on horizontal branch with female perched crosswise, lasting 6–9 seconds. Calls and head swinging may precede copulation. Engages in flight display during pair formation.

Southern and coastal populations breed earlier than northern and high-elevation birds; nesting may be postponed by wet weather. Some adult pairs maintain close contact in winter, others are rarely together.

#### *Parenting*

Known to relocate eggs if active cavity is destroyed by weather. Parents brood nestlings up to 10 days old. Male returns to nest up to 2 hours before sunset to brood young for night. Young are fed about every hour when small, frequency decreasing to every 2 hours after about 1 week.

### *Young Birds*

In days before leaving nest, young vocalize, flex wings, and teeter on edge of cavity entrance. Older nestlings will leave cavity if nest tree is disturbed. Nests typically fledge 2–3 young, up to 3 days apart. After fledging, family may stay together or parents may divide young between them. Young are dependent until at least Sept.; until then, adults feed young directly and teach them how to acquire food.

### *Dispersal*

Young stay in adult territory 3–5 months, then wander until spring. Frequent drumming and calling accompany dispersal of young in fall. Ultimate dispersal distance of juveniles depends on habitat quality; up to 3.7 mi. (6 km) in first winter, later nesting up to 20 mi. (32 km) from natal site.

Home ranges are generally much larger in West (up to 1,200 ac. [485 ha]) than in East (up to 400 ac. [162 ha]); largest ranges occur with lower log and stump volume and more open canopy.

## NONBREEDING BEHAVIOR

### *Feeding*

***Preferred Foods:*** Favors carpenter ants throughout range and all year. Also consumes thatching ants and various beetle larvae; takes berries and wild nuts when available; rarely feeds at large fruits such as persimmon. Throughout year and range, consumes 40–97 percent ants (3:2 carpenter ants to thatching ants in West), 22 percent beetle larvae, 11 percent other insects, and 27 percent plant matter.

Diet changes seasonally, varied by region; eats carpenter ants all year but prefers thatching ants June–Sept. in East, fruit in fall, carpenter ants in winter, wood-boring beetle larvae in early spring, and insects in summer.

Up to one-quarter of Pileated Woodpecker stomachs are found to contain termites. Will regularly patronize suet feeders when available in appropriate habitats. May rarely eat carrion.

***Feeding Behaviors:*** Most often excavates into rotten cambium (53–77 percent of feeding time); also gleans from branches and trunks, pecks at bark, and strips bark off trees. Large rectangular or oval excavations in trees and logs are characteristic; can

An adult female Pileated Woodpecker at the base of a chasm excavated by this individual and its mate (see image to the right). *Marie Read; Mar. 2006, Tompkins Co., NY.*

An adult male Pileated Woodpecker excavates at the base of the chasm shown at left. *Marie Read; Mar. 2006, Tompkins Co., NY.*

A persistent carpenter ant continues to remove woody material despite the fact that a Pileated Woodpecker has gutted the log and eaten most of the ant's colony. *Stephen Shunk; May 2009, Jefferson Co., OR.*

be 1 ft. (0.3 m) or more long and extensive enough to weaken tree (although tree is usually dead or dying before Pileated begins working it); large furrows expose ant galleries behind long slivers of wood. Occasionally forages on ground, often on downed logs. Rarely forages on ponderosa pine cones, presumably for seeds.

While feeding in rotten log or snag, may remove single wood chips up to 5 in. (13 cm) long and 1 in. (2.5 cm) wide; may toss smaller chips a distance of up to 4 ft. (1.2 m).

Extremely long, barbed tongue and sticky saliva assist in catching and extracting ants from tunnels. Occasionally hops on ground or hangs awkwardly on small branches or vines while feeding.

### *Territory Defense and Sociality*

Estimates of population density range widely owing to varied survey methods; range of 1 pair per 106 ac. (43 ha) in parts of LA to 1 pair per 880 ac. (356 ha) in ne. OR; mean density reported from ne. OR as 5 birds per 247 ac. (100 ha).

Defends territory all year, more aggressively in breeding season; occasionally tolerates wanderers in winter. Both sexes call and drum when intruders approach territory boundaries. Most often, pair members chase off intruders of same sex and sometimes opposite sex, while mate observes.

Will chase, call, strike with wings, and jab with bill during encounters. Frequently raises and spreads wings to show white patch; swings head or waves bill, often pointing bill upward; commonly raises crest during interactions. Birds may posture opposite one another while circling trunk, periodically alternating direction, which may be followed by chasing in flight.

New pairs often formed following death of one pair member. Surviving bird remains in original territory and seeks new mate from territory periphery. New pair defends territory by drumming, calling, and chasing intruders.

### *Interactions with Other Species*

No general territory defense is reported against other species, but will chase away potential nest competitors from nest cavity, including waterfowl, passerines, and other woodpeckers. However, will nest in same tree, in different cavity, as other species; well known to roost, but not necessarily to nest, in same cavity with nesting Vaux's Swifts. Will use wings to strike squirrels approaching nest tree, or will lunge at them on ground with spread-wing display. May rarely usurp passerines occupying cavity used in previous season, removing young and nest materials and enlarging cavity before nesting.

Rarely antagonized by other birds. Northern Mockingbird observed chasing Pileated from holly berries. Known to nest in same tree with Yellow-bellied Sapsucker, but also to be chased by Red-breasted Sapsucker. Ivory-billed Woodpecker rarely observed chasing a pair of Pileateds.

Extensive excavations often attract other species to forage. In fall and winter, observed foraging on logs with Northern Flicker and Williamson's Sapsucker. Well known to enlarge Red-cockaded Woodpecker cavity entrances beyond what Red-cockadeds will use. Known to depredate nestlings of Red-bellied Woodpecker.

## MISCELLANEOUS BEHAVIORS

Roosts in hollow trees or vacated nest cavities at night and during inclement weather. Individuals in some populations may use as many as seven different roost sites in a given year. Roost trees may have up to 16 entrance holes, possibly providing alternate escape routes from potential predators; some roost cavities may be very low on tree.

Typically only one bird roosts in a particular tree, but mates occasionally use same tree. Fledglings roost in protected sites on large, live trees, but will not roost in cavities until 2–5 weeks after fledging. Rarely adult may roost with juvenile.

Flight is rather slow, vigorous, and direct; rarely undulating like that of smaller woodpeckers, with

**OPPOSITE:** A pair of adult Pileated Woodpeckers performs a nest exchange. *Dick Tipton; June 2011, Cariboo Co., BC.*

An adult Pileated Woodpecker takes a vigorous dust bath at the Archbold Biological Station in south-central Florida. *Stephen Shunk; Highlands Co., FL.*

labored wingbeats occasionally resembling that of large corvid flight.

Preens frequently, occasionally associated with drumming. Bathes in dirt or moving water, and may sunbathe for 5 minutes or more. Drinks from streams and ponds at dusk before roosting.

## CONSERVATION

**HABITAT THREATS:** Timber harvest and forest fragmentation currently have the most profound impacts on habitat; also large-scale stand-replacement fire, especially in West. Removal of large-diameter live and dead trees and downed woody material is especially detrimental; prescribed burns destroy foraging substrate by removing downed wood. A decrease in canopy cover, for whatever reason, may increase risk to predation.

**POPULATIONS CHANGES:** Declined rapidly and became rare in e. U.S. after wholesale clearing of forests before 1900. Recovery began in 1920s and 1930s, as forests reclaimed abandoned farms and state and federal legislation provided protection. Increase and expansion of populations continue, especially in NY, OH, OK, and WI.

Spread of Dutch elm disease in e. N. Am. since 1950s increased supply of invertebrate foods and nesting substrate, and may have contributed to increases in local populations.

**CONSERVATION STATUS AND MANAGEMENT:** Despite rangewide population increases, listed as Species of Concern in ND, RI, WA, and AB. Focal species in coniferous forests of CA; listed as critically dependent on late successional/old-growth forest in Sierra Nevada. Until 1994, considered a Management Indicator Species on 16 of 19 national forests in Pacific Northwest, until this provision was removed from Northwest Forest Plan.

In OR and WA, U.S. Forest Service designates Pileated management areas of 300 ac. (120 ha) in old-growth forests for nesting, and an additional 300 ac., with greater than two snags per acre (five snags per ha), for foraging.

Manual thinning of forest in Pileated Woodpecker habitats retains more productive foraging substrates than prescribed burns. Snags and any remaining downed wood should be retained after stand-replacement fire or bug-kill episodes, especially when known territories are reduced or eliminated.

## REFERENCES

Allert 1925; American Ornithologists' Union 1923, 1947, 1949; Aubrey and Raley 1994; Bangs 1898; Beckwith and Bull 1985; Bedell 1924; Brewster 1890; Bridge 1905; Brooks 1944; Bryant 1916; Bryens 1926; Bull 2001; Bull and Holthausen 1993; Bull and Jackson 1995; Bull and Meslow 1977; Bull et al. 1992; Bull et al. 2005; Bull et al. 2007; Bush 2001; Carriger 1919; Carter 1942; Conner 1973; Conner et al. 2001, 2004a; Cook 1917; Cox 1902; Duncan 2003; Eastwood 1930; Eifrig 1927, 1944; Fisher 1904; Forsyth 1988; Fox 1956; Ginther 1916; Grater 1936; Grinnell 1935; Hartwig 1999; Hartwig et al. 2004; Henderson 1934; Hofslund 1958; Houghton 1924; Hoyt 1944, 1952, 1953; Hutchins 1908; Jackson et al. 1998; Janisch 2008; Job 1901; Johnson and McGarigal 1984; Kilham 1958d, 1959b, 1959f, 1975, 1976, 1979a; Kinsey 1955; Mannan 1984; Marshall 1992c; Maxon and Maxon 1981; Mazlowski 1937; McClelland 1977; McClelland and McClelland 1999; McCovey 2012; Michael 1921, 1928; Michael 1925; Miller 1918; Nice 1927; Nielsen-Pincus 2005; Nolan 1959; Rathcke and Poole 1974; Renken and Wiggers 1989, 1993; Rosen 1925; Saenz et al. 1998; Savignac et al. 2000; Schemnitz 1924; Schwab et al. 2006; Scoville 1923; Servin et al. 2001; Short 1965a; Soelner 1904; Southgate et al. 1951; Springarn 1924; Tanner 1942; Thayer 1911; Thornburgh 1937; Vickers 1914; Wallace 1926; Warren 1916; Wimpfheimer, D., pers. comm.

# IVORY-BILLED WOODPECKER
*Campephilus principalis*
L: 19–20 in. (48.0–50.1 cm)
WS: 30–32 in. (76.2–81.3 cm)
WT: 15.9–20.1 oz. (450–570 g)

**In the winter of 2006,** I walked through the Big Woods of Arkansas, across the first and second bottoms of the White River, many miles upstream from the mighty Mississippi. It was a dry winter, making foot travel easy, though I could easily envision paddling under the leafless canopy of massive oak, cypress, and tupelo trees. I held no expectations when I ventured into these woods, but as I trod deeper into the forest an ineffable feeling crept over me. Just being there and knowing that Ivory-billed Woodpeckers once coursed up and down the White River gave me a sense of discovery. Are there still Ivory-billeds in the Big Woods? Maybe; maybe not. But I learned that I do not require their presence to experience their majesty and mystery.

Nearly everything we know about the natural history of the Ivory-billed Woodpecker comes from the meticulous work of ornithologist James Tanner. In 1937–1939, Tanner was commissioned to search for and study the species throughout its range. He visited habitats across the Southeast, traveling over 45,000 miles (72,420 km) by car—and many more miles by horseback, canoe, and foot—documenting what was believed to be the "last stand" of the Ivory-billed Woodpecker in the United States.

Recent reports of purported Ivory-billeds have rekindled hopes that the "Lord God Bird" may still persist in the southeastern swamps. The following account draws primarily from Tanner's observations, along with recent writings on the Ivory-billed's plight and reports from recent search efforts. Whether or not the Ivory-billed Woodpecker persists, it will always be a part of North America's fascinating ecological history.

## DISTRIBUTION

Historically, occurred in se. U.S. along Gulf coastal plain from upper TX coast to near tip of FL peninsula; up Atlantic coastal plain to s. NC; inland along major Gulf Coast river systems, especially Mississippi R. and tributaries as far north as lower Ohio R.; also Mobile Bay delta, and swamps of interior FL peninsula. Distinct subspecies from Cuba—also possibly extinct—was last documented in mature forests.

**KNOWN SEASONAL MOVEMENTS:** Generally sedentary and no evidence of migration, but prone to wandering with fluctuations in food supply and habitat quality; thrived at least partly on catastrophic

As a graduate student at Cornell University in the late 1930s, James Tanner published the only detailed field studies on the Ivory-billed Woodpecker. *Nancy Tanner, courtesy of David Tanner; 1940, Saranac Lake, NY.*

disturbance such as flooding, fire, and hurricanes, shifting populations regionally with changing mosaic of suitable habitats.

Long-distance wandering up and down narrow bottomland corridors could at least partly explain birds occurring for only short periods in suitable or even marginal habitats.

**RECENT REPORTS:** Sporadic but unconfirmed and unresearched sightings since middle of 20th century throughout Southeast. Most recent sightings determined valid enough to warrant systematic search efforts include Pearl R. swamp in ne. LA; bottomlands of Cache and White Rs. in se. AR; and Choctawhatchee R. in FL panhandle.

## HABITAT

Virgin bottomland forests almost always below 100 ft. (30 m) elevation. May also have occurred in uplands, but by 1900s restricted to areas downstream of pine-bald cypress interface. Requires large tracts of contiguous forest with very large-diameter trees

**OPPOSITE:** An adult male Ivory-billed Woodpecker approaches the nest cavity as its mate peers out. *James Tanner, courtesy of Nancy & David Tanner; Apr. 1935, Singer Refuge, Madison Par., LA.*

An adult female Ivory-billed Woodpecker at its nest cavity in the Singer Tract of Louisiana. *James Tanner, courtesy of Nancy & David Tanner; Apr. 1935, Singer Refuge, Madison Par., LA.*

and adequate dying and dead trees to provide forage and nest sites; ideal habitat characterized by wide range of age classes produced by high mortality from flooding, wind, fire, and lightning. An abundance of other woodpecker species may indicate high-quality Ivory-billed habitat.

Tanner described three primary regional habitat types, each with its own predominant tree associations and all with bald cypress and tupelo gum in the wettest swamps: river bottoms outside Mississippi delta—sweetgum, water oak, and laurel oak in East, with sweetgum and willow oak in West; bottomlands of Mississippi River—primarily higher parts of first (lowest) bottoms, with sweetgum, Nuttall's or bottomland red oak, and green ash; Florida region—nesting primarily among river-bottom swamp with cypress, black gum, and green ash; also creek swamps with bald cypress, red maple, laurel oak, black gum, and cabbage palmetto; feeding in pine woods bordering swamps, especially if fire-killed.

## DETECTION

**VOCAL SOUNDS AND BEHAVIOR:** Vocalizations are known from early historic firsthand accounts, as well as from sound-recording expedition of Allen, Kellogg, and Tanner in 1935.

One basic call is described, with up to three variations. Typical call note is likened to sound of blowing on clarinet mouthpiece, generally given only once or twice, then often repeated after pause of 3–4 seconds; frequently referred to as a nasal *kent, kent*. Also an up-slurred *kient, kient, kient* and a courtship contact call like *yent, yent, yent*.

Call note is frequently compared to that of Red-breasted Nuthatch, but much less repetitive; higher pitched and longer than that of southeastern White-breasted Nuthatch. Call possibly audible for ¼ to ½ mi. (0.4–0.8 km), especially in winter, when deciduous trees are bare.

Considered quite vocal; reported calling shortly after leaving nightly roost and chatting softly, often continuously, with mate and offspring while feeding and traveling together; also upon lighting and while shimmying up a trunk. Typically silent in flight.

**NONVOCAL SOUNDS:** Delivers a loud, resonant double-knock, like other *Campephilus* species but unlike rolling drum of other North American woodpeckers; second note slightly softer and suggestive of an echo of first note. Lone male may also give single raps. Rarely reported giving typical woodpecker drum.

## VISUAL IDENTIFICATION

**ADULT:** Largest woodpecker north of Mex., and third largest in world. Boldly black and white with light iris and light-colored bill. Black parts, especially wing coverts, occasionally show bluish gloss above; tail and primary tips duller. Forehead and crown black. Adult male has bright scarlet red crest, extending forward and tapering to a point on cheek. Female has all-black crest, longer and occasionally recurved slightly at tip.

White facial stripe extends down side of neck and onto back; perched bird shows these two stripes as white V on upper back above large white shield-shaped patch formed by white secondaries. Long tail feathers strongly decurved.

Bill is long, broad-based, and ivory-toned, similar in color to aged ivory piano keys (museum specimens show strongly yellowish bill); boldly chisel-tipped. Nasal tufts pale white. Reports of adult eye color vary widely, from creamy white to vivid yellow.

To prepare for the extremely remote possibility of seeing an extant Ivory-billed Woodpecker, birders should be very familiar with the flight profile of the Pileated Woodpecker—ideally, under a wide range of conditions. And always remember the birders' parsimony rule: Before making a positive identification of a rare bird, consider all the reasons why the bird you are seeing is not the more likely species to be present. *Originals and composite image by Stephen Shunk; June 2015, Cold Lake, AB.*

"Sonny Boy," the nestling Ivory-billed Woodpecker and the only Ivory-bill known to have been banded; here on the arm of J.J. Kuhn, warden of the Singer Refuge. *James Tanner, courtesy of Nancy & David Tanner; Mar. 1937, Singer Refuge, Madison Par., LA.*

In flight, rarely seen from above. Upperwing shows striking two-tone pattern, with leading edge and nearly all primaries dark black, hindwing bold white.

Black stripe through center of underwing, broadening and extending to black wingtips, sandwiched between bold white hindwing and paler white wing coverts.

**JUVENILE:** Browner and less iridescent than adult, with more rounded bill tip, short black crest, and white tips on primaries. Male crest may show a few red feathers, otherwise sexes alike. Tail is square-tipped because of slowly developing central feathers. Iris of nestling and fledgling is dark brownish, changing to bright yellow after a short time.

**INDIVIDUAL VARIATION:** Females are slightly smaller than males. Leucism rarely reported. Red on crest rarely reported as extending down back of neck.

**PLUMAGES AND MOLTS:** Molt is difficult to determine with certainty. May experience a single annual molt, primarily in summer–fall. Lack of field data from transitioning individuals leads to much confusion in describing molts, plumages, and body colors.

**DISTINCTIVE CHARACTERISTICS:** Massive feet are easily seen when perched. Third toe is longest; claws are large, sharp, and strongly curved. Toe arrangement is considered a modified pamprodactyl foot, the most specialized among woodpeckers for climbing. In their most extreme positions, toes 1 and 4 rotate into a more or less forward position with toes 2 and 3; typically observed in a less extreme rotation, with the long hallux extended laterally and next to toe 4, allowing bird to counter outward gravitational force. Strongly decurved and very stiff tail feathers assist with tree climbing.

**SIMILAR SPECIES:** Ivory-billed Woodpecker shares habitat with Pileated Woodpecker, and both have bold patches of black and white; much debate has occurred over interpretation of recent photos and video of purported Ivory-billeds. Observers should obtain photographs of any bird even remotely believed to be an Ivory-billed.

As with any difficult identification, factors such as shadowing, light angle, and amount of ambient light should be carefully considered. Reports of questionable birds or those believed to be Ivory-billeds should be sent (with photographs if possible) to U.S. Fish and Wildlife Service (contact information at www.fws.gov).

**GEOGRAPHIC VARIATION:** Size gradation is similar to that of Pileated Woodpecker; larger birds in cooler and more northern areas, smaller birds in warmer, more southern areas.

### SUBSPECIES

Two subspecies are recognized: *C. p. principalis* of se. U.S. and *C. p. bairdii* of old-growth forests (upland pines) on main island of Cuba. There is broad variation and overlap in plumage characters between *bairdii* and FL *principalis*. Future DNA research may show these two forms to be genetically distinct species, with some possibility that Cuban Ivory-billed may be more closely related to Imperial Woodpecker of Mex. than to nominate Ivory-billed.

## BEHAVIOR

### BREEDING BIOLOGY

Few data exist on breeding biology, hence much of information below is either postulated or based on small sample sizes.

#### *Nest Site*

Uses variety of tree species for nest cavities, especially bald cypress and large oaks; palmetto trunks also reported. Cuban birds nest and roost almost exclusively in pines. Cavity excavation is always in wood rotted by fungi characteristic of high-humidity swamps. Less than one-quarter of recorded nests are in dead trees, but frequently uses dead, though not rotten, branches in live trees.

Favored cavity site is generally in or below snapped-off dead branch, close to main trunk; downward-facing location under overhanging branch protects cavity entrance from weather, shades it, and makes detection by predator more difficult than with exposed cavity.

Unlike other woodpeckers, may not be deterred by foliage close to cavity entrance. Occasionally nests in same tree in successive years, but in newly excavated cavity.

#### *Courtship and Parenting*

Presumed to be monogamous and mated for life; pairs likely remain together year-round. Mates may fly to same snag, male landing above female, each giving soft courtship call and preening itself; female then climbs up below male, male bends his head down, and both clasp bills. Male also observed flying in below waiting female, then climbing to upper position. Mates may follow closely together on the wing, alternately excavating in different trees.

Both parents share most care duties. Fledglings are adept at flight and begin roosting independently after about 2 weeks. Young are fed by parents for more than 2 months after fledging; by 3 months they can typically feed independently. Some young may depend on adults for a year or more, possibly because of long period required for bill to fully develop and hence be optimally efficient for foraging.

#### *Dispersal*

Young birds are nomadic until they establish their own territories. Individuals travel 0.75–2.5 mi. (1.2–4 km) from nest.

### NONBREEDING BEHAVIOR

#### *Feeding*

***Preferred Foods:*** Strongly favors large long-horned beetle larvae. Also takes larvae of other wood-boring and bark beetles; animal foods may represent 40–50 percent of diet. Feeds opportunistically on pine-bark beetles during outbreaks. Vegetable foods include cherries, various berries and nuts, wild persimmons, and wild grapes. May bore into galls, and known to ingest small amounts of gravel, presumably to aid in digestion of vegetable foods.

***Foraging Behaviors:*** Often feeds among dying, beetle-infested trees at ecotone between swamp and pine forest. Primarily found stripping bark (scaling) of recently dead, large trees. Very rarely forages on ground; observed feeding among palmetto roots in a recent burn.

Primarily feeds until midmorning. Uses bill like chisel, stabbing it horizontally into bark crevices and prying off huge sheets of bark to expose beetle larvae. Occasionally bores into standing trunks, removing large splinters of wood with each chopping

Researcher David Luneau fixes a motion-detecting camera on a large patch of bark scaling in the Big Woods of Arkansas. *Stephen Shunk; Jan. 2006, Arkansas Co., AR.*

motion. Some deeper excavations may resemble diamond shape, as opposed to long ovals of Pileated Woodpecker, but this is not diagnostic.

### *Territory Defense and Sociality*

No descriptions are available of Ivory-billed Woodpecker defending its territory.

Maintains immense home range and may wander outside range into areas of recently burned or bug-killed forests. Winter ranges believed to be much larger; maximum density estimated at about one pair in 6 sq. mi. (15.4 sq. km).

Up to 11 individuals observed at once, in winter, and 4 of which foraged in same bald cypress tree. Other observations include five individuals feeding together and two males and two females feeding together.

Small group sociality may have declined as availability of large timber decreased; if such behavior helped facilitate survival of a group, then deterioration of group size in general may have contributed to decline of the species as a whole. (Though these are reasonable postulations, they are purely speculative.)

### *Interactions with Other Species*

Observed chasing Pileateds, but both species also seen foraging in close proximity. Species in range of Ivory-billed that may benefit from its excavation activities include more than 40 species of cavity-nesting birds as well as flying squirrels, opossums, and tree-nesting bats.

Separation in nesting chronology between Ivory-billed and Pileated Woodpeckers, the former nesting earlier, may have decreased competition for nest cavity sites in an undisturbed forest ecosystem; similar observations have been made in Mex. and Cen. and S. Am. where a *Campephilus* and a *Dryocopus* species overlap in habitat.

Found to host a species of feather mite endemic to Ivory-billed.

### MISCELLANEOUS BEHAVIORS

Flight is unlike that of most woodpeckers except Pileated: strong and direct, crowlike, with steady wingbeats and little undulation. Flies longer distances above canopy; shorter distances, especially over water, occasionally involve long, single undulations, ending with deep swoop onto a trunk.

Each individual likely excavates its own roost cavity. Birds typically emerge late in morning, first climbing up roost tree to preen and stretch at top, then calling for and joining mate; return to roost cavity in late afternoon. Pair members observed roosting in different cavities in same tree.

## CONSERVATION

Multiple factors precipitated decline of Ivory-billed Woodpecker, including collection for sport, food, curiosity, and science, but habitat loss limited species' ability to expand beyond carrying capacity of local territories. Destruction of virgin bottomland forests through early 1900s restricted Ivory-billed to tiny islands of habitat, too small to sustain healthy populations.

**HABITAT THREATS:** Final overlying cause of species' decline was destruction of its native habitats. Initially, settlement itself led to gradual clearing of forests and woodlands. Broad demand included need for railroad and utility lines, steamboat fuel, and homes. By 1880s, northeastern forests were cleared, and forests and swamps of southern states harbored the nation's primary wood supply.

Successive wars waged within and outside U.S. required timber for airplanes, ships, crates, barracks, and bridges, and wood pulp supplied the key ingredient in nitrated cellulose explosives. Opposing cutting of timber for these purposes was seen as unpatriotic.

Current threats include diversion of surface water from federally protected waterways for agricultural uses in adjacent communities. Some areas that once supported Ivory-billed (such as swamps) now generate considerable tourism revenue as productive fishing lakes and hunting preserves, especially waterfowl habitats. Suitable habitats may remain only in SC, FL, and LA.

**POPULATION CHANGES:** Tanner reported the species was historically "most abundant" in lowest bottoms of Mississippi R. and rivers of SC, GA, and FL swamps. Probably more common in FL than in other states.

In early 1800s, Audubon observed the species "frequently" on upper Mississippi R. delta and as "common" in TX and LA, but "rare" on Ohio R. to Henderson. He also reported the species as "very abundant" along Buffalo Bayou in se. TX. In late 1800s, reported by Wayne as "once very common" in nw. FL, but "now rapidly becoming extinct."

By 1885, gone from most of western portions of range, including AR (especially Ouachita and Red Rs.), and upper reaches of se. TX rivers; also gone from especially northeast corner of range, ne. SC and se. NC.

At least 21 confirmed nesting records from 1892 to 1935. Tanner estimated that 24 individual birds remained in 1939, including 5 he observed in Singer Tract, ne. LA. Last definitive Ivory-billed Woodpecker photos in U.S. were taken by Tanner in ne. LA in 1939; last confirmed sighting by Don Eckelberry in 1944. May have persisted through at least 1942 in Okefenokee Swamp, GA. By 1956 considered rarest bird in N. Am. Last reported from Cuba in 1986.

**RECENT SEARCH EFFORTS:** Minor search efforts followed a reported 1999 sighting in Pearl R. swamp, ne. LA, followed by major expedition in 2002. No irrefutable proof was obtained. Sight reports and a cryptic video obtained from e. AR in 2004 catalyzed formation of Big Woods Conservation Partnership, launching unprecedented search efforts in 2004–2007 and resulting in 142 sq. mi. (367 sq. km) being searched. Some reports of putative Ivory-billed sightings and sound recordings have been acquired, but still no irrefutable proof.

Reported sighting in 2005 from Choctawhatchee R. of FL panhandle precipitated systematic search that led to multiple purported sightings and sound recordings; no photographs were obtained, and no evidence was accepted as irrefutable. Regional search efforts in 2006–2007 from SC to TX ended with same results. Following reports from LA, AR, and FL, organized searches ensued throughout se. U.S., including aerial searches in LA and GA and ground searches in MS and TX. Thousands of cumulative search hours failed to produce irrefutable proof that the species remains.

**CONSERVATION STATUS AND MANAGEMENT:** FL law passed in 1901 protected all nongame birds; federal Lacey Act of 1900 criminalized interstate shipment of birds. With confirmation of remaining pairs in Singer Tract, 1930s in ne. LA, state of LA issued a moratorium on all collection permits.

Following reports in 1950 from Chipola R. swamp in FL panhandle, 1,300-ac. (526-ha) Chipola River Wildlife Sanctuary was established; ab-

Bayou DeView in the Big Woods of eastern Arkansas, the site of a flurry of Ivory-billed Woodpecker reports in 2004. *Stephen Shunk; Jan. 2006, Monroe Co., AR.*

Steinhagen Reservoir on the Neches River in eastern Texas, where Ivory-billed Woodpeckers likely once roamed. Today, many of the Ivory-billed Woodpecker's former bottomland habitats have been flooded for reservoirs and are managed for fishing, watersports, and other public recreation. *Stephen Shunk; May 2014, Jasper Co., TX.*

sence of further sightings ended sanctuary status in 1952.

Following late 1960s reports of purported Ivory-billed Woodpeckers in Big Thicket of e. TX, Senator Ralph W. Yarborough of Beaumont, TX, authored legislation to designate portions of Big Thicket as national monument; designated as national preserve in 1974. Species was officially listed as endangered by U.S. Fish and Wildlife Service (USFWS) in 1967 and 1970. Red Listed on 2007 WatchList.

State of LA established 2-year moratorium on timber cutting following reports from Pearl R. area in 1999. Following 2005 announcement of prior year's sightings in AR, Big Woods region was closed to public; later reopened by permit only. In May 2005, AR formed task force on Ivory-billed Woodpecker conservation, and USFWS formed Ivory-billed Woodpecker Recovery Team. USFWS also designated $10 million for Corridor of Hope Conservation Plan in Big Woods, AR.

In 2007, USFWS issued Draft Recovery Plan, estimating cost of recovery at $27.785 million between 2006 and 2010. Among the proposed actions were habitat inventory and monitoring, landscape conservation design, and management of rediscovered populations.

Current emphasis is on purchasing land for habitat restoration and conservation. AR conservation issues include balancing ecological water needs with proposed diversions for local rice farming.

## REFERENCES

Allen 1924; Allen and Kellogg 1937; American Association for the Advancement of Science 1939, 1942, 1956; American Ornithologists' Union 1976; Argus 2005; Bailey 1927; Bailey 1939; Berchok 2005; Beyer 1900; Chapman 1930; Charif 2005; Chu 2005; Collinson 2007; Cornell Lab of Ornithology, no date; Dittman 2003; Driscoll 2005; Fenwick 2005; Fitzpatrick 2006a, 2006b; Gallagher 2005a, 2005b; Graham 2000; Harrison 2005, 2006a, 2006b; Harwood 1986; Hasbrouck 1891; Hill et al. 2006; Hoose 2004; Jackson 2002a, 2002b, 2004, 2006; Jacques 2005; Koenig 2005; LaBranche 2005; Lammertink, M., pers. comm.; Lammertink 2005; Loftin 1991; Luneau, D., pers. comm.; McAtee 1942a; McIlhenney 1941; McKinley 1958; Mirinov 2005; Nature Publishing Group 1941; Niskanen 2005; Oberholser 1930; Oberle 1970; Osier 2007; Parr 2005; Peterson 1957; Powers 2005; Rohrbaugh 2005, 2007; Rosenberg 2005; Rosenberg et al. 2005; Schoch 2005; Schorger 1949; Simon 2005; Snyder 2007; Spahr 2005; Stefferud 1966; Stolzenburg 2002; Swarthout and Rohrbaugh 2005; Tanner 1942a, 1942b; Tanner, N., pers. comm.; The Nature Conservancy of Arkansas 2005; Thompson 1889; Thone 1936; U.S. Fish and Wildlife Service 2005, 2007; Wayne 1893, 1895, 1905; Wetmore 1943; Williams 2005.

# ACKNOWLEDGMENTS
# APPENDICES
# GLOSSARY
# BIBLIOGRAPHY
# INDEX

# ACKNOWLEDGMENTS

**Any effort of this magnitude** requires teamwork, mentoring, and unconditional generosity. Eighteen years of woodpecker study have given me a measure of enthusiasm and confidence in the subject material, but I could not have prepared this volume without accessing the results from many cumulative decades of fieldwork and other research performed by the biologists, birders, colleagues, and clients who have contributed to my work.

As so many authors have said before me, there is not room to thank everyone who had something to do with this book. Also, I will undoubtedly forget someone who deserves special mention. I apologize and hope that I expressed my gratitude earlier; your support has not gone unappreciated.

The American Birding Association first provided a public audience for my woodpecker chronicles at its annual convention in Eugene in 2003. My field studies in central Oregon would not have been possible without the ongoing support of the East Cascades Bird Conservancy and the biology team at the Sisters Ranger District, Deschutes National Forest. Thanks also to the Deschutes Land Trust and to Wolftree, Inc., for their financial support of field studies in the Metolius River basin.

The ornithological community rallied around my requests to review both the narrative chapters and species accounts. My esteemed review team included Carl Bock, Monica Bond, Evelyn Bull, Richard Conner, Troy Corman, Rita Dixon, Danielle Dodenhoff, Chris Forrestal, Kimball Garrett, Dan Gleason, Jocelyn Hudon, Michael Husak, Jerome Jackson, Walt Koenig, Jeff Kozma, Martjan Lammertink, Mark Lockwood, Dawn McCovey, Will McDearman, Joe Neal, Rob Neyer, Nicole Nielsen-Pincus, A. Townsend Peterson, Peter Pyle, John Rakestraw, James Van Remsen, Vicki Saab, Ivan Schwab, Cliff Shackelford, Rodney Siegel, Kim Smith, Robert Stark, Nancy Tanner, Eric Walters, and Karen Wiebe.

Special thanks to David Luneau for showing me around the Big Woods of Arkansas; Walt Koenig for a tour of the Hastings Reservation; Tim Gallagher, Martjan Lammertink, and Nancy Tanner for personal and professional insights into the quest for the Ivory-billed Woodpecker; Tanya Brummit of the Sabine and Angelina National Forests for early field trips into Red-cockaded Woodpecker clusters; the Black-backed Woodpecker crew at the Institute for Bird Populations for welcoming me into their family; and Joe Neal for special wisdom on woodpecker conservation, and for showing me more Yellow-bellied Sapsuckers in one day than I had ever seen before.

There is no comparison for the hands-on learning afforded by exploring bird collections in natural history museums. For access and support, thanks to Eugene Cardiff, San Bernardino County Museum; Carla Cicero, Museum of Vertebrate Zoology, University of California at Berkeley; Krista Fahy, Santa Barbara Museum of Natural History; Moe Flannery, California Academy of Sciences; Kimball Garrett, Natural History Museum of Los Angeles County; Karen Morton, Dallas Museum of Nature and Science; and Phil Unitt, San Diego Natural History Museum.

A handful of dedicated volunteers dove into special tasks that enhanced the final manuscript and would not have been completed without their help: Rod Bonacker, Scott Carpenter, Christopher Ciccone, Scott Cronenweth, Kris Falco, Ron Halvorson, Laura Kammermeier, Rob Neyer, Nicole Nielsen-Pincus, and Skye Richardson.

Special thanks to Lisa White of Houghton Mifflin Harcourt for taking a chance with this first-time book author, and for displaying the patience it took awaiting repeated revisions of the manuscript. And I may never have finished the final manuscript without the loving support and planning assistance of Christine Elder. Thanks to my mother and late father for believing in and encouraging me and for their financial support during the years of negative cash flow. And my greatest thanks and perpetual gratitude to Kris Falco for managing the farm while I spent endless hours at the computer; for editing, proofreading, rereading, and otherwise reviewing nearly every page of the manuscript; and for putting her dreams on hold while she helped me realize my own.

# APPENDIX 1: MEASUREMENTS OF NORTH AMERICAN WOODPECKERS

| | LENGTH | | WINGSPAN | | WEIGHT | | BILL | WING | TAIL | TARSUS |
|---|---|---|---|---|---|---|---|---|---|---|
| | IN. | CM | IN. | CM | OZ. | G | MM | MM | MM | MM |
| Lewis's Woodpecker | 10-10.8 | 25.3-27.3 | 20-21 | 50.8-53.3 | 3.0-4.9 | 85-138 | 26-33 | 153-180 | 86-106 | 22-27 |
| Red-headed Woodpecker | 7.6-9.5 | 19.4-24.0 | 16-18 | 40.6-45.7 | 2.0-3.4 | 56-97 | 21-30 | 125-150 | 63-85 | 19-25 |
| Acorn Woodpecker | 8.3-9.3 | 21.1-23.6 | 17-17.5 | 43.2-44.5 | 2.4-3.1* | 67-89* | 24-33 | 131-154 | 70-88 | 20-25 |
| Gila Woodpecker | 8-10 | 20.3-25.4 | 15-18 | 38.1-45.7 | 1.8-2.9* | 51-81* | 26-35 | 130-140 | 74-88 | 22-25 |
| Golden-fronted Woodpecker | 8.5-10 | 21.6-25.4 | 16-18 | 40.6-45.7 | 2.2-3.5* | 63-100* | 30-35 | 127-144 | 71-85 | 21-25 |
| Red-bellied Woodpecker | 9-10.5 | 22.9-26.7 | 15-18 | 38.1-45.7 | 2.0-3.2 | 56-91 | 25-33 | 122-139 | 68-85 | 19-23 |
| Williamson's Sapsucker | 8.5-9 | 21.6-22.9 | 17 | 43.2 | 1.6-2.3 | 44-64 | 22-29 | 128-143 | 71-90 | 20-24 |
| Yellow-bellied Sapsucker | 7.5-9 | 19.0-22.9 | 16-18 | 40.6-45.7 | 1.4-2.2 | 40-62 | 21-26 | 115-131 | 59-78 | 19-22 |
| Red-naped Sapsucker | 7.5-9 | 19.0-22.9 | 16-18 | 40.6-45.7 | 1.3-2.2 | 37-61 | 20-26 | 118-133 | 67-79 | 19-22 |
| Red-breasted Sapsucker | 8-9 | 20.3-22.9 | 16-18 | 40.6-45.7 | 1.4-1.9 | 40-55 | 22-27 | 118-134 | 71-85 | 19-22 |
| Ladder-backed Woodpecker | 6-8 | 15.2-20.3 | 11-13 | 27.9-33.0 | 0.7-1.6* | 21-48* | 20-27 | 101-106 | 60-68 | 16-20 |
| Nuttall's Woodpecker | 6.7-7.5 | 17.0-19.0 | 13-13.5 | 33.0-34.3 | 1.0-1.7 | 28-47 | 18-24 | 96-108 | 59-72 | 18-19 |
| Downy Woodpecker | 6.8-7 | 17.1-17.8 | 11-12 | 27.9-30.5 | 0.7-1.2 | 21-33 | 14-19 | 86-106 | 48-72 | 15-21 |
| Hairy Woodpecker | 8.5-10.5 | 21.6-26.7 | 15-17.5 | 38.1-44.5 | 1.5-2.8 | 42-80 | 21-39 | 97-138 | 49-92 | 18-25 |
| Arizona Woodpecker | 7-8 | 17.8-20.3 | 14-15 | 35.6-38.1 | 1.2-1.8* | 34-51* | 24-28 | 109-121 | 60-71 | 19-21 |
| Red-cockaded Woodpecker | 8.5 | 21.6 | 14 | 35.6 | 1.4-1.9 | 40-54 | 19-23 | 95-126 | 70-82 | 18-21 |
| White-headed Woodpecker | 9-9.3 | 22.9-23.5 | 16-17 | 40.6-43.2 | 1.8-2.8 | 50-79 | 24-32 | 118-131 | 74-90 | 20-24 |
| Am. Three-toed Woodpecker | 8-9 | 20.3-22.9 | 14-16 | 35.6-40.6 | 1.7-2.3 | 47-65 | 22-31 | 109-129 | 65-83 | 19-23 |
| Black-backed Woodpecker | 8.1-9.8 | 20.5-24.8 | 16-17 | 40.6-43.2 | 2.2-3.1 | 61-88 | 29-35 | 119-134 | 73-85 | 21-25 |
| Northern Flicker | 10-14 | 25.4-35.6 | 19-21 | 48.3-53.3 | 3.7-5.9 | 105-167 | 28-43 | 137-178 | 87-124 | 26-32 |
| Gilded Flicker | 10-12 | 25.4-30.5 | 18 | 45.7 | 3.2-4.6 | 92-129 | 34-41 | 141-153 | 86-101 | 25-29 |
| Pileated Woodpecker | 16.5-19.5 | 42.0-49.5 | 27-30 | 68.6-76.2 | 8.5-12.0 | 240-341 | 41-60 | 210-253 | 136-174 | 31-36 |
| Ivory-billed Woodpecker | 19-20 | 48.0-50.1 | 30-32 | 76.2-81.3 | 15.9-20.1* | 450-570* | 61-73 | 240-263 | 147-166 | 40-46 |

**STANDARD MEASUREMENTS**

Unless noted, figures include males and females of all subspecies occurring north of Mexico; see Species Accounts for individual variation.

All decimals rounded to the nearest tenth; (g) and (mm) rounded to the nearest whole number.

Conversions calculated from http://www.onlineconversion.com/.

*These weights include ranges and subspecies south of the U.S., though most southern forms are generally on the lighter end of the scale.

LENGTH - bill tip to tail tip (with head tilted up so that bill is in line with body)

BILL - length of exposed culmen, or centerline on dorsal surface of maxilla

WING - chord length, or distance between the bend in the wing and the tip of the longest primary, with wing folded and pressed flat

TAIL - length of central tail feathers

TARSUS - length of upper section of foot, between heel and toes

**REFERENCES**

Alsop 2001; Birds of North America species accounts; Burt 1930; Dickenson 2002; Howell & Webb 1995; Pyle 1997; Ridgway 1914; Sibley 2000; Winkler et al. 1995.

# APPENDIX 2: NEST SITE DATA FOR NORTH AMERICAN WOODPECKERS

| | NEST TREE | | | NEST CAVITY | | | | | | | | EGG | |
|---|---|---|---|---|---|---|---|---|---|---|---|---|---|
| | dbh (CM) | Diam. at cavity height (CM) | Tree height (M) | Avg. height from ground to bottom of entrance (M) | Min., max. cavity height (M) | Avg. elliptical area of cavity entrance (SQ. CM) | Avg. vert. depth of cavity from upper to lower apex (CM) | Min., max. cavity depth (CM) | Avg. interior diam. of cavity at widest point (CM) | Min., max. interior diam. (CM) | Compass direction of cavity entrance (DEG.) | Avg. diam. at widest point x avg. diam. at narrowest point (MM) | Avg. weight (G) |
| Lewis's Woodpecker | 20–120 | 52 | 1.6–28 | 7.5–24.5 | 0.6, 51.8 | 21.1 | 32–34 | 23, 76 | 16–17 | 13, 20 | ? | 26 x 20–22 | 5.6 |
| Red-headed Woodpecker | 58–59 | 16–26 | 14.6–18.3 | 7–12.4 | 2, 24.5 | 18.5 | 38 | 20, 60 | 8–11 | <8, 14 | 67.5–225 | 25 x 19 | ? |
| Acorn Woodpecker | ? | 17–100 | ? | 8.3 | 2.3, 18 | ? | 40 | 22, 70 | 15 | ? | 90 | 25 x 19 | 5.3 |
| Gila Woodpecker | ? | ? | 4–12 | 5.8–9.6 | 1, 10.7 | 18.8 | 23–40 | 18, 42 | 13–16 | 9, 24 | 351 | 25 x 19 | ? |
| Golden-fronted Woodpecker | ? | ? | ? | 2–9 | <1, 11 | 16.3 | 32 | 15, 46 | ? | ? | ? | 22–26 x 17–20 | ? |
| Red-bellied Woodpecker | ? | 27 | 5–31 | 8.4 | ? | 18.2 | ? | 22, 32 | ? | ? | 225 | 25 x 19 | ? |
| Williamson's Sapsucker | 23–118 | ? | 10–20 | 13–28 | 2, 18 | 13.1 | ? | 20, 27 | ? | 9, 13 | 157.5–202.5 | 23–24 x 17–18 | 3.8 |
| Yellow-bellied Sapsucker | 12–43 | 23 | 3–33 | 8.6 | 2, 13.7 | 12.9 | 27 | ? | 7 | ? | 45–270 | 22–23 x 17 | 3.4 |
| Red-naped Sapsucker | 17–119 | ? | 1.0–50.6 | 2.7–21.5 | 0.7, 42.7 | 13.1 | 17 | ? | 12 | ? | 90–225 | 23 x 17 | 3.8 |
| Reb-breasted Sapsucker | 30–117 | 32–125 | 16.8–47.1 | 10–17.2 | 1.8, 50 | 14.6 | 25 | 14, 34 | * | ? | * | * | * |
| Ladder-backed Woodpecker | ? | 14 | 9 | 0.6–9 | 0.6, 9 | 11.0 | ? | 18, 36 | ? | 4 | ? | 21 x 16 | 2.9 |
| Nuttall's Woodpecker | ? | ? | ? | 5.2 | 0.7, 13.7 | 15.7 | 28 | ? | ? | ? | ? | 22 x 16 | 2.9 |
| Downy Woodpecker | 15–66 | 14–21 | 8–10.2 | 6.1–7.75 | 1, 13.5 | 10.1 | ? | 15, 30 | ? | 6, 9 | 157.5–337.5 | 19–20 x 14–15 | 2.1 |
| Hairy Woodpecker | 25–53 | 25 | 12–17 | 9–10 | 1, 18.3 | 15.3 | 25 | 20, 41 | ? | ? | 157.5–337.5 | 21–25 x 18–20 | 4.4 |
| Arizona Woodpecker | ? | 18 | ? | 4.6–6.1 | 2.4, 15.2 | 16.8 | ? | 30, 38 | ? | 9, 11 | ? | 23–24 x 17–18 | ? |
| Red-cockaded Woodpecker | 28–64 | 38 | 13.5–29.4 | 10–13 | <1, 30 | 20.1 | ? | 17, 18 | ? | 9, 10 | 180–270 | 24 x 17–18 | ? |
| White-headed Woodpecker | 14–152 | 15–152 | 0.6–44 | 2.9–5 | 0.4, 22 | 15.1 | 25 | 19, 40 | 9–12 | 5, 14 | ? | 24–25 x 18–19 | 4.4 |
| Am. Three-toed Woodpecker | 20–33 | ? | 10–23.1 | 5.2–8.9 | 3, 25 | 12.9 | ? | ? | ? | ? | ? | 22–25 x 17–19 | 4.4 |
| Black-backed Woodpecker | 16–40 | ? | 4–32.7 | 4.7–9 | 1, 11 | 12.2 | ? | 24, 30 | ? | 8, 12 | ? | 21–25 x 18 | 4.8 |
| Northern Flicker | 8–84 | 15–61 | 3–14.9 | 1.9–18 | 0.4, 22.1 | 20.4 | 34–48 | 13, 156 | 15–17 | 7, 28 | 67.5–337.5 | 28–29 x 22 | 7.2 |
| Gilded Flicker | ? | ? | 5–8 | 6.2 | ? | 23.9 | 38 | ? | 13 | ? | ? | 29 x 21 | 7.1 |
| Pileated Woodpecker | 35–91 | 25–60 | 18.3–28.8 | 10–17.5 | 5, 17.5 | 24.0 | 37–43 | ? | 13 | ? | 45–180 | 32–34 x 24–25 | 10.0 |
| Ivory-billed Woodpecker | ? | 33–56 | ? | 14.3 | 7.6, 21 | 38.4 | 52 | ? | 17 | ? | 0–360 | 35–37 x 25–28 | ? |

**NOTES**

dbh, or diameter-at-breast-height, is a standard measurement of the diameter of a tree, generally considered to be approx. 4 ft. (1.2 m) above ground level.

Where ranges of figures are given in a column labeled as averages, this is the range of averages from multiple studies.

? = unknown or not enough data.

* = presumed to be similar to above, but no direct evidence.

Measurements: (cm) and (mm) are rounded to the nearest whole number; (g), (sq. cm), and (m) are rounded to the nearest tenth.

Egg weights are based on recently laid eggs (generally within 4 days of laying).

**REFERENCES:** Aitken and Martin 2004; Bate 2001; Benson and Arnold 2001; Birds of North America accounts; Carriger and Wells 1919; Conner 1975, 1977; Cooper and Beauchesne 2000; Cooper and Gillies 2000; Dixon 1995; Fisher and Wiebe 2006a; Frenzel 2004; Graham 1940; Gutzwiller and Anderson 1987; Hill et al. 2006; Inouye 1976; Jackson 1977a; Joy 2000; Kerpez and Smith 1990a, 1990b; Ligon 1968b; Linder and Anderson 1998; McAuliffe and Hendricks 1988; McClelland 1977; McClelland and McClelland 2000; Newlon 2005; Nielsen-Pincus 2005; Rodewald et al. 2005; Saab et al. 2002; Sedgwick and Knopf 1991; Shunk 2008; Stauffer and Best 1982; Tanner 1942; USFWS 1999, 2003; Van Oort 2006; Vierling et al. 2008; Walter and Maguire 2005; Wiebe 2001; Wood 1905a.

# APPENDIX 3: PARENTING DATA FOR NORTH AMERICAN WOODPECKERS

| | NEST | | | | CLUTCH | | | INCUBATION | | | | |
|---|---|---|---|---|---|---|---|---|---|---|---|---|
| | Sex that selects tree and location on tree for nest site[1] | Sex that excavates nest cavity[1] | Number of days it takes to excavate nest cavity | Frequency that a pair reuses previously excavated nest cavity (%) | Avg. number of eggs laid in a single clutch | Range of clutch size | Number of broods raised in a single season | Avg. number of days in incubation period | Range of days in incubation period[3] | Sex that incubates clutch in daytime hours[13] | Sex that broods hatchlings in daytime hours[13] | Egg hatching period |
| Lewis's Woodpecker | ? | fm | ? | 41–100 | 6.0 | 3–9 | 1 | 13 | 12–16 | ? | fm | ? |
| Red-headed Woodpecker | fm | f/m+ | 12–17 | 14 | 4.7 | 4–7 | 1–2[7], 3? | 12 | 12–14 | ? | fm | asynch |
| Acorn Woodpecker | ? | ? | ? | 57 | 4.3 | 3–7 | 1,2ss | 11 | 11–14 | m/f+[8] | br/f+[8] | <24 hr |
| Gila Woodpecker | ? | fm | ? | 85 | 4.1 | 2–7 | 2,3oc | 13 | 12–14 | fm | ? | ? |
| Golden-fronted Woodpecker | f/m+ | fm | 6–12 | 85 | 5.4 | 4–7 | 1,2oc | 13 | 12–14 | ? | fm | <48 hr |
| Red-bellied Woodpecker | f/m+ | fm | 14 | 15 | 4.6 | 2–8 | 1–2,3r | 13 | 11–14 | fm | m/f+ | <48 hr |
| Williamson's Sapsucker | f/m+ | f/m+ | 21–28 | 21–23 | 5.3 | 3–7 | ? | 13 | 12–14 | fm | fm | ? |
| Yellow-bellied Sapsucker | fm | m73%[2] | 7–28 | 20[4] | 4.3 | 4–7 | 1[7] | 12 | 10–13 | m/f+ | fm | 24 hr |
| Red-naped Sapsucker | fm | m+/f[2] | 7–28 | 9–624 | 4.8 | 3–7 | 1[7] | 10 | 8–14 | fm | fm | * |
| Red-breasted Sapsucker | fm | m+/f[2] | 7–28 | 0–254 | 4.5 | 4–7 | 1[7] | 13 | 12–15 | * | fm | * |
| Ladder-backed Woodpecker | ? | f/m+ | ? | ? | 4.3 | 2–7 | 1 | 13 | ? | ? | ? | ? |
| Nuttall's Woodpecker | f/m+? | f/m+ | ? | 0 | 4.5 | 3–6 | 1[7] | 14 | ? | ? | fm | ? |
| Downy Woodpecker | m/f+ | fm | 7–21 | 0–21 | 4.8 | 3–8 | 1, 2r[7] | 12 | ? | fm | ? | 12–48 hr |
| Hairy Woodpecker | f | f/m+ | 7–21 | 2–16 | 3.8 | 3–7 | 1[7] | 14 | 11–15 | ? | fm | synch |
| Arizona Woodpecker | ? | ? | ? | 65 | 4.1 | 2–5 | 1,2r[7] | 11 | 10–11 | fm | fm | <24 hr |
| Red-cockaded Woodpecker | ? | m/f? | ? | 64–66 | 3.4 | 2–4 | 1[7], 2r | 14 | ? | fm[8] | f/m+[8] | 12–24 hr |
| White-headed Woodpecker | f/m+ | ? | ? | 0–10 | 4.3 | 4–7; 9[5] | 1 | 14 | ? | fm | fm | asynch |
| Am. Three-toed Woodpecker | ? | m/f+ | 7+ | 0 | 3.4 | 3–7 | 1? | 14 | 11–14 | fm | fm | synch? |
| Black-backed Woodpecker | ? | f/m+ | <7–28 | 0–7 | 3.6 | 2–6 | 1[7] | 13 | 10–14 | f/m+ | m/f+ | asynch |
| Northern Flicker | ? | f/m+[12] | 14 | 5–65 | 7.0 | 3–13 | 1[7] | 12 | 11–16 | m/f+ | f+ then m+ | asynch |
| Gilded Flicker | ? | ? | ? | 40 | 4.2 | 3–5 | ? | 11 | 11–14 | ? | ? | ? |
| Pileated Woodpecker | m? | f/m+ | ? | 1–2 | 3.9 | 1–8 | 1[7] | 18 | 15–18 | ? | f/m+ | asynch |
| Ivory-billed Woodpecker | ? | m/f+? | ? | 15 | 2.9 | 1–6[6] | ? | 20? | 14–20? | m/f+ | f/m+ | ? |

**KEY**

| | |
|---|---|
| fm | Female and male generally share duty equally. |
| + | Plays a larger role than the other (also applies to breeders vs. helpers in Red-cockaded and Acorn Woodpeckers). |
| ++ | Plays a much larger role than the other. |
| ss | Generally has a second nesting season later in summer. |
| oc | Occasionally double (2) or triple (3) brooded. |
| br | Adults of the breeding pair, i.e., does not include helpers from previous years' broods that are of adult age. |
| r | Rarely double (2) or triple (3) brooded. |
| synch | Eggs hatch more or less at the same time. |
| asynch | Eggs generally hatch in the order they were laid (over a period, where given). |
| ad | Adult. |
| ? | Uncertain, not enough data. |
| * | Presumed to be same as above, but not confirmed. |

| FEEDING | SANITATION | | | FLEDGING[10] | | | | |
|---|---|---|---|---|---|---|---|---|
| Frequency that each sex (m/f) spends feeding nestlings (%), or division of duties where frequency figures are unavailable | Practice of nest sanitation during the first week of the nestling period | Sex responsible for nest sanitation in 2d and 3d weeks of nestling period | Number of days at end of nestling period during which nest is not cleaned[11] | Avg. number of days in between hatching and fledging | Range of days in nestling period | Nest success, or frequency that at least 1 nestling fledges (%) | Time that fledglings are dependent on parents for feeding | |
| ? | ? | f/m++ | ? | 31 | 28–34 | 30–91 | ? | Lewis's Woodpecker |
| 50/50 (<12d); 25/75 (>12d) | ad eats | f/m+ | ? | 28 | ? | 16–78 | 21–25 d | Red-headed Woodpecker |
| br/f+ | ad eats | m/br+ | 7 | 31 | ? | up to 88 | 3–4 mo | Acorn Woodpecker |
| 44/55 | ? | f/m++ | ? | 28 | ? | ? | extended per. | Gila Woodpecker |
| fm | ? | fm? | ? | 30 | >14 | ? | 3–4 wks | Golden-fronted Woodpecker |
| 52/48(<12d); 42/58 (>12d) | ad eats? | m | ? | 25 | 24–27 | 47–82 | 2–10 wks | Red-bellied Woodpecker |
| 50/50 (2 wks) | ? | f/m+ | ? | 24 | 24–33 | 92–100 | 1+ wk | Williamson's Sapsucker |
| 50–55/45–50 | occ. eaten[14] | f/m++ | 4–8 | 27 | 23–29 | up to 100 | 6–8 wks | Yellow-bellied Sapsucker |
| 50/50 | * | f/m+ | * | * | 23–32 | 75–100 | * | Red-naped Sapsucker |
| 50/50 | * | m | * | * | 20–28 | 70 | * | Red-breasted Sapsucker |
| ? | ? | ? | ? | ? | ? | ? | ? | Ladder-backed Woodpecker |
| 47/53 | ? | f/m+ | ? | 29 | ? | ? | 2 wks | Nuttall's Woodpecker |
| 54/46 | eaten[14] | m | few days | 21 | 19–22 | 33–100 | ? | Downy Woodpecker |
| 40–50/50–60 | ad eats | fm | ? | 29 | 28–30 | 38–100 | 3–5 wks | Hairy Woodpecker |
| fm, br+ | eaten[14] | fm | nestlings eat? | 25 | 26–29 | ? | ? | Arizona Woodpecker |
| 82/18 | ? | m+/f?[8] | ? | 26 | 24–29 | 50–90 | up to 5 mo | Red-cockaded Woodpecker |
| 50/50 | ? | fm | ? | 26 | ? | 23–100 | ? | White-headed Woodpecker |
| fm, m+later | ? | m | after 18 | 24 | ? | 33–58 | ? | Am. Three-toed Woodpecker |
| fm, m+later | fm+ | f/m++ | ? | 25 | 19–25 | 50–100 | ? | Black-backed Woodpecker |
| fm | ad eats | f/m+ | ? | 28 | 24–27 | 20–100 | ? | Northern Flicker |
| ? | ad eats | f/m+ | ? | 24 | ? | ? | ? | Gilded Flicker |
| m/f+ | ad eats | fm | 2–5 | 27 | 24–31 | 45–83 | ? | Pileated Woodpecker |
| fm | m | m | ? | 36 | ? | 50 | 4–6 mos | Ivory-billed Woodpecker |

1 In many cases, one mate may propose a site (occ. with multiple starts) and then seek other mate's approval.

2 If first attempt fails, female increases her contribution on second attempt.

3 Inclement weather may extend period.

4 Often to occasionally in same tree but different cavity.

5 Large clutch may indicate intraspecific brood parasitism.

6 Early-season nests may have smaller clutches; 6 eggs reported once but not substantiated.

7 May renest, occasionally in second cavity, if first brood fails early enough in season.

8 Helpers occasionally, mostly males and rarely females.

9 For most species, 1 or 2 eggs may hatch later than the others, but these young usually do not survive.

10 Most nests that fledge > 1 young do so asynchronously over < 24 hours; figures shown apply to first fledgling only.

11 When young are too big for adults to enter nest, sanitation ceases for the remaining nestling periods shown.

12 Female's contribution increases and may exceed male's as nest nears completion.

13 Male does most night incubation and brooding.

14 Unknown whether fecal material may be eaten by adult(s) or young.

**REFERENCES**

Bate 2001; Birds of North America accounts; Blewett and Marzluff 2005; Bull 2001; Burns 1915; Fisher and Wiebe 2006a; Forestall et al. 2004; Hannah 1924; Johnson and Kermott 1994; Kilham 1966, 1979; Koenig 1984, 1986, 1987; Koenig et al. 2005; Li and Martin 1991; Newlon 2005; Saab 2001; Saab et al. 2004; Tanner 1942; Vasquez 2005; Vierling et al. 2008; Wiebe et al. 2006; Yom-Tov and Ar 1993.

# APPENDIX 4: The Woodpecker "Family Tree"

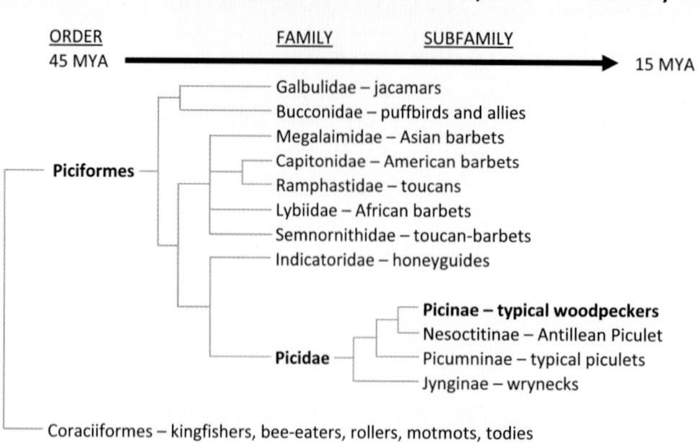


SUMMARY

Relationships depicted here are based on the author's interpretations of the sources below. Study results vary, and studies are ongoing, so this chart should not be considered a definitive reference. Instead, it represents just one possible evolutionary scenario.

Brackets represent divergence into newly evolved forms. *Parent* taxa diverge into two or more closely related *sister* taxa. For example, in this chart, the ancestral **Picidae** diverged into the ancestral **Jynginae** and its sister taxon, which was ancestral to the remaining three subfamilies.

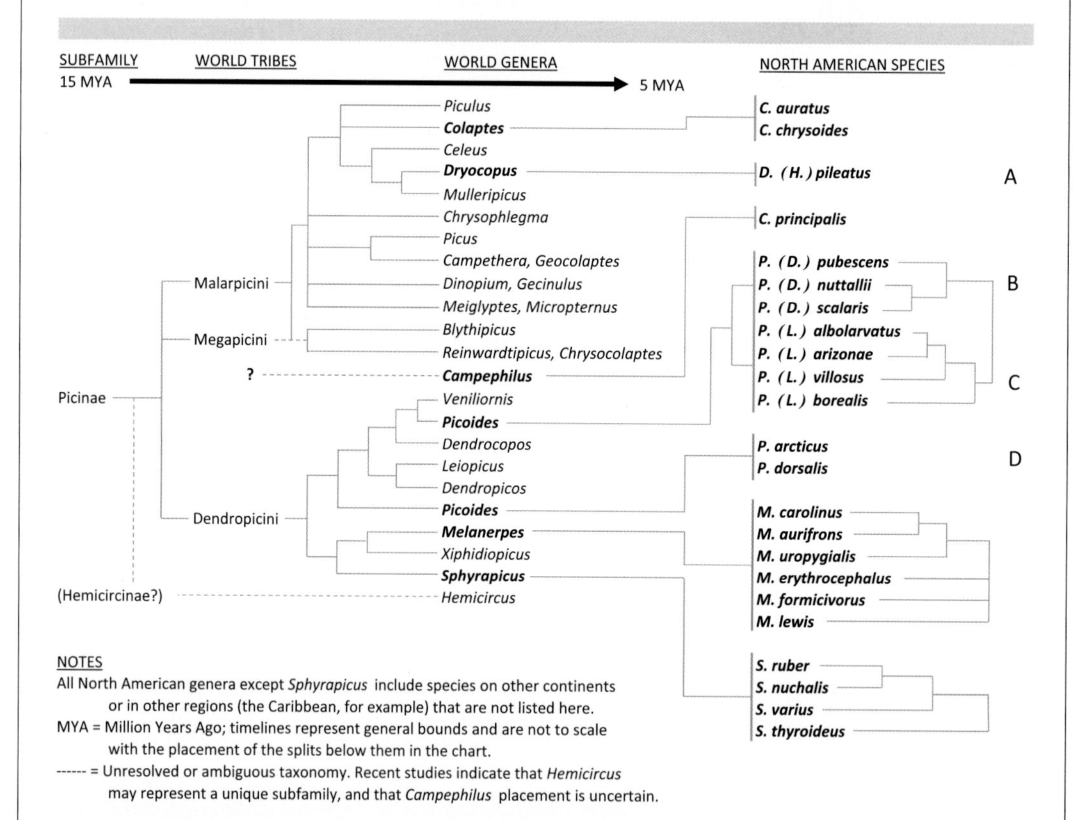


NOTES

All North American genera except *Sphyrapicus* include species on other continents or in other regions (the Caribbean, for example) that are not listed here.

MYA = Million Years Ago; timelines represent general bounds and are not to scale with the placement of the splits below them in the chart.

------ = Unresolved or ambiguous taxonomy. Recent studies indicate that *Hemicircus* may represent a unique subfamily, and that *Campephilus* placement is uncertain.

A – Pileated Woodpecker has been proposed for reversion to the former genus *Hylascopus*.

B – Downy, Nuttall's, and Ladder-backed Woodpeckers have been proposed for placement in the former genus *Dryobates*.

C – White-headed, Arizona, Hairy, and Red-cockaded Woodpeckers have been proposed for placement in a new genus, *Leuconotopicus*.

D – Under the proposals above, American Three-toed and Black-backed Woodpeckers would remain in the genus *Picoides*.

*Note that the taxonomic arrangement above differs markedly from the current AOU order, which is that used in the text.*
*Note also that the taxonomic changes discussed above have neither been accepted, nor are they currently under consideration, by the AOU North American Classification Committee (AOU 2015,* http://www.aou.org/committees/nacc/, retrieved 23 Dec 2015).
*As of Dec 2015, these changes are formally recognized only in del Hoyo et al 2015, which also lists the Northern Flicker as two species:*
del Hoyo, J., Elliott, A., Sargatal, J., Christie, D.A. & de Juana, E. (eds.), 2015. Handbook of the Birds of the World Alive. Lynx Edicions, Barcelona. (http://www.hbw.com/, retrieved 23 Dec 2015).

SOURCES: AOU 2015; Benz et al 2006; DeFilippis and Moore 2000; del Hoyo et al 2015; Fuchs et al 2006, 2007; Garcia-Trejo et al 2009; Gorman 2014; Hackett et al 2008; Johansson and Ericson 2003; Laybourne et al 1994; Manegold and Topfer 2012; Overton and Rhodes 2006; Short and Horne 2001; Weibel and Moore 2001a,b, 2005.

**Woodpeckers are among** the most recognized of all birds, and their recognition often results from conflicts with humans. Drumming on resonant artificial surfaces and excavating in artificial structures are the two most commonly reported nuisance behaviors. Both behaviors result from the loss of native habitats.

Habitat loss due to urban and agricultural (including timber) development has plagued North American bird populations for centuries, and woodpeckers are no exception. One particular human activity falls under the category of habitat loss and profoundly affects all communities with cavity-dwelling wildlife: the removal and exclusion of snags, or dead trees. Lack of snag availability leads woodpeckers to pursue their livelihoods in whatever places most closely resemble their native habitats. And this leads to conflict with humans.

## DRUMMING

On early spring mornings and often with frustrating persistence, the sound of a woodpecker's drum resonates into your home like a machine gun targeting your metal chimney cover. Prior to our introduction of wonderfully resonant artificial surfaces, woodpeckers likely performed most of their drumming on loud resonant snags. Snag removal on the landscape forced woodpeckers to search for alternate drumming posts.

Recruitment and retention of snags may provide alternative drumming posts, but only the conscious exclusion of resonant surfaces in building practices will permanently eliminate these favored communication sites. Rest assured that drumming behavior is seasonal and does not damage your home. This assurance does not stop the broadcast of your natural alarm clock, but it may help you accept this fascinating woodpecker behavior.

## EXCAVATIONS

Excavations in undesired places represent an entirely different level of nuisance. Opportunistic woodpeckers have probably exploited artificial structures for cavity excavation since the advent of artificial structures. Excavations in utility poles have plagued engineers since the earliest telegraph lines were strung into the pioneer countryside. Today, property owners spend millions of dollars annually to mitigate or prevent woodpecker damage.

Very little formal research adequately summarizes the costs and overall effects of undesirable woodpecker excavations. Instead, we have invented a plethora of products and practiced many futile behaviors to deter woodpeckers. These have included:

- *Visual scaring devices,* such as compact disks hanging on fishing line or fake predators (e.g., owls) installed on rooftops;
- *Sound deterrents,* such as recorded predator calls triggered by motion-detection devices, or even the banging of metal pans;
- *Tactile repellants,* such as sticky commercial products that prevent effective footing;
- *Forced exclusion,* including the prompt covering of newly excavated holes with tin can lids, or plastic garden netting hung across favored excavation sites;
- *Woodpecker-proof siding materials,* such as synthetic siding; and the last resort,
- *Lethal control,* which violates federal laws when conducted without a permit.

None of these deterrents provides a permanent solution to woodpecker nuisance behavior. While we are "cultivating" snags in our yards, forests, and woodlands, there is one mitigation effort that will keep most woodpeckers away from our homes: the installation of artificial nest structures.

The most ubiquitous and most conspicuous of all North American woodpeckers is the Northern Flicker. Because flickers are weak excavators, they will often reuse old cavities from prior years, many of which were once excavated by other woodpecker species. Given a lack of suitable nest cavities—or snags that are easy to excavate—flickers will readily excavate in the side of your building. Thankfully, flickers are also the only North American woodpeckers that will consistently use an artificial nest structure. If flickers are attempting to excavate in your home, install a constructed nest box in a nearby tree, and you may quickly solve the problem. Search online for nest-box specifications, or visit your local backyard bird store and buy one.

In the meantime, advocate for the conservation and restoration of healthy forest and woodland habitats. Cavity-dwelling wildlife communities deserve their own chance for survival.

# GLOSSARY

**albinism.** Complete absence of melanin pigmentation resulting in white or translucent body parts; can occur in single feathers, sets of feathers (often bilaterally), and unfeathered parts of the head. See leucism.

**altricial.** Referring to birds that are featherless, blind, and essentially immobile when they hatch.

**anisodactyl.** Toe arrangement with toes 2, 3, and 4 pointing forward and toe 1 (the hallux) pointing backward; the most common toe arrangement in birds. See Legs and Feet, p. 7.

**anvil site.** Any substrate used as a hard surface on which prey items can be easily pulverized prior to ingestion; bark furrows and jagged tops of snapped-off trees or large limbs are frequently used.

**basic plumage.** The freshest plumage acquired through the annual molt, typically late summer through fall for most species.

**bast.** Fibrous material of phloem tissue that is often ingested when woodpeckers forage into the trunk of a tree.

**bolus.** Tightly packed mass of food containing multiple individual food items that have been packed together for efficient digestion.

**cambium.** Tissues of a tree located outside the phloem (cork cambium) and between the phloem and xylem (vascular cambium); references to "cambium" in this book refer to cork cambium, unless otherwise noted.

**clinal.** Referring to a gradual change in a morphological feature, clutch size, or other element, over a geographic distance.

**confirmed nesting/breeding.** Observed behavior(s) providing proof that a bird has nested in a given area; such observations may include but are not limited to nest with eggs, nest with young, and adults feeding fledglings.

**convergent evolution.** Acquisition of similar behaviors or plumage characters despite a distant phylogenetic relationship between organisms.

**corvid.** A bird of the family Corvidae, including jays, crows, ravens, magpies, and nutcrackers.

**dorsal.** Referring to the back or upper surface.

**double-brooded.** Referring to raising two broods of young in the same season, usually in succession.

**ecotone.** Transitional boundary between two different habitat types or ecosystems.

**ectropodactyl.** Toe arrangement in woodpeckers that positions toes 2 and 3 forward, toe 1 (the hallux) backward, and toe 4 movable through an arc between the hallux and the forward toes; typical of woodpeckers that spend the majority of their time climbing rather than perching. See Legs and Feet, p. 7.

**gallery.** Pathway excavated by larval and adult beetles between the bark and cambium (bark beetles) or inside the xylem and phloem of a tree (wood-boring beetles).

**genus.** The highest level of binomial nomenclature (*genus—species*) used for classifying organisms within the same taxonomic family, subfamily, or tribe (plural, genera).

**granary.** Location used by Acorn Woodpeckers to store acorns, often in the bark of a standing tree, in which the woodpeckers bore many single holes to store individual acorns; the communal center of activity for a family group of Acorn Woodpeckers, and often the substrate for a nest cavity.

**gynandromorphic.** Possessing both male and female characteristics, often so that the left side of the bird expresses one sex and the right side the opposite sex.

**hallux.** The first toe of all birds, typically pointing to the rear, and in woodpeckers often laid flat when an individual is climbing a vertical surface; absent in Black-backed and American Three-toed Woodpeckers.

**hardwood.** Usually referring to trees other than conifers.

**harvesting.** Collecting foods (usually fruits) from a live plant, such as berries or corn kernels off the cob.

**hybrid.** Offspring of parents that represent one or both of two closely related species, i.e., the parents can both be "pure" species, or one or both parents could themselves be hybrids. See intergrade.

**intergrade.** Offspring of parents that represent one or both of two subspecies; the parents can both be "pure" subspecies, or one or both parents could themselves be intergrades. See hybrid.

**introgression.** Interbreeding between multiple subspecies.

**keystone.** Referring to a species whose removal from the ecosystem may precipitate deleterious impacts on other species or the entire ecosystem.

**leucism.** Deficiency in melanin expression causing feathers to appear lighter than usual. See albinism.

**maxilla.** Upper jaw bone and bill structure.

**nasal tufts.** Two bunches of barbless feathers that extend from the base of the maxilla to cover the nares (nasal openings in the bill), and which prevent debris from entering the nasal passages.

**Neotropics.** Region of the Western Hemisphere between the Tropics of Cancer and Capricorn (roughly 23.5 degrees north and south of the equator), including southern Florida, Mexico, Central America, the Caribbean, and South America to approximately southern Brazil.

**nomenclature.** The system of names assigned to organisms.

**nominate.** Referring to the first form of a taxon to be described by taxonomists.

**pamprodactyl.** Toe arrangement with all four toes pointing forward of a lateral midline through the leg; typical of *Campephilus* woodpeckers (such as the Ivory-billed). See Legs and Feet, p. 7.

**phloem.** Thin outer layer of a tree trunk between the cork cambium and the sapwood. See cambium.

**polyandry.** Breeding system in which a female has more than one mate.

**polygynandry.** Breeding system in which two or more females each breed with two or more males; eggs are laid in the same nest, and any or all of the individuals may be involved in incubation and brood rearing.

**polygyny.** Breeding system in which a male mates with multiple females, usually in different nests.

**prebasic molt.** The molt by which an adult acquires its definitive basic plumage.

**preformative molt.** The molt by which a juvenile bird acquires its first basic plumage.

**rectrices.** Longest tail feathers extending beyond upper- and undertail coverts; generally the feathers we typically think of as the tail (singular, rectrix); with the remiges, form the flight feathers.

**remiges.** Feathers forming the hindwing, including the primaries, secondaries, and tertials; with the rectrices, comprise the flight feathers.

**retina.** Layered light-sensitive tissue at the back of the eye that contains the rods and cones and is responsible for vision.

**scaling.** Removal or "flaking" of bark plates in search of prey items.

**scansorial.** Specialized for climbing.

**sexual dichromatism.** Exhibition of different plumage colors or patterns between males and females of the same species.

**sexual dimorphism.** Exhibition of different plumage, shape, or size between males and females of the same species; dichromatism (above) is just one type of dimorphism.

**sexual niche partitioning.** Division of a microhabitat whereby mates feed in different areas or on different substrates, avoiding competition for food.

**species.** Second taxonomic level of binomial nomenclature; taxon within a genus.

**stand-replacement fire.** Fire in which most trees and all layers of the canopy perish; also known as high-severity fire.

**sternal ribs.** Ribs extending distally from the sternum, each connecting to a thoracic rib.

**subspecies.** Closely related taxa within a single species for which the behavioral, morphological, or genetic differences between them are insufficient to separate them into distinct species.

**subspecies group.** One or more subspecies that are more closely related to each other than to the other subspecies of a single species.

**superspecies.** A group of multiple species that are closely related, probably descended relatively recently from a common ancestor, and each of which typically has separate geographic ranges.

**systematics.** The study of phylogenetic relationships among organisms.

**taxon.** A single unit at any level in the taxonomical hierarchy (plural, taxa).

**taxonomy.** The application of systematics to classify organisms in a hierarchical system.

**tertials.** Innermost feathers of the remiges, considered the innermost secondaries, typically seen as covering the "outer" secondaries in a folded wing.

**tribe.** The highest level of phylogentic classification within a subfamily.

**type genus/species/specimen.** The representative taxon or individual bird that inspired the original taxonomic description of that genus or species.

**vestibular window.** Passage to the inner ear covered by a membrane that creates motion waves in the cochlear fluid; smaller in woodpeckers than in other birds.

**xylem.** Parts of a tree trunk inward from the phloem, including the vascular cambium (see cambium), sapwood, and heartwood.

**zygodactyl.** Toe arrangement with toes 2 and 3 pointing forward and toes 1 and 4 pointing backward; generally considered a perching adaptation, and typical of some "true" woodpeckers, especially of the genus Colaptes (flickers); also found in cuckoos, owls, most parrots, and some other birds. See Legs and Feet, p. 7.

# BIBLIOGRAPHY

**Abbreviations for Organizations Referenced Below**
ABC—American Bird Conservancy
AMNH—American Museum of Natural History
AOU—American Ornithologists' Union
PIF—Partners in Flight
USAEC—United States Army Environmental Command
USDA—United States Department of Agriculture
USFS—United States Forest Service
USFWS—United States Fish and Wildlife Service


Abbott, C. E. 1937. Birds in western Texas. *Wilson Bull.* 49:44–45.

Abbott, C. G. 1929. Woodpecker perching on a wire. *Condor* 31:252.

———. 1930. Wire-perching woodpeckers. *Condor* 32:129–130.

Abele, S. C., V. A. Saab, and E. O. Garton. 2004. Lewis's Woodpecker (*Melanerpes lewis*): a technical conservation assessment. Bozeman, MT: USDA Forest Service, Rocky Mtn. Region.

Adams, L. 1941a. Aberrant mating activities of the California Woodpecker. *Condor* 43:68–69.

———. 1941b. Lewis Woodpecker migration. *Condor* 43:119.

Adamus, P. R. 1985. *Atlas of Breeding Birds of Maine, 1978–1983*. Augusta: Maine Dept. of Inland Fisheries and Wildlife.

Adamus, P. R., K. Larsen, G. Gillson, and C. R. Miller. 2001. *Oregon Breeding Bird Atlas*. CD-ROM. Eugene, OR: Oregon Field Ornithithologists.

Adkins, T. R. 1926. The Red-headed Woodpecker occasionally wintering in Alabama. *Wilson Bull.* 38:161.

Aitken, A. E. H., and K. Martin. 2004. Nest cavity availability and selection in aspen-conifer groves in a grassland landscape. *Can. J. For. Res.* 34:2099–2109.

Alaska Department of Fish and Game. 2006. Our Wealth Maintained: A Strategy for Conserving Alaska's Diverse Wildlife and Fish Resources. Juneau: Alaska Dept. of Fish and Game.

Alcorn, J. R. 1988. *The Birds of Nevada*. Fallon, NV: Fairview West Publ.

Alexander, M. P., and K. J. Burns. 2006. Intraspecific phylogeography and adaptive divergence in the White-headed Woodpecker. *Condor* 108:489–508.

Allen, A. A. 1924. Vacationing with birds. *Bird-Lore* 26:208–213.

Allen, A. A., and P. P. Kellogg. 1937. Recent observations on the Ivory-billed Woodpecker. *Auk* 54:164–184.

Allen, A. A., and A. H. Wright. 1918. Sap drinking by sapsuckers and hummingbirds. *Auk* 35:79–80.

Allen, J. A. 1892. The North American species of the genus *Colaptes*, considered with special reference to the relationships of *C. auratus* and *C. cafer*. *Bull. of AMNH* 4:21–44.

Allen, O., B. Barkus, and K. Bennett. 2006. Delaware Wildlife Action Plan: 2007–2117. Dover: Delaware Natural Heritage and Endangered Species Program, Delaware Div. of Fish and Wildlife.

Allert, O. P. 1925. Northern Pileated Woodpecker in Clayton Co., Iowa. *Auk* 42:269–270.

Alsop, F. J., III. 2001. *Birds of North America: Western Region*. Smithsonian Handbooks. New York: DK Publishing.

American Association for the Advancement of Science. 1880. General notes: instrumental substitute for singing in birds. *Science* 1(2):24.

———. 1929. Acorn-storing woodpeckers. *Science* 70:xii.

———. 1939. The Bishop Ornithological Collection. *Science* 90:347.

———. 1942. The Singer Wildlife Refuge. *Science* 96:7–8.

———. 1956. News briefs. *Science* 123:718.

———. 1957. Cedar Creek Forest. *Science* 126:691–692.

American Automobile Association. 2001. North American Road Atlas, 2002 ed. Heathrow, FL: American Automobile Assoc.

American Bird Conservancy. n.d. Partners in Flight Bird Conservation Plan for the East Gulf Coastal Plain (Physiographic Area 4). The Plains, VA: ABC.

———. 2008. The United States WatchList of Birds of Conservation Concern. www.abcbirds.org/abcprograms/science/watchlist/.

American Ornithologists' Union. 1889. Check-list of North American Birds, Abridged Edition, Revised. Washington, D.C.: AOU.

———. 1902. Eleventh supplement to the American Ornithologists' Union check-list of North American birds. *Auk* 19(3):315–342.

———. 1903. Twelfth supplement to the American Ornithologists' Union check-list of North American birds. *Auk* 20(3):331–368.

———. 1908. Fourteenth supplement to the American Ornithologists' Union check-list of North American birds. *Auk* 25(3):343–399.

———. 1912. Sixteenth supplement to the American

Ornithologists' Union check-list of North American birds. *Auk* 29(3):380–387.
———. 1916. Changes in the AOU check-list of North American birds proposed since the publication of the sixteenth supplement. *Auk* 33(4):425–431.
———. 1923. Eighteenth supplement to the American Ornithologists' Union check-list of North American birds. *Auk* 40(3):513–525.
———. 1931. Check-list of North American Birds. 4th ed. Lancaster, PA: AOU.
———. 1947. Twenty-second supplement to the American Ornithologists' Union check-list of North American birds. *Auk* 64(3):445–452.
———. 1949. Twenty-fourth supplement to the American Ornithologists' Union check-list of North American birds. *Auk* 66(3):281–285.
———. 1951. Twenty-sixth supplement to the American Ornithologists' Union check-list of North American birds. *Auk* 68(3):367–369.
———. 1957. Check-list of North American Birds. 5th ed. Washington, D.C.: AOU.
———. 1973. Thirty-second supplement to the American Ornithologists' Union check-list of North American birds. *Auk* 90(2):411–419.
———. 1976. Thirty-third supplement to the American Ornithologists' Union check-list of North American birds. *Auk* 93(4):875–879.
———. 1983. Check-list of North American Birds. 6th ed. Lawrence, KS: AOU.
———. 1985. Thirty-fifth supplement to the American Ornithologists' Union check-list of North American birds. *Auk* 102(3):680–686.
———. 1995. Fortieth supplement to the American Ornithologists' Union check-list of North American birds. *Auk* 112(3):819–830.
———. 1998. Check-list of North American Birds. 7th ed. Lawrence, KS: AOU.
———. 2000. Forty-second supplement to the American Ornithologists' Union check-list of North American birds. *Auk* 117(3):847–858.
———. 2003. Forty-fourth supplement to the American Ornithologists' Union check-list of North American birds. *Auk* 120(3):923–931.
Anderson, A. H. 1934. Food of the Gila Woodpecker. *Auk* 51:84–85.
Anderson, J., coord. 2005. The Arkansas Comprehensive Wildlife Conservation Strategy, Wildlife Action Plan. Rev. 2006. Little Rock: Arkansas Game and Fish Commission.
Anderson, S., and M. McGee. 2001. Region 2 Sensitive Species Evaluation Form: Three-toed Woodpecker *in* USFS 2009, Forest Service Manual—Supplement no. 2600-2009-1, Rocky Mountain Region (Region 2), Denver, CO.
Andres, B. A. 1999. Landbird Conservation Plan for Alaska Biogeographic Regions. Ver. 1.0. Anchorage, AK: Boreal PIF Working Group.
Andrews, R., and R. Righter. 1992. *Colorado Birds.* Denver: Denver Museum of Natural History.
Andrle, R. F., and J. R. Carroll, eds. 1988. *The Atlas of Breeding Birds in New York State.* Ithaca, NY: Cornell Univ. Press.
Antevs, A. 1947. Behavior of the Gila Woodpecker, Ruby-crowned Kinglet, and Broad-tailed Hummingbird. *Condor* 50:91–92.
Anthony, A. W. 1886. Field notes on the birds of Washington County, Oregon. *Auk* 3(2):161–172.
———. 1895a. Birds of San Fernando, lower California. *Auk* 12:134–143.
———. 1895b. New races of *Colaptes* and *Passerella* from the Pacific Coast. *Auk* 12:347–349.
———. 1896. A new subspecies of the genus *Dryobates.* *Auk* 13:31–34.
———. 1897. A new name for *Dryobates v. montanus.* *Auk* 4:54.
Apfelbaum, S., and A. Haney. 1981. Bird populations before and after fire in a Great Lakes pine forest. *Condor* 83:347–354.
Arizona Game and Fish Department. 2006. Arizona's Comprehensive Wildlife Conservation Strategy: 2005–2015. Phoenix: Arizona Game and Fish Dept.
Armstrong, R. H. 1995. *Guide to the Birds of Alaska.* 4th ed. Anchorage: Alaska Northwest Books.
Askins, R. A. 2000. *Restoring North America's Birds: Lessons from Landscape Ecology.* 2d ed. New Haven, CT: Yale Univ. Press.
Attwater, H. P. 1892. List of birds observed in the vicinity of San Antonio, Bexar County, Texas. *Auk* 9(3):229–238.
Aubry, K. B., and C. M. Raley. 1994. Landscape- and stand-level studies of Pileated Woodpeckers: design constraints and stand-level results. *Northwest Sci.* 68:113.
Audubon, J. J., and J. B. Chevalier. 1840. *The Birds of America.* Vol. 4. Reprint, New York: Dover, 1967.
Aududon, M. R. 1897. *Audubon and His Journals.* Reprint, New York: Dover, 1960.
Austin, G. T. 1976. Sexual and seasonal differences in foraging of Ladder-backed Woodpeckers. *Condor* 78:317–323.
Austin, O. L., Jr. 1971. *Families of Birds.* New York: Golden Press.
Avise, J. C., and C. F. Aquadro. 1987. Malate dehydrogenase isozymes provide a phylogenetic marker for the Piciformes (woodpeckers and allies). *Auk* 104:324–328.
Backhouse, F. 2005. *Woodpeckers of North America.* New York: Firefly Books.
Bagg, A. C. 1919. Arctic Three-toed Woodpecker at Southampton, Mass. *Auk* 36:421–422.
Baicich, P. J., and C. J. O. Harrison. 2005. *A Guide to the Nest, Eggs, and Nestlings of North American Birds.* 2d ed. San Diego, CA: Academic Press.
Bailey, A. L. 1953. Eastern race of Yellow-bellied Sapsucker in Colorado. *Condor* 55:219.
Bailey, A. M. 1939. Ivory-billed Woodpecker's beak in an Indian grave in Colorado. *Condor* 41:164.
Bailey, A. M., and R. J. Niedrach. 1938. Nelson's Downy Woodpecker from Colorado. *Auk* 55: 672-673.
Bailey, F. M. 1914. *Handbook of Birds of the Western*

*United States*. 2d ed. Cambridge, MA: Riverside Press.
———. 1923. Birds recorded from the Santa Rita Mountains in southern Arizona. *Pacific Coast Avifauna* no. 15.
Bailey, H. H. 1927. The Ivory-billed Woodpecker in Florida. *Oologist* 44(2):19–20.
Bailey, M. A. 1999. Woodpeckers of Alabama. *Alabama's Treasured Forests*. Winter 1999:28–31.
Bailey, S. W. 1912. Red-headed Woodpecker at Newburyport, Mass. *Auk* 29:541.
Baird, S. F. 1858. Reports of explorations and surveys, to ascertain the most practicable and economical route for a railroad from the Mississippi River to the Pacific Ocean. Vol. IX, Part II. Birds. Washington, D.C.: Government Printing Office.
Baker, J. N. 1975. Egg-carrying by a Common Flicker. *Auk* 92:614–615.
Baker, M. F., and G. O. Hendrickson. 1937. The wintering of a Northern Flicker in central Iowa. *Bird-Banding* 8:114–117.
Baldwin, P. H., and R. W. Schneider. 1963. Flight in relation to form of a wing in the Lewis's Woodpecker. *J. of Col.-Wyo. Acad. of Sci.* 5:58–59.
Baldwin, R. M. 1910. A hybrid flicker in eastern Missouri. *Auk* 27:340–341.
Bales, B. R. 1989. Another unusual laying of the Flicker. *Wilson Bull.* 29(4):188–191.
Ballard, G., R. Burnett, D. Burton, A. Chrisney, L. Comrack, G. Elliott, et al. 2004. The Riparian Bird Conservation Plan: A Strategy for Reversing the Decline of Riparian Associated Birds in California. Ver. 2.0. California PIF and Riparian Habitat Joint Venture. Stinson Beach, CA: Point Reyes Bird Observatory.
Bangs, O. 1898. Some new races of birds from eastern North America. *Auk* 15:173–183.
———. 1900. A review of the Three-toed Woodpeckers of North America. *Auk* 17:126–142.
Barlow, C. 1900. Brewer's Blackbird nesting in cavities. *Auk* 2:18.
Barron, A. D. 2001. *A Birdfinding Guide to Del Norte County, California*. Crescent City, CA: Redwood Economic Development Inst.
Batchelder, C. F. 1889. An undescribed subspecies of *Dryobates pubescens*. *Auk* 6:253–255.
Bate, L. J. 2001. Woodpeckers and their habitat use in the Columbia River Basin Region: A Literature Review. Kalispell, MT: Interior Columbia Basin Ecosystem Management Science Team.
Batts, H. L. 1953. Siskin and goldfinch feeding at sapsucker tree. *Wilson Bull.* 65:198.
Baumel, J. J., A. S. King, J. E. Breazile, H. E. Evans, and J. C. Vanden Berge. 1993. *Handbook of Avian Anatomy: Nomina Anatomica Avium*. 2d ed. Publication of the Nuttall Ornithological Club no. 23.
Bayer, R. 1995a. Menu of June–December 1995 bird field notes. *The Sandpiper* 95:81. Yaquina Birders and Naturalists, Newport, OR.
———. 1995b. Semimonthly bird records through 1992 for Lincoln County, Oregon; Part II: Records sorted by species. *J. of Ore. Ornith.* 4:395–543.
Beal, F. E. L. 1895. Preliminary report on the food of woodpeckers. *USDA Division of Ornithology and Mammalogy Bull.* no. 7: 7–33.
———. 1911. Food of the Woodpeckers of the United States. *USDA Biol. Survey Bull.* no. 37.
Beaton, G., P. W. Sykes, Jr., and J. W. Parrish, Jr. 2003. *Annotated Checklist of Georgia Birds*. 5th ed. Occasional Publ. no. 14. Georgia Ornithological Society.
Beckwith, R. C., and E. L. Bull. 1985. Scat analysis of the arthropod component of Pileated Woodpecker diet. *Murrelet* 66:90–92.
Bedell, E. 1924. Pileated Woodpecker in Helderberg Mts. N.Y. *Auk* 41:602–603.
Beedy, E. C., and S. L. Granholm. 1985. *Discovering Sierra Birds*. Yosemite Natural History Assoc. and Sequoia Natural History Assoc., San Francisco, CA, and Three Rivers, CA.
Behle, W. H. 1976. Mohave Desert avifuna in the Virgin River valley of Utah, Nevada, and Arizona. *Condor* 78:40–48.
Beidleman, C. A. 2000. Partners in Flight Landbird Conservation Plan: Colorado. Ver. 1.0. Estes Park, CO: PIF.
Bender, S., S. Shelton, K. C. Bender, and A. Kalmbach, eds. 2005. Texas Comprehensive Wildlife Conservation Strategy: 2005–2010. Austin: Texas Parks and Wildlife Dept.
Bendire, C. E. 1888. Notes on the habits, nests, and eggs of the genus *Sphyrapicus* Baird. *Auk* 5:225–240.
———. 1889. *Sphyrapicus ruber* breeding in coniferous trees. *Auk* 6:71.
Benítez-Díaz, H. 1993. Geographic variation in coloration and morphology of the Acorn Woodpecker. *Condor* 95:63–71.
Benson, K. L. P., and K. A. Arnold. 2001. *The Texas Breeding Bird Atlas*. College Station and Corpus Christi: Texas A&M Univ. System.
Bent, A. C. 1939. Life Histories of North American Woodpeckers. *U.S. National Museum Bull.* no. 174.
Benz, B. W., M. R. Robbins, and A. T. Peterson. 2006. Evolutionary history of woodpeckers and allies (Aves: Picidae): placing key taxa on the phylogenetic tree. *Molec. Phylogen. and Evol.* 40:389–399.
Berchok, C. 2005. Never a dull moment. *BirdScope* 19(3):13.
Bevier, L. R. 1990. Eleventh report of the California Bird Records Committee. *Western Birds* 21:145–176.
———. 1994. *The Atlas of Breeding Birds of Connecticut*. State Geological and Natural History Survey of Connecticut bull. 113. Hartford: Connecticut Dept. of Environmental Protection.
Beyer, G. E. 1900. The Ivory-billed Woodpecker in Louisiana. *Auk* 17:97–99.
Bezener, A. M., K. De Groot, W. Easton, I. Hartasanchez, and S. Pelech. 2003. Canada's Great Basin Landbird Conservation Plan, Version 1.0. Delta, BC: PIF British Columbia and Yukon.
Bird Studies Canada. 2008. *Maritimes Breeding Bird Atlas*. Sackville, NB: Bird Studies Canada.
Blackford, J. L. 1955. Woodpecker concentration in burned forest. *Condor* 57:28–30.

Blake, F. G., and M. C. Blake. 1902. A winter record for the Flicker (*Colaptes auratus luteus*) in Berkshire County. *Auk* 19:199.

Blancher, P., and J. Wells. 2005. The Boreal Forest Region: North America's Bird Nursery. Ottawa, ON: Canadian Boreal Initiative, and Seattle, WA: Boreal Songbird Initiative.

Blewett, C. M., and J. M. Marzluff. 2005. Effects of urban sprawl on snags and the abundance and productivity of cavity-nesting birds. *Condor* 107:678–693.

Block, W. M. 1991. Foraging ecology of Nuttall's Woodpecker. *Auk* 108:303–317.

Blood, D. A. 1997. White-headed Woodpecker: Restricted range and dramatic habitat change makes this bird threatened in British Columbia. Victoria: British Columbia Ministry of Environment, Lands, and Parks, Wildlife Branch.

Blow, K. L., L. W. Clark, A. G. Gubanich, and D. McNinch, eds. 2000. *A Birding Guide to Reno and Beyond.* Reno, NV: Lahontan Audubon Society.

Bock, C. E. 1970. The ecology and behavior of the Lewis Woodpecker. *Univ. Calif. Publ. Zool.* no. 92.

———. 1971. Pairing in hybrid flicker populations in eastern Colorado. *Auk* 88:921–924.

———. 1999. Functional and evolutionary morphology of woodpeckers. *Ostrich* 70(1):23–31.

Bock, C. E., and W. M. Block. 2005a. Fire and birds in the southwestern United States. Pp. 1–19, in V. A. Saab and D. W. Powell, eds., Fire and avian ecology in North America. *Studies in Avian Biology* no. 30.

———. 2005b. Response of Birds to Fire in the American Southwest. Gen. Tech. Rept. PSW-GTR-191:1093–1099. Albany, CA: USDA Forest Service, Pacific Southwest Research Station.

Bock, C. E., and J. Bock. 1974. On the geographic ecology and evolution of the Three-toed Woodpeckers, *Picoides tridactylus* and *P. arcticus*. *Am. Midland Naturalist* 92:397–405.

Bock, C. E., H. H. Hadow, and P. Somers. 1971. Relations between Lewis' and Red-headed Woodpeckers in southeastern Colorado. *Wilson Bull.* 83:237–248.

Bock, C. E., and D. L. Larson. 1986. Winter habitats of sapsuckers in southeastern Arizona. *Condor* 88:246–247.

Bock, C. E., and J. F. Lynch. 1970. Breeding populations of burned and unburned conifer forest in the Sierra Nevada. *Condor* 72:182–189.

Bock, W. J., and W. D. Miller. 1959. The Scansorial Foot of the Woodpeckers, with Comments on the Evolution of Perching and Climbing Feet in Birds. American Museum Novitates no. 1931.

Bolander, L. P. 1914. The Lewis Woodpecker nesting in Alameda County, California. *Condor* 16:183.

———. 1930. Is the Lewis Woodpecker a regular breeder in the San Francisco region? *Condor* 32:263–264.

Bolles, F. 1891. Yellow-bellied Woodpeckers and their uninvited guests. *Auk* 8:256–270.

———. 1892. Young sapsuckers in captivity. *Auk* 9:1–11.

———. 1893. Scars on apple tree trunks. *Science* 22:217.

Bond, R. M. 1957. A second record of the Yellow-bellied Sapsucker from St. Croix, Virgin Islands. *Condor* 59:211–212.

Borelli, A. 1680. *De Motu Animalium*. Opus Postbumum, Pars Prima. Rome: Superiorum Permissu.

Borror, D. J. 1960. *Dictionary of Word Roots and Combining Forms.* Mountain View, CA: Mayfield Publ. Co.

Bower, A. R., and D. J. Ingold. 2004. Intraspecific brood parasitism in the Northern Flicker. *Wilson Bull.* 116:94–97.

Boyne, N. 2008. Army installations a refuge for struggling woodpecker. *Environmental Update* [USAEC] Summer 2008. http://www.army.mil/article/9199/.

Brackbill, H. 1942. Dusting Flickers. *Wilson Bull.* 54:250.

———. 1953. Notes on the drumming of some woodpeckers. *J. of Field Ornith.* 24(1):18.

———. 1955. Possible function of the flicker's black breast crescent. *Auk* 72:205.

———. 1957. Observations on a wintering flicker. *Bird-Banding* 28:40–41.

———. 1969a. Red-bellied taking bird's eggs. *Bird-Banding* 40:323–324.

———. 1969b. Reverse mounting by the Red-headed Woodpecker. *Bird-Banding* 40:255–256.

Brackett, F. H. 1896. The Redheaded Woodpecker in eastern Massachusetts. *Auk* 13:258.

Brauning, D. W. 1992. *Atlas of Breeding Birds of Pennsylvania.* Pittsburgh, PA: Univ. of Pittsburgh Press.

Brenowitz, G. L. 1978a. An analysis of Gila Woodpecker vocalizations. *Wilson Bull.* 90:451–455.

———. 1978b. Gila Woodpecker agonistic behavior. *Auk* 95:49–58.

Brewer, R., G. A. McPeek, and R. J. Adams, Jr. 1991. *The Atlas of Breeding Birds of Michigan*. East Lansing: Michigan State Univ. Press.

Brewer, T. M. 1853. *Wilson's American Ornithology*. New York: H. S. Samuels.

Brewster, W. 1884. Recent occurrence of the Black-backed Three-toed Woodpecker in Massachusetts. *Auk* 1:93.

———. 1886. An ornithological reconnaissance in western North Carolina. *Auk* 3(1):94–112.

———. 1889. *Melanerpes carolinus* eating oranges. *Auk* 6:337–338.

———. 1890. Breeding of the Pileated Woodpecker in Worcester County, Massachusetts. *Auk* 7:400–401.

———. 1893. A brood of young Flickers (*Colaptes auratus*) and how they were fed. *Auk* 10:231–236.

———. 1897. On the nomenclature of certain forms of the Downy Woodpecker (*Dryobates pubescens*). *Auk* 14:80–82.

———. 1898. Lewis's Woodpecker storing acorns. *Auk* 15:188.

Bridge, L. E. 1905. Northern Pileated Woodpecker in Massachusetts. *Auk* 22:414.

Bright, D. E. 2014. A Catalog of Scolytidae and Platypodidae (Coleoptera), supplement 3 (2000–2010), with notes on subfamily and tribal reclassifications. *Insecta Mundi* paper 861. http://digitalcommons.unl.edu/insetcamundi/861.

Brimble, L. J. F., and A. J. V. Gale, eds. 1941. News and Views, Uses of wood in warfare. *Nature* 148(3745):161.

British Columbia Conservation Data Centre. 2008. BC Species and Ecosystems Explorer. http://a100.gov.bc.ca/pub/eswp/.

Brock, K. J. 2006. *Brock's Birds of Indiana*. CD-ROM. Chesterton, IN: Indiana Audubon Society.

Brodkorb, P. 1928. Flicker trapped by resin. *Auk* 45:503–504.

———. 1942. The subspecific status of Michigan Flickers. *Wilson Bull.* 54:50–51.

Brooks, A. 1944. A deplumed Pileated Woodpecker. *Condor* 46:124.

Brooks, M. 1934. An unusual Red-headed Woodpecker accident. *Auk* 51:379.

Brown, C. 1987. Avian Sociobiology. Review of *Population Ecology of the Cooperatively Breeding Acorn Woodpecker*, by W. D. Koenig and R. L. Mumme. *Science* 238:4833.

Brown, H. 1902. Unusual abundance of Lewis's Woodpecker near Tucson, Arizona, in 1884. *Auk* 19:80–83.

———. 1904. The Elf Owl in California. *Condor* 6:45–47.

Brown, S., J. Crocker, C. Dillingham, T. Mickel, J. Rogers, B. Stotz, et al. 1996. *Birding the Southern Oregon Coast*. Coos Bay, OR: So. Coast Printing.

Brown, W. S., W. S. Kordek, K. Leo, B. McDonald, B. Sargent, and J. Wykle. 2005. It's About Habitat: West Virginia Wildlife Conservation Action Plan. South Charleston: West Virginia Div. of Natural Resources, Wildlife Resources Sect. www.wvdnr.gov/Wildlife/PDFFiles/wvwcap.pdf.

Browning, M. R. 1977. Interbreeding members of the *Sphyrapicus varius* group (Aves: Picidae) in Oregon. *Bull. of So. Cal. Acad. of Sci.* 76(1):38–41.

———. 1995. Do Downy Woodpeckers migrate? *J. of Field Ornith.* 66:12–21.

———. 2003. The generic distinction of pied woodpeckers. *Western Birds* 34:97–107.

Browning, M. R., and S. P. Cross. 1994. Third specimen of Nuttall's Woodpecker (*Picoides nuttalli*) in Oregon from Jackson County and comments on earlier records. *Oregon Birds* 20:119–120.

Browning, M. R., and W. English. 1967. Possible Yellow-shafted Flicker in southwestern Oregon. *Condor* 69:210.

Bryan, A., and M. Sarell. 2006. Restoring habitat for the White-headed Woodpecker. Penticton, BC: So. Okanagan-Similkameen Stewardship Prog.

Bryan, W. A. 1899. *Melanerpes erythrocephalus* wintering in Chicago. *Auk* 16:272–273.

Bryant, H. C. 1916. A note on the food of the Northern Pileated Woodpecker. *Condor* 18:32.

———. 1921. California Woodpecker steals eggs of Wood Pewee. *Condor* 23:33.

Bryant, W. L. 1929. Lewis' Woodpecker in Rhode Island. *Auk* 46:113–114.

Bryens, O. M. 1926. Actions of the Northern Pileated Woodpecker. *Auk* 43:98.

———. 1929. The American Three-toed Woodpecker in Luce County, Michigan. *Auk* 46:239–240.

Buchanan, J. B., R. E. Rogers, D. J. Pierce, and J. E. Jacobson. 2003. Nest-site habitat use by White-headed Woodpeckers in the eastern Cascade Mountains, Washington. *Northwestern Naturalist* 84:119–128.

Buckelew, A. R., Jr., and G. A. Hall. 1994. *The West Virginia Breeding Bird Atlas*. Pittsburgh, PA: Univ. of Pittsburgh Press.

Bull, E. L. 1987. Ecology of the Pileated Woodpecker in northeastern Oregon. *J. of Wildl. Mgmt.* 51:472–481.

———. 2001. Survivorship of Pileated Woodpeckers in northeastern Oregon. *J. of Field Ornith.* 72:131–136.

Bull, E. L., A. A. Clark, and J. F. Shepherd. 2005. Short-term effects of fuel reduction on Pileated Woodpeckers in northeastern Oregon: a pilot study. Research paper PNW-RP-564. Portland, OR: USDA Forest Service, Pacific Northwest Res. Sta.

Bull, E. L., J. W. Deal, and J. E. Hohmann. 2001. Avian and amphibian use of fenced and unfenced stock ponds in northeastern Oregon forests. Research paper PNW-RP-539. Portland, OR: USDA Forest Service, Pacific Northwest Res. Sta.

Bull, E. L., and R. S. Holthausen. 1993. Habitat use and management of Pileated Woodpeckers in northeastern Oregon. *J. of Wildl. Mgmt.* 57:335–345.

Bull, E. L., R. S. Holthausen, and M. G. Henjum. 1992. Roost trees used by Pileated Woodpeckers in northeastern Oregon. *J. of Wildl. Mgmt.* 56:786–793.

Bull, E. L., and J. E. Jackson. 1995. Pileated Woodpecker (*Dryocopus pileatus*). No. 148 in *The Birds of North America*, A. Poole and F. Gill, eds. Philadelphia, PA: Acad. of Nat. Sci., and Washington, D.C.: AOU.

Bull, E. L., and C. E. Meslow. 1977. Habitat requirements of the Pileated Woodpecker in northeastern Oregon. *J. Forestry* 75:335–337.

Bull, E. L., N. Nielsen-Pincus, B. C. Wales, and J. L. Hayes. 2007. The influence of disturbance events on Pileated Woodpeckers in northeast Oregon. *Forest Ecol. and Mgmt.* 243:320–329.

Bull, J. 1964. *Birds of the New York Area*. New York: Harper and Row.

Bunn, D., A. Mummert, M. Hoshovsky, K. Gilardi, and S. Shanks. 2007. California Wildlife: Conservation Challenges. California's Wildlife Action Plan. Sacramento: California Dept. of Fish and Game.

Bunnell, F. L., E. Wind, and R. Wells. 2002. Dying and Dead Hardwood: Their Implications and Management. Gen. Tech. Rept. PSW-GTR-181. Albany, CA: USDA Forest Service, Pacific Southwest Research Station.

Burchsted, A. E. 1987. Downy Woodpecker caches food. *Wilson Bull.* 99:136–137.

Bureau of Land Management. 2002. Wee Thump Joshua Tree Wilderness Fact Sheet. BLM, Southern Nevada Dist. Office, Las Vegas, NV.

Burleigh, T. D. 1972. *Birds of Idaho*. Caldwell, ID: Caxton Printers.

Burns, F. L. 1900. A monograph of the flicker. *Wilson Bull.* 7(2):1–82.

———. 1901. A few additional notes on the Flicker. *Wilson Bull.* 42:24–26.

———. 1910. Additional vernacular names of the Flicker. *Wilson Bull.* 22:55.
———. 1915. Comparative periods of deposition and incubation of some North American birds. *Wilson Bull.* 27(1):275–286.
———. 1916. One hundred and thirty-two vernacular names for the flicker. *Wilson Bull.* 95:90–91.
Burns, J. 2004. Arizona's Special Species: Gilded Flicker. *Cactus Wren-Dition* 52(1):6. Maricopa Audubon Society.
Burr, F. 1926. Woodpeckers and the automobile. *Science* 63(1638):524.
Burrows, R. 1988. *Birding in Atlantic Canada: Nova Scotia.* St John's, NL: Jesperson Press, Ltd.
Burt, W. H. 1930. Adaptive modifications in the woodpeckers. *Univ. Calif. Publ. Zool.* 32:455–524.
Burtch, V. 1923. Red squirrel eating young Hairy Woodpecker. *Auk* 40:340–341.
Burton, P. J. K. 1984. Anatomy and evolution of the feeding apparatus in the avian orders Coraciiformes and Piciformes. *Bull. Br. Mus. Nat. Hist. Zool.* 47(6):331–439.
Busby, W. H., and J. L. Zimmerman. 2001. *Kansas Breeding Bird Atlas.* Lawrence: Univ. Press of Kansas.
Bush, P. G. 2001. *Influence of Landscape-Scale Forest Structure on the Presence of Pileated Woodpeckers,* Dryocopus pileatus, *in Central Ontario Forests.* Ottawa, ON: National Library of Canada.
Butcher, L. R., S. A. Fleury, and J. M. Reed. 2002. Orientation and vertical distribution of Red-naped Sapsucker (*Sphyrapicus nuchalis*) nest cavities. *Western North American Naturalist* 62:365–369.
Cadman, M. D., P. F. J. Eagles, and F. M. Helleiner, eds. 1987. *Atlas of the Breeding Birds of Ontario 1981–1985.* Toronto and Port Rowan: Federation of Ontario Naturalists and Long Point Bird Observatory.
Cadman, M. D., D. A. Sutherland, G. G. Beck, D. Lepage, and A. R. Couturier, eds. 2008. *Atlas of the Breeding Birds of Ontario 2001–2005.* Toronto: Ontario Nature.
California Automobile Association. 1991. Lake Tahoe Region, Map. Emeryville and Los Angeles: California State Automobile Assoc. and Automobile Club of Southern California.
Campbell, R. W., N. K. Dawe, I. McTaggart-Cowan, J. M. Cooper, G. W. Kaiser, and M. C. E. McNall. 1990. *The Birds of British Columbia.* Vol. 2. Victoria: Royal British Columbia Museum.
Canadian Wildlife Service. 2006. NWT/Nunavut Bird Checklist Survey Newsletter, May 2006. Yellowknife, NT: NWT/Nunavut Bird Checklist Survey.
———. 2008a. Northwest Territories/Nunavut Bird Checklist Program. Yellowknife, NT: NWT/Nunavut Bird Checklist Survey.
———. 2008b. NWT/Nunavut Bird Checklist Survey Newsletter, Spring 2008. Yellowknife, NT: NWT/Nunavut Bird Checklist Survey.
Cannings, R. J. 1995. Status of the White-headed Woodpecker in British Columbia. Wildlife bull. no. B-80. Victoria: British Columbia Ministry of Environment, Lands, and Parks.
———. 1998. The birds of British Columbia: a taxonomic catalogue. Wildlife bull. no. B-86. Victoria: Ministry of Environment, Lands, and Parks.
Capen, F. M. 1926. Arctic Three-toed Woodpecker in Winchendon, Mass. *Auk* 43:545–546.
Cardiff, E. A. 1972. Red-headed Woodpecker in the Imperial Valley of California. *California Birds* 3:23–24. San Diego, CA: Western Field Ornithologists.
Carpenter, N. K. 1919. Luck. *Condor* 21:235–236.
Carriger, H. W., and G. Wells. 1919. Nesting of the Northern Pileated Woodpecker. *Condor* 21:153.
Carter, T. D. 1942. Northern Pileated Woodpecker eating salmon. *Auk* 59:585.
Casey, D. 2000. Partners in Flight Draft Bird Conservation Plan: Montana. Ver. 1. Kalispell, MT: Montana Fish, Wildlife and Parks.
Cassin, J. 1858. *United States Exploring Expedition during the Years 1838, 1839, 1840, 1841, 1842 under the Command of Charles Wilkes, U.S.N.: Mammalogy and Ornithology.* Philadelphia, PA: J. B. Lippencott.
Cassirer, E. F. 1993. Cavity nesting by Harlequin Ducks in the Pacific Northwest. *Wilson Bull.* 105(4):691–694.
Cecil, R. 2007. The regional reports: Iowa and Missouri. *North American Birds* 61(2):271.
Cerovski, A. O., M. Grenier, B. Oakleaf, L. Van Fleet, and S. Patla. 2004. *Atlas of Birds, Mammals, Amphibians, and Reptiles in Wyoming.* Lander: Wyoming Game and Fish Dept.
Chamberlain, M. 1884. A woodpecker destroying cocoons. *Auk* 1:93.
———. 1891. *A Popular Handbook of the Ornithology of the United States and Canada, Based on Nuttall's Manual: Volume I. The Land Birds.* Boston: Little, Brown.
Chambers, J. A. 1979. Unusual feeding behavior of a Hairy Woodpecker. *Bird-Banding* 50:365.
Chapin, J. P. 1921. The abbreviated inner primaries of nestling woodpeckers. *Auk* 38:531–552.
Chapman, F. M. 1891. On the color-pattern of the upper tail-coverts in *Colaptes auratus. Bull. of AMNH* 3:311–314.
———. 1930. Notes on the plumage of North American birds. *Bird-Lore* 32:265–267.
Charif, R. 2005. Listening for Ivory-bills, the high-tech way. *BirdScope* 19(3):11–13.
Choate, E. A. 1985. *The Dictionary of American Bird Names.* Rev. ed. Harvard, MA: Harvard Common Press.
Chu, M. 2005. Ivory-bill evidence from sound and video. *BirdScope* 19(4):4–5.
Cicero, C., and N. K. Johnson. 1995. Speciation in sapsuckers (*Sphyrapicus*): III. Mitochondrial-DNA sequence divergence at the cytochrome-B locus. *Auk* 112:547–563.
Clabaugh, B. D. 1928. Bird casualties due to automobiles. *Condor* 30:157.
Claypole, B. 2004. *Klamath River Bird Finder.* Klamath River, CA: Living Gold Press.
Clement, R. C. 1943. American Three-toed Woodpecker in Massachusetts. *Auk* 60:106.

Clements, J. F. 2007. *The Clements Checklist of Birds of the World.* 6th ed. Ithaca, NY: Cornell Univ. Press.
Clevenger, S. V. 1881. Contributions to comparative psychology. II. Language. *Science* 2(56):342–344.
Cole, F. J. 1949. *A History of Comparative Anatomy: From Aristotle to the Eighteenth Century.* Reprint, New York: Dover, 1976.
Collinson, J. M. 2007. Video analysis of the escape flight of Pileated Woodpecker *Dryocopus pileatus*: does the Ivory-billed Woodpecker *Campephilus principalis* persist in continental North America? *BMC Biology* 5:1741–1748.
Colorado Division of Wildlife. 2006. Colorado's Comprehensive Wildlife Conservation Strategy and Wildlife Action Plans. Denver: Colorado Div. of Wildlife.
Committee on the Status of Endangered Wildlife in Canada. 2001. COSEWIC assessment and status report on the Lewis's Woodpecker *Melanerpes lewis* in Canada. Ottawa, ON: Comm. on the Status of Endangered Wildlife in Canada.
———. 2005. COSEWIC assessment and status report on the Williamson's Sapsucker *Sphyrapicus thyroideus* in Canada. Ottawa, ON: Comm. on the Status of Endangered Wildlife in Canada.
Confer, J. L., and P. Paicos. 1985. Downy Woodpecker predation of goldenrod galls. *J. of Field Ornith.* 56:56–64.
Connecticut Department of Environmental Protection. 2005. Connecticut's Comprehensive Wildlife Conservation Strategy. Hartford: Connecticut Bureau of Natural Resources.
Conner, R. N. 1973. Eastern Screech Owl displaces nesting Pileated Woodpeckers. *J. of Field Ornith.* 44:316.
———. 1974. Red-bellied predation on nestling Carolina Chickadees. *Auk* 91:836.
———. 1975. Orientation of entrances to woodpecker nest cavities. *Auk* 92:371–374.
———. 1976. Nesting habitat for Red-headed Woodpeckers in southwestern Virginia. *Bird-Banding* 47:40–43.
———. 1977. The effect of tree hardness on woodpecker nest orientation. *Auk* 94:369–370.
Conner, R. N., and C. S. Adkisson. 1977. Principal component analysis of woodpecker nesting habitat. *Wilson Bull.* 89:122–129.
Conner, R. N., and J. C. Kroll. 1979. Food-storing by Yellow-breasted Sapsuckers. *Auk* 96:195.
Conner, R. N., and D. C. Rudolph. 1989. Red-cockaded Woodpecker colony status and trends on the Angelina, Davy Crockett, and Sabine National Forests. Res. Paper so-250. New Orleans, LA: USDA Forest Service, Southern Forest Exp. Sta.
Conner, R. N., D. C. Rudolph, and J. R. Walters. 2001. *The Red-cockaded Woodpecker. Surviving in a Fire-Maintained Ecosystem.* Austin: Univ. of Texas Press.
Conner, R. N., D. Saenz, and D. C. Rudolph. 2004a. The Red-cockaded Woodpecker: interactions with fire, snags, fungi, rat snakes and Pileated Woodpeckers. *Tex. J. Sci.* 56:415–426.
———. 2006. Population trends of Red-cockaded Woodpeckers in Texas. *Bull. Texas Ornith. Soc.* 39:42–48.
Conner, R. N., D. Saenz, R. R. Schaefer, J. R. McCormick, D. C. Rudolph, and D. B. Burt. 2004b. Group size and nest success in Red-cockaded Woodpeckers in the West Gulf coastal plain: helpers make a difference. *J. of Field Ornith.* 75:74–78.
Conner, R. N., C. E. Shackelford, D. Saenz, and R. R. Schaefer. 2001. Interactions between nesting Pileated Woodpeckers and Wood Ducks. *Wilson Bull.* 113:250–253.
Constanz, G. D. 1974. Robbing of breeding Lewis's Woodpecker food stores. *Auk* 91:171.
Contreras, A. 1998. *Birds of Coos County, Oregon: Status and Distribution.* Oregon Field Ornithologists Special Publication no. 12.
———. 2006. *Birds of Lane County Oregon.* Corvallis: Oregon State Univ. Press.
Cook, O., and A. R. Phillips. 1917. Pileated Woodpecker. *Wilson Bull.* 99:106–107.
Cooper, J. M. 1995. Status of the Williamson's Sapsucker in British Columbia. Wildlife Working Rept. no. WR-69. Victoria: British Columbia Ministry of Environment, Lands, and Parks.
Cooper, J. M., and S. Beauchesne. 2000. Inventory of Lewis's Woodpecker breeding population and habitat in the East Kootenay. Wildlife Working Rept. no. WR-100. Victoria: British Columbia Ministry of Environment, Lands, and Parks.
Cooper, J. M., and C. Gillies. 2000. Breeding Distribution of the Lewis's Woodpecker in the East Kootenay Trench in relation to fire history. Pp. 423–428 in L. M. Darling, ed., Proceedings of a Conference on the Biology and Management of Species and Habitats at Risk, Kamloops, BC, February 1999. Vol. 1. Victoria: British Columbia Ministry of Environment, Lands, and Parks, and Kamloops, BC: Univ. College of the Cariboo.
Cooper, J. M., C. Siddle, and G. Davidson. 1998. Status of the Lewis' Woodpecker (*Melanerpes lewis*) in British Columbia. Wildlife Working Rept. no. WR-91. Victoria: British Columbia Ministry of Environment, Lands, and Parks.
Copeland, H. W. 1926. Red-headed Woodpecker nesting in Maine. *Auk* 43:544–545.
Coppa, J. B. 1960. Sapsuckers breeding in the Hualapai Mountains, Arizona. *Condor* 62:294.
Corben, C. 1999. Hybrid woodpecker. Frontiers of field identification. National Birding Hotline Cooperative. University of Arizona Listserve.
Cordle, S. 2006. Voice of the naturalist. Hotline report Apr. 4, 2006. Audubon Naturalist Soc. of the Atlantic States, Chevy Chase, MD. www.freelists.org/archives/va-bird/04-2006/msg00057.html.
Corman, T. E., and C. Wise-Gervais. 2005. *Arizona Breeding Bird Atlas.* Albuquerque: Univ. of New Mexico Press.
Cornell Lab of Ornithology. n.d. Identifying and Reporting an Ivory-billed Woodpecker. [Brochure.] Ithaca, NY: Cornell Lab of Ornithology.

Costa, R. 2002. Red-cockaded Woodpecker. [Brochure.] Clemson, SC: USFWS.
———. 2006. Red-cockaded Woodpecker (*Picoides borealis*) 5-Year Review: Summary and Evaluation. Clemson, SC: USFWS.
Costa, R. N., C. S. Collins, D. Saenz, T. Trees, R. R. Schaefer, and D. C. Rudolph, eds. 2004. *Red-cockaded Woodpecker: Road to Recovery*. Blaine, WA: Hancock House Publishers.
Coston, C. 1999. Changing Bay Area bird populations II—woodpeckers. *Stilt* 18(1):4–5.
Coues, E. 1872. *Key to North American Birds*. Boston: Estes and Lauriat.
———. 1884. *Key to North American Birds*. 2d ed. Boston: Estes and Lauriat.
———. 1892. Original description of Lewis's Woodpecker. *Auk* 9:394.
Covel, P. F. 1936. Occurrence of the Red-naped Sapsucker in Santa Cruz County. *Condor* 38:87.
Covert, K. A. 2003. Hairy Woodpecker Winter Ecology Following Wildfire: Effects of Burn Severity and Age. Master's thesis, Northern Arizona Univ.
Cowan, I. M. 1938. Distribution of the races of the Williamson Sapsucker in British Columbia. *Condor* 40:128–129.
Cox, U. O. 1902. The Pileated Woodpecker. *Auk* 19:288–290.
Cracraft, J., and J. J. Morony, Jr. 1969. A New Pliocene Woodpecker, with Comments on the Fossil Picidae. American Museum Novitates no. 2400.
Crenshaw, F. 1954. Red-cockaded Woodpecker banded in Rome, Ga. *J. of Field Ornith.* 25:112.
Crewe, T., and D. Badzinski. 2006. Red-shouldered Hawk and spring woodpecker survey, 2005 final report. Port Rowan, ON: Ontario Ministry of Natural Resources and Bird Studies Canada.
Crockett, A. B., and H. H. Hadow. 1975. Nest site selection by Williamson's and Red-naped sapsuckers. *Condor* 77:365–368.
Crockett, A. B., and P. L. Hansely. 1978. Apparent response of *Picoides* woodpeckers to outbreaks of the pine bark beetle. *Western Birds* 9:67–70.
Cruz, A., and D. W. Johnston. 1979. Occurrence and feeding ecology of the Common Flicker on Grand Cayman Island. *Condor* 81:370–375.
Culbertson, A. E. 1936. Abnormal mandible in a Flicker. *Condor* 38:124.
Currier, E. S. 1928. Lewis's Woodpeckers nesting in colonies. *Condor* 30:356.
Curry, J. R. 1969. Red-bellied Woodpecker feeds Tufted Titmouse. *Wilson Bull.* 81:470.
Daily, G. C. 1993. Heartwood decay and vertical distribution of Red-naped Sapsucker nest cavities. *Wilson Bull.* 105:674–679.
Daily, G. C., P. R. Ehrlich, and N. M. Haddad. 1993. Double keystone bird in a keystone species complex. *Proc. Nat. Acad. Sci.* 90:592–594.
Dales, L. 1956. Mobbing of an attacking Scrub Jay by a Mockingbird and Red-shafted Flicker. *Condor* 58:446–447.
Danforth, C. G. 1938. Some feeding habits of the Red-breasted Sapsucker. *Condor* 40:219–224.
Daniel, J. W. 1901. Nesting of the Hairy Woodpecker near Washington, D.C. *Auk* 18:272.
Davidson, G. S. 1999. B. C. field ornithologists bird records committee report for 1996–1997. *BC Birds* vol. 9.
Davis, B. L. 1997. *A Field Guide to Birds of the Desert Southwest*. Houston, TX: Gulf Publ. Co.
Davis, J. 1965. Natural history, variation, and distribution of the Strickland's Woodpecker. *Auk* 82:537–590.
Davis, J., and B. S. Davis. 1959. Red-naped Sapsucker in Monterey County California. *Condor* 61:59.
Davis, J., and T. R. Howell. 1951. A record of *Sphyrapicus varius varius* for California. *Condor* 53:102.
Dawson, R. W. 1921. Fall migration in northwestern Nebraska in 1920. *Wilson Bull.* 33(1):35–37.
Dawson, W. L. 1923. *The Birds of California: A Complete, Scientific and Popular Account of the 580 Species and Subspecies of Birds Found in the State.* Vol. 2. San Diego, CA: South Moulton Co.
Delacour, J. 1951. The significance of the number of toes in some woodpeckers and kingfishers. *Auk* 68:49–51.
DeLotelle, R. S., and R. J. Epting. 1992. Reproduction of the Red-cockaded Woodpecker in central Florida. *Wilson Bull.* 104:285–294.
DeLotelle, R. S., R. J. Epting, and J. R. Newman. 1987. Habitat use and territory characteristics of Red-cockaded Woodpeckers in central Florida. *Wilson Bull.* 99:202–217.
Dennis, J. V. 1964. Woodpecker damage to utility poles: with special reference to the role of territory and resonance. *Bird-Banding* 35:225–253.
———. 1967. Damage by Golden-fronted and Ladder-backed Woodpeckers to fence posts and utility poles in south Texas. *Wilson Bull.* 79:75–88.
———. 1969. The Yellow-shafted Flicker (*Colaptes auratus*) on Nantucket Island, Massachusetts. *Bird-Banding* 40:290–308.
Denny, M. 2001. Female Three-toed Woodpecker utilizes "conking" behavior in the Blue Mountains of Umatilla County, Oregon. *Oregon Birds* 27(3):72–73.
Dettmers, R., and K. Rosenberg. 2000. Partners in Flight Landbird Conservation Plan: Physiographic Area 9, Southern New England. Ver. 1.0. The Plains, VA: ABC.
———. 2003. Partners in Flight Landbird Conservation Plan, Lower Great Lakes Plain (Physiographic Area 15) Ver. 1.1. The Plains, VA: ABC.
Diamond, J. R. 1971. Comparison of faunal equilibrium turnover rates on a tropical island and a temperate island. *Proc. Nat. Acad. Sci.* 68:2742–2745.
Dickinson, M. B., ed. 2002. *Field Guide to the Birds of North America*. 4th ed. Washington, D.C.: National Geographic Society.
Dill, H. R. 1926. Is the automobile exterminating the woodpecker? *Science* 63:69–70.

Dinsmore, J. J., T. H. Kent, D. Koenig, P. C. Petersen, and D. M. Roosa. 1984. *Iowa Birds.* Ames: Iowa State Univ. Press.

Dixon, R. D. 1995. Ecology of White-headed Woodpeckers in the Central Oregon Cascades. Master's thesis, University of Idaho.

Dixon, R. D., and V. A. Saab. 2000. Black-backed Woodpecker (*Picoides arcticus*). No. 509 in *The Birds of North America*, A. Poole and F. Gill, eds. Philadelphia, PA: Birds of North America, Inc.

Dobbs, M., and D. Hedeen. 2006. Distribution of birds in Georgia: kingfisher and woodpeckers. Georgia Birds County Distribution Maps. Georgia Ornithological Society. www.gos.org/species_maps/king_wood/maps.htm.

Dobbs, R. C., T. E. Martin, and C. J. Conway. 1997. Williamson's Sapsucker (*Sphyrapicus thyroideus*). No. 285 in *The Birds of North America*, A. Poole and F. Gill, eds. Philadelphia, PA: Acad. of Nat. Sci., and Washington, D.C: AOU.

Dobkin, D. S., A. C. Rich, J. A. Pretare, and W. H. Pyle. 1995. Nest-site relationships among cavity-nesting birds of riparian and snow pocket aspen woodlands in the northwestern Great-Basin. *Condor* 97:694–707.

Dodenhoff, D., R. D. Stark, and E. V. Johnson. 2001. Do woodpecker drums encode information for species recognition? *Condor* 103:143–150.

Doherty, P. F., Jr., T. C. Grubb, Jr., and C. L. Bronson. 1996. Territories and caching-related behaviors of Red-headed Woodpeckers wintering in a beech grove. *Wilson Bull.* 108:740–747.

Driscoll, M. 2005. A life-altering glimpse of an Ivory-billed Woodpecker. *BirdScope* 19(3):10.

du Plessis, M., W. W. Weathers, and W. D. Koenig. 1994. Energetic benefits of communal roosting by Acorn Woodpeckers in the non-breeding season. *Condor* 96:631–637.

Dudley, J. G. 2005. Home Range Size and Foraging Habitat of Black-backed Woodpeckers. Master's thesis, Boise State Univ.

DuMont, P. A. 1933. An old specimen of hybrid flicker from central Arkansas. *Auk* 50:362.

———. 1935. A Red-shafted Flicker secured at Des Moines, Iowa. *Wilson Bull.* 47:158.

Duncan, D. 1933. The White-headed Woodpecker in Marin County, California. *Condor* 35:123–124.

Duncan, S. 1961. Wasp attack on a Flicker. *Auk* 79:277.

———. 2003. Coming home to roost: the Pileated Woodpecker as ecosystem engineer. *Science Findings* 57:1–6.

Dunham, D. W. 1963. Head-scratching in the Hairy Woodpecker, *Dendrocopos villosus. Auk* 80:375.

Dykstra, B. L., M. A. Rumble, and L. D. Flake. 1997. Effects of timber harvesting on birds in the Black Hills of South Dakota and Wyoming, USA. First Biennial North American Forest Ecology Workshop. Raleigh: North Carolina State Univ.

Eagle, A. C., E. M. Hay-Chmielewski, K. T. Cleveland, A. L. Derosier, M. E. Herbert, and R. A. Rustem, eds. 2005. Michigan's Wildlife Action Plan. Lansing: Michigan Dept. of Natural Resources. www.michigan.gov/dnrwildlifeactionplan.

Eastwood, S. K. 1930. The Northern Pileated Woodpecker in Westmoreland County, Pennsylvania. *Wilson Bull.* 42:54.

Eberhardt, L. S. 1997. A test of an environmental advertisement hypothesis for the function of drumming in Yellow-bellied Sapsuckers. *Condor* 99:798–803.

———. 2000. Use and selection of sap trees by Yellow-breasted Sapsuckers. *Auk* 117:41–51.

Eckstorm, F. H. 1901. *The Woodpeckers.* Boston: Houghton, Mifflin.

Edwards, H. H., and G. D. Schnell. 2000. Gila Woodpecker (*Melanerpes uropygialis*). No. 532 in *The Birds of North America*, A. Poole and F. Gill, eds. Philadelphia, PA: Birds of North America, Inc.

Eifrig, C. W. G. 1927. Some notes on Pileated Woodpeckers. *Wilson Bull.* 39:174–175.

———. 1944. The Southern Pileated Woodpecker: an unusual victim of the automobile. *Auk* 61:299.

Elchuk, C. L., and K. L. Wiebe. 2002. Food and predation risk as factors related to foraging locations of Northern Flickers. *Wilson Bull.* 114:349–357.

———. 2003. Ephemeral food resources and high conspecific densities as factors explaining lack of feeding territories in Northern Flickers (*Colaptes auratus*). *Auk* 120:187–193.

Elert, G. 2008. Acceleration: the physics hypertext book. www.hypertextbook.com/physics/mechanics/acceleration/.

Elzroth, M. 1987. Red-naped Sapsucker in Western Oregon. *Oregon Birds* 13(1):36–37.

England, E. G. 1940. A nest of the Arctic Three-toed Woodpecker. *Condor* 42:242–245.

———. 1941. A record of the Northern Flicker in Butte County, California. *Condor* 43:198.

Environment Yukon. 2008. Yukon Species at Risk. Government of Yukon. http://environmentyukon.gov.yk.ca/wildlifebiodiversity/speciesrisk.php.

Erlich, P. R., D. S. Dobkin, and D. Wheye. 1988. *The Birder's Handbook: A Field Guide to the Natural History of North American Birds.* New York: Simon and Schuster.

Errington, P. L. 1936. Winter-killing of Flickers in central Iowa. *Auk* 53:334–335.

Erskine, A. J. 1992. *Atlas of Breeding Birds of the Maritime Provinces.* Halifax, NS: Nimbus Publ. Ltd. and Nova Scotia Museum.

Erskine, A. J., and W. D. McLaren. 1976. Comparative nesting biology of some hole-nesting birds in the Cariboo Parklands, British Columbia. *Wilson Bull.* 88:611–620.

Ertep, S. A., and G. W. Lee. 1994. Use of GRASS to facilitate Red-cockaded Woodpecker management at Fort Benning Military Reservation. Pp. 628–633 in 1994 Annual Conference Proceedings, Urban and Regional Information Systems Association, Washington, D.C.

Erwin, C. A., K. B. Rozell, and L. H. DeCicco. 2004. Update on the status and distribution of Wilson's

Phalarope and Yellow-bellied Sapsucker in Alaska. *Western Birds* 35:42–44.
Eubanks, T. 2005. *The Roads of Texas*. Addison, TX: MAPSCO Inc.
Ewins, P. J. 1994. Birds breeding in or beneath Osprey nests in the Great Lakes basin. *Wilson Bull.* 106(4):743–749.
Fajer, E. D., K. J. Schmidt, and J. G. Eschler. 1987. Acorn Woodpecker predation on Cliff Swallow nests. *Condor* 89:177–178.
Farner, D. S. 1952. *The Birds of Crater Lake National Park*. Lawrence: Univ. of Kansas Press.
Feduccia, A. 1999. *The Origin and Evolution of Birds*. 2d ed. New Haven, CT: Yale Univ. Press.
Fenn, R. C. 1940. Red-headed Woodpecker nesting in New Hampshire. *Auk* 20:254.
Fenwick, G. H. 2005. First step. *Birder's World* 19(4):24–25.
Fisher, G. C. 1910. Additional vernacular names for the Flicker. *Wilson Bull.* 22:127.
Fisher, J. 1954. *A History of Birds*. Boston: Houghton Mifflin.
Fisher, R. J., and K. L. Wiebe. 2006a. Nest site attributes and temporal patterns of Northern Flicker nest loss: effects of predation and competition. *Oecologia* 147:744–753.
———. 2006b. Effects of sex and age on survival of Northern Flickers: a six-year field study. *Condor* 108:193–200.
———. 2006c. Breeding dispersal of Northern Flickers *Colaptes auratus* in relation to natural nest predation and experimentally increased perception of predation risk. *Ibis* 148:772–781.
Fisher, W. H. 1902. The Downy Woodpeckers of California. *Condor* 4:68–70.
———. 1903. Nesting of the Red-bellied Woodpecker in Harford County, Maryland. *Auk* 20:305–306.
———. 1904. The Pileated Woodpecker in Anne Arundel County, Md. *Auk* 21:278.
Fisher, W. K. 1905. Northern Flickers at Auburn, California. *Condor* 6:172.
———. 1906. An acorn store-house of the California Woodpecker. *Condor* 8:1.
Fitzgerald, J. A., B. Busby, M. Howery, R. Klataske, D. Reinking, and D. Pashley. 2000. Partners in Flight Bird Conservation Plan, Osage Plains (Physiographic Area 33). Ver. 1.0. The Plains, VA: ABC.
Fitzgerald, J. A., J. R. Herkert, and J. D. Brawn. 2000. Partners in Flight Bird Conservation Plan, Prairie Peninsula (Physiographic Area 31). Ver. 1.0. The Plains, VA: ABC.
Fitzgerald, J. A., and D. N. Pashley. 2000a. Partners in Flight Bird Conservation Plan, Dissected Till Plains (Physiographic Area 32). The Plains, VA: ABC.
———. 2000b. Partners in Flight Bird Conservation Plan, Ozark/Ouachitas (Physiographic Area 19). The Plains, VA: ABC.
Fitzgerald, J. A., D. N. Pashley, S. J. Lewis, and B. Pardo. 1998. Partners in Flight Bird Conservation Plan, Northern Tallgrass Prairie (Physiographic Area 40). Ver. 1.0. The Plains, VA: ABC.
Fitzgerald, J. A., D. N. Pashley, and B. Pardo. 1999. Partners in Flight Bird Conservation Plan, Northern Mixed-grass Prairie (Physiographic Area 37). Ver. 1.0. The Plains, VA: ABC.
Fleming, J. H. 1902. American Avocet and American Three-toed Woodpecker in Toronto. *Auk* 19:79.
Fletcher, S. D., and W. S. Moore. 1992. Further analysis of allozyme variation in the Northern Flicker, in comparison with mitochondrial DNA variation. *Condor* 94:988–991.
Florida Fish and Wildlife Conservation Comm. 2003. *Florida's Breeding Bird Atlas: A Collaborative Study of Florida's Birdlife*. Tallahassee: Florida Fish and Wildlife Conservation Commission.
———. 2005. Florida's Wildlife Legacy Initiative. Florida's Comprehensive Wildlife Conservation Strategy. Tallahassee: Florida Fish and Wildlife Conservation Commission.
Floyd, C. B. 1937. A Northern Flicker's unusual manner of bathing. *J. of Field Ornith.* 8:176.
Floyd, T. 2000. Posting to Birding the Americas listserve regarding Gilded Flicker in Nevada. http://maybank.tripod.com/USA/NV-06-2000.htm.
Floyd, T., C. S. Elphick, G. Chisholm, K. Mack, R. G. Elston, E. M. Ammon, and J. D. Boone. 2007. *Atlas of the Breeding Birds of Nevada*. Reno: Univ. of Nevada Press.
Ford, B., S. Carr, C. Hunter, J. York, and M. Roedel. 2000. Partners in Flight Bird Conservation Plan, Interior Low Plateaus (Physiographic Area 14). The Plains, VA: ABC.
Forristal, C., V. Saab, and A. Markus. 2005. Examining the Influence of Post-wildfire Timber Harvest on Sensitive Woodpecker Reproduction. 2005 Combined Progress Report: Birds and Burns Network. USFS. http://www.rmrs.nau.edu/lab/4251/birds burns.
Forsyth, A. 1988. *The Nature of Birds*. Rochester, NY: Camden House Publishing.
Fournier, A. 2005. Benning woodpeckers picked to repopulate parks. *Environmental Update* [USAEC] Winter 2005.
Fowler 1903. Stray notes from southern Arizona. *Condor* 5:106–107.
Fox, W., Jr. 1956. Pileated Woodpecker south of the Sierra Nevada in California. *Condor* 58:74.
Franken, R. J., and C. S. Gillies. 2001. Second confirmed occurrence of a Red-headed Woodpecker, *Melanerpes erythrocephalus*, in British Columbia. *BC Birds* 11:17–20.
Franzreb, K. E., and A. E. Higgins. 1975. Possible bear predation on a Yellow-breasted Sapsucker nest. *Auk* 92:817.
Freer, R. S. 1933. Yellow-bellied Sapsucker breeding in the Virginia Blue Ridge. *Auk* 50:437–438.
Freer, R. S., and J. J. Murray. 1935. The Yellow-bellied Sapsucker and the Ruby-throated Hummingbird—commensals? *Auk* 52:187–188.
Frenzel, R. W. 2004. Nest-site occupancy, nesting success, and turnover-rates of White-headed Woodpeckers in the Oregon Cascade Mountains in 2004.

Unpubl. report. Portland: Oregon Natural Heritage Program, The Nature Conservancy of Oregon.
Frisch, R. 1987. *Birds by the Dempster Highway*. Rev. ed. Victoria, BC: Robert Frisch.
Fritz, E. 1937. Sapsuckers on redwoods. *Condor* 39:36–37.
Fuchs, J., J. I. Ohlson, P. G. P. Ericson, and E. Pasquet. 2006. Molecular phylogeny and biogeographic history of the piculets (Piciformes: Picumninae). *J. of Avian Biology* 37: 487–496.
Gabrielson, I. N. 1924. Notes on the birds of Wallowa County, Oregon. *Auk* 41:552–565.
Gabrielson, I. N., and S. G. Jewett. 1940. *Birds of Oregon*. Corvallis: Oregon State College.
Gadd, S. W. 1941. Range of the Texas Woodpecker in Colorado. *Condor* 43:201.
Gaines, D. 1992. *Birds of Yosemite and the East Slope*. Rev. and updated. Lee Vining, CA: Artemisia Press.
Galen, C. 1989. A preliminary assessment of the status of the Lewis' Woodpecker in Wasco County, Oregon. Tech. Rept. 88-3-01. Salem: Oregon Dept. of Fish and Wildlife, Nongame Wildlife Program.
Gallagher, T. 2005a. *The Grail Bird*. Boston: Houghton Mifflin.
———. 2005b. Back to the bayou. *BirdScope* 19(3):20.
Ganier, A. F. 1926. An unusual Flicker's nest. *Wilson Bull.* 38:116.
Gardner, L. L. 1925. The adaptive modifications and the taxonomic value of the tongue in birds. *Proc. U.S. Nat. Mus.* 67(19):1–49.
———. 1959. Gila Woodpecker in San Diego County, California. *Condor* 61:435.
Garrett, K. L., and J. Dunn. 1981. *Birds of Southern California: Status and Distribution*. Los Angeles: Los Angeles Audubon Society.
Garrett, K. L., M. G. Raphael, and R. D. Dixon. 1996. White-headed Woodpecker (*Picoides albolarvatus*). No. 252 in *The Birds of North America,* A. Poole and F. Gill, eds. Philadelphia, PA: Acad. of Nat. Sci., and Washington, D.C.: AOU.
Garry Oak Ecosystems Recovery Team. 2002. Recovery Strategy for Garry Oak and Associated Ecosystems and Their Associated Species at Risk in Canada, 2001–2006. Victoria, BC: Garry Oak Ecosystems Recovery Team.
Gebauer, M. 2004. White-headed Woodpecker, *Picoides albolarvatus*. Accounts and Measures for Managing Identified Wildlife Version 5:1–10.
Gehlbach, F. R. 1993. *Mountain Islands and Desert Seas: A Natural History of the U.S.-Mexican Borderlands*. 2d ed. College Station: Texas A&M Univ. Press.
George, W. G. 1972. Age determination of Hairy and Downy Woodpeckers in eastern North America. *Bird-Banding* 43:128–135.
Georgia Department of Natural Resources. 2005. A Comprehensive Wildlife Conservation Strategy for Oregon. Social Circle; Georgia Dept. of Natural Resources, Wildlife Resources Division.
Gibbs, M. 1892. Acorn-eating birds. *Science* 20:133.
Gibson, D. D., and B. Kessel. 1992. Seventy-four new avian taxa documented in Alaska 1976–1991. *Condor* 94:454–467.
Gignoux, C. 1921a. The storage of almonds by the California Woodpecker. *Condor* 23:118.
———. 1921b. Speed of flight of the Red-shafted Flicker. *Condor* 23:33–34.
Giles, L. W. 1958. Occurrence of the Yellow-shafted Flicker in northern California. *Condor* 60:193.
Gill, F., and D. Donsker, eds. 2015. IOC World Bird List (v 5.1). doi:10.14344/IOC.ML.5.1.
Gilman, M. F. 1915. Woodpeckers of the Arizona lowlands. *Condor* 17:151–163.
Gilmer, A. n.d. Virginia Avian Records Committee Review List. Virginia Society of Ornithology. http://www.virginiabirds.net/varcom_reviewlist.html.
Gilmore, A. E. 1930. The Red-shafted Flicker in Tulsa County, Oklahoma. *Wilson Bull.* 42:133.
Ginther, H. J. 1916. Northern Pileated Woodpecker in Ashtabula County, Ohio. *Wilson Bull.* 94:40.
Godfrey, W. E. 1986. *The Birds of Canada*. Rev. ed. Ottawa, ON: National Museum of Natural Science.
Goggans, R., R. D. Dixon, and L. C. Seminara. 1988. Habitat use by Three-toed and Black-backed woodpeckers, Deschutes National Forest, Oregon. Nongame Wildlife Program report 87-3-02. Salem: Oregon Dept. of Fish and Wildlife.
Goodge, W. R. 1972. Anatomical evidence for phylogenetic relationships among woodpeckers. *Auk* 89:65–85.
Gorman, G. 2014. *Woodpeckers of the World, The Complete Guide*. London: Christopher Helm.
Gorrell, J. V., M. E. Anderson, K. D. Bunnell, M. F. Canning, A. G. Clark, D. E. Dolsen, and F. P. Howe. 2005. Utah Comprehensive Wildlife Conservation Strategy. Salt Lake City: Utah Div. of Wildlife Resources.
Graber, J. W., R. R. Graber, and E. L. Kirk. 1977. *Illinois Birds: Picidae*. Urbana, IL: Illinois Natural History Survey.
Graham, J. D. 1940. An unusually low nest of the Nuttall Woodpecker. *Condor* 42:223–224.
Grater, R. K. 1936. Pileated Woodpecker in Grand Canyon National Park. *Auk* 53:218.
———. 1939. Nesting records of the Red-shafted Flicker from Charleston Mountain, Nevada. *Condor* 41:125.
Great Basin Bird Observatory. 2005. Landbirds of Nevada and the Habitats They Need: A Resource Manager's Guide to Conservation of Priority Species. Great Basin Bird Observatory Tech. Rept. no. 05-01. Reno, NV: Great Basin Bird Observatory.
———. 2008. Nevada Bird Records Committee (NBRC) Complete List of All Records as of 9/18/2008. Reno, NV: Great Basin Bird Observatory.
Green, J. 1998. Breeding Birds—NE Minnesota, Landscape Region 1, N. Superior Uplands and S. Superior Uplands. Minneapolis: Minnesota Ornithologists Union.
Greenway, J. C. 1978. Type specimens of birds in the American Museum of Natural History, Part 2. *Bull. of AMNH* 161(1).

Gregory, S. S. 1923. Another Three-toed Woodpecker in Michigan. *Auk* 40:534.
Gremillion-Smith, C., proj. coord. 2006. Indiana Comprehensive Wildlife Strategy. Mishawaka, IN: D. J. Case and Assoc.
Grinnell, J. 1901. Two races of the Red-breasted Sapsucker. *Condor* 3:12.
———. 1902. The southern White-headed Woodpecker. *Condor* 4:89–90.
———. 1915. A distributional list of the birds of California. *Pacific Coast Avifauna* no. 11.
———. 1923. The status of the Rocky Mountain Downy Woodpecker in California. *Condor* 25:30–31.
———. 1927. A new race of Gila Woodpecker. *Condor* 29:168–169.
———. 1928. Do Willow Downy Woodpeckers ever drill in tree-bark? *Condor* 30:253–254.
———. 1935. Winter Wren and Pileated Woodpecker on the Greenhorn Mountains, California. *Condor* 37:44–45.
Grinnell, J., and A. H. Miller. 1944. The distribution of the birds of California. *Pacific Coast Avifauna* no. 27.
Grinnell, J., and H. S. Swarth. 1926. A new race of acorn-storing woodpecker, from lower California. *Condor* 28:176–178.
Grinnell, J., and M. W. Wythe. 1927. Directory to the bird-life of the San Francisco Bay region. *Pacific Coast Avifauna* no. 18.
Griscom, L. 1924. Arctic Three-toed Woodpecker in New Jersey. *Auk* 41:343–344.
Grudzien, T. A., and W. S. Moore. 1986. Genetic differentiation between the Yellow-shafted and Red-shafted subspecies of the Northern Flicker. *Biochemistry Syst. and Ecology* 14(4):451–453.
Guldager, N., and M. Bertram. 1997. Yukon Flats National Wildlife Refuge Bird Checklist. Fairbanks, AK: USFWS.
Gullion, G. W. 1949. A heavily parasitized flicker. *Condor* 51:232.
Gutzwiller, K. J., and S. H. Anderson. 1987. Multiscale associations between cavity-nesting birds and features of Wyoming streamside woodlands. *Condor* 89:534–548.
Hackett, S. J., R. T. Kimball, S. Reddy, R. C. K. Bowie, E. L. Braun, M. J. Braun, et al. A phylogenomic study of birds reveals their evolutionary history. *Science* 320:1763–1768.
Hadow, H. H. 1973. Winter ecology of migrant and resident Lewis's Woodpeckers in southeastern Colorado. *Condor* 75:210–224.
Hagen, S. K., P. T. Isakson, and S. R. Dyke. 2005. North Dakota Comprehensive Wildlife Conservation Strategy. Bismarck: North Dakota Game and Fish Dept.
Hailman, J. P. 1959. Drumming by female Hairy Woodpecker. *Auk* 30:47.
Hall, G. A. 2008. West Virginia bird review list. Brooks Bird Club, WV. www.brooksbirdclub.org/WVREC09.pdf.
Hall, R. E. 1938. Broad-tailed Hummingbird attracted to food of the Red-naped Sapsucker. *Condor* 40:264.
Halsey, L. 2008. Posting to listserve regarding Northern x Gilded Flicker hybrids. BIRDWG05 Archives — June 2008, week 3 (#19).
Hamilton, R., M. Billings, R. Carmona, and G. Ruiz-Campos. 2007. The regional reports: Baja California peninsula. *North American Birds* 61(2):332.
Hamilton, R. A., and J. L. Dunn. 2002. Featured photo: Red-naped and Red-breasted Sapsuckers. *Western Birds* 33:128–130.
Hanna, W. C. 1924. Weights of about three thousand eggs. *Condor* 26:146–153.
Hanson, J. D. 2003. Seasonal and sexual differences in foraging behavior of Ladder-backed Woodpeckers (*Picoides scalaris*) in an altered riparian habitat in west-central Texas. San Angelo, TX.
Hardie, K. 2005. Lewis's Woodpecker Nesting Study Report 2005. Bend, OR: East Cascades Bird Cons.
Harrington, A. B. 1924. Red-headed Woodpecker in Lincoln, Mass. *Auk* 41:480.
Harris, R. D. 1983. Albinistic Red-breasted Sapsucker. *Western Birds* 14:168.
Harrison, B. R. 2005. Updates from Arkansas. *WildBird* Sept./Oct. 2005: 6–7.
———. 2006a. Updates from Arkansas, part 3. *WildBird* Jan./Feb. 2006: 6–7.
———. 2006b. Updates from Arkansas, part 8. *WildBird* Nov./Dec. 2006: 6–7.
Harrison, C., and H. Loxton. 1993. *The Bird: Master of Flight*. New York: Barron's Educational Series, and Auckland, New Zealand: David Bateman, Ltd.
Harrison, G. 2002. The friendly little jackhammer. *Birder's World* Oct. 2002: 62–65.
Harrison, H. H. 1979. *A Field Guide to Birds' Nests*. Boston: Houghton Mifflin.
Hartwig, C. L. 1999. Effect of Forest Age, Structural Elements, and Prey Density on the Relative Abundance of Pileated Woodpecker (*Dryocopus pileatus abieticola*) on South-eastern Vancouver Island. Master's thesis, Univ. of Victoria.
Hartwig, C. L., D. S. Eastman, and A. S. Harestad. 2004. Characteristics of Pileated Woodpecker (*Dryocopus pileatus*) cavity trees and their patches on southeastern Canada. *Forest Ecol. and Mgmt.* 187:225–234.
Hasbrouck, E. M. 1889. Summer birds of Eastland County, Texas. *Auk* 6(3):236–241.
———. 1890. *Picoides arcticus* in central New York. *Auk* 7:206.
———. 1891. The present status of the Ivory-billed Woodpecker (*Campephilus principalis*). *Auk* 8:174–186.
Hauser, D. C. 1957. Some observations on sun-bathing in birds. *Wilson Bull.* 69:78–90.
Hay, O. P. 1887. The Red-headed Woodpecker a hoarder. *Auk* 4:193–196.
Haydock, J., and W. D. Koenig. 2002. Reproductive skew in the polygynandrous acorn woodpecker. *Proc. Nat. Acad. Sci.* 99:7178–7183.
Hayward, C. L., C. Cottam, A. M. Woodbury, and H. H. Frost. 1976. *Birds of Utah*. Great Basin Naturalist Memoirs no. 1. Provo, UT: Brigham Young Univ.
Hazler, K. R., D. E. W. Drumtra, M. R. Marshall, R. J. Cooper, and P. B. Hamel. 2004. Common, but

commonly overlooked: Red-bellied Woodpeckers as songbird nest predators. *Southeastern Naturalist* 3:467–474.
Hebard, F. V. 1949. Sexual selection in woodpeckers. *Auk* 66:90–91.
Heindel, T., and J. Heindel. 2001. Woodpeckers of Inyo County. Eastern Sierra Audubon Society newsletter, Mar. 2001.
———. 2005. Another exciting fall in Inyo County: 2004. Eastern Sierra Audubon Society newsletter, Jan. 2005.
Heintzselman, D. S. 1981. *A World Guide to Whales, Dolphins, and Porpoises.* Tulsa, OK: Winchester Press.
Helmuth, W. T. 1937. Arctic Three-toed Woodpecker on Long Island. *Auk* 54:102.
Hempel, K. M. 1922. The whistled call of the Hairy Woodpecker. *Auk* 39:259–260.
Henderson, G. 1931. An icebound Woodpecker. *Wilson Bull.* 43:310–311.
———. 1934. The Pileated Woodpecker in Decatur County, Indiana. *Wilson Bull.* 46:117.
Henshaw, H. W. 1921. Storage of acorns by the California Woodpecker. *Condor* 23:109–118.
Herbert, R. J., J. J. Hickey, and I. Kassoy. 1926. The Arctic Three-toed Woodpecker in New Jersey. *Auk* 43:99.
Herrick, F. H. 1884. Rare Vermont birds. *Science* 3:216.
Hess, G. K., R. L. West, M. V. Barnhill III, and L. M. Flemming. 2000. *Birds of Delaware.* Greenville, DE: Delmarva Ornithological Society.
Hickey, M. 2003. Posting to listserve regarding possible hybrid Gila x Golden-fronted Wodpecker. BIRDWG05 Archives—February 2003, week 2.
Hickman, G. C. 1970. Egg transport recorded for the Red-bellied Woodpecker. *Wilson Bull.* 82:463.
Hicks, L. E. 1939. Southern Downy Woodpecker in Ohio. *Auk* 56:83–84.
Hill, G. E., D. J. Mennill, B. W. Rolek, T. L. Hicks, and K. A. Swiston. 2006. Evidence suggesting that Ivory-billed Woodpeckers (*Campephilus principalis*) exist in Florida. *Avian Cons. and Ecol.* 1:1–34.
Hine, A. 1924. Lewis's Woodpecker visits Chicago. *Auk* 41:156–157.
Hlady, D. A. 1990. South Okanagan Conservation Strategy 1990–1995. Victoria: British Columbia Ministry of the Environment.
Hodgman, T. P., and K. Rosenberg. 2000. Partners in Flight Landbird Conservation Plan, Northern New England (Physiographic Area 27). Ver. 1.0. The Plains, VA: ABC.
Hoffman, E.C. 1928. Red-headed Woodpecker recovery. *Bulletin of the Northeastern Bird-Banding Association* 5(2):81–82.
Hoffman, R. 1927. The Gila Woodpecker at Holtville, Imperial County, California. *Condor* 29:162.
———. 1931. Saw-whet Owl and California Woodpecker on Santa Cruz Island. *Condor* 33:171.
Hofslund, P. B. 1958. Chimney Swift nesting in an abandoned Pileated Woodpecker hole. *Wilson Bull.* 70:192.
Holdstein, O. 1899. A musical woodpecker. *Auk* 16:353.
Holst, L., M. Schiavone, and T. Tomajer. 2005. New York State Comprehensive Wildlife Conservation Plan: A Strategy for Conserving New York's Fish and Wildlife Resources. Final Submission Draft. Albany: New York State Dept. of Environmental Conservation.
Hooge, P. N., M. T. Stanback, and W. D. Koenig. 1999. Nest-site selection in the Acorn Woodpecker. *Auk* 116:45–54.
Hooper, E. T. 1936. Red-shafted Flicker foraging on a cement pillar. *Condor* 38:43–44.
Hooper, R. G., and M. R. Lennartz. 1982. Roosting behavior of Red-cockaded Woodpecker clans with insufficient cavities. *J. of Field Ornith.* 54:72–76.
Hoose, P. 2004. *The Race to Save the Lord God Bird.* New York: Farrar, Straus and Giroux.
Houghton, C. 1924. Pileated Woodpecker in Washington County, N.Y. *Auk* 41:157.
Houston, R. 2007. Fort Benning sends woodpeckers to repopulate storm-hit forest. *Environmental Update* [USAEC] Winter 2007.
Howell, S. N. G., and S. Webb. 1995. *A Guide to the Birds of Mexico and Northern Central America.* Oxford, UK: Oxford Univ. Press.
Howell, T. R. 1952. Natural history and differentiation in the Yellow-breasted Sapsucker. *Condor* 54:237–282.
———. 1953. Racial and sexual differences in migration in *Sphyrapicus varius. Auk* 70:118–126.
Howitt, H. 1925. Habits of the Flicker. *Auk* 42:143.
———. 1927. Arctic Three-toed Woodpecker in Guelph, Ontario. *Auk* 44:252–253.
Hoyt, J. S. Y. 1944. Preliminary notes on the development of nestling Pileated Woodpeckers. *Auk* 61:376–384.
Hoyt, S. F. 1952. An additional age record of a Pileated Woodpecker. *Bird-Banding* 23:29.
———. 1953. Forehead color of the Pileated Woodpecker (*Dryocopus pileatus*). *Auk* 70:209–210.
Hubbard, J. P. 1963. Noteworthy records from New Mexico. *Condor* 65:236–239.
Hudon, J. 2000. On with Project Sapsucker. *Alberta Naturalist* 30(1):22–23.
———. 2001. Effect of Human Disturbance on Genetic Diversity: Case Study using Sapsuckers in Kananaskis Country. Final Report to the Alberta Conservation Association, Project 030-50-60-002. Edmonton: Provincial Museum of Alberta.
———. 2005. The official list of the birds of Alberta: now 400 species and counting. *Nature Alberta* Spring 2005:10–18.
Huey, L. M. 1932. Boreal Flicker in San Diego County, California. *Condor* 34:140.
Hunter, C., R. Katz, D. Pashley, and B. Ford. 1999. Partners in Flight Bird Conservation Plan, Southern Blue Ridge (Physiographic Area 23). Ver. 1.0. The Plains, VA: ABC.
Hunter, C., L. Peoples, and J. Collazo. 2001. Partners in Flight Bird Conservation Plan, South Atlantic Coastal Plain (Physiographic Area 03). Ver. 1.0. The Plains, VA: ABC.

Husak, M. S. 2000. Seasonal variation in territorial behavior of Golden-fronted Woodpeckers in west-central Texas. *Southwestern Naturalist* 45:30–38.
Husak, M. S., and J. F. Husak. 2002. Low frequency of site fidelity by Golden-fronted Woodpeckers. *Southwestern Naturalist* 47:110–114.
———. 2005. Atypical pair-bonding behavior among Golden-fronted Woodpeckers (*Melanerpes aurifrons*). *Southwestern Naturalist* 50:85–88.
Husak, M. S., and T. C. Maxwell. 1998. Golden-fronted Woodpecker (*Melanerpes aurifrons*). No. 373 in *The Birds of North America*, A. Poole and F. Gill, eds. Philadelphia, PA: Acad. of Nat. Sci., and Washington, D.C: AOU.
———. 2000. A review of 20th century range expansion and population trends of the Golden-fronted Woodpecker (*Melanerpes aurifrons*): historical and ecological perspectives. *Tex. J. Sci.* 52:275–284.
Hutchins, J. 1908. Pileated Woodpecker near Litchfield, Conn. *Auk* 25:475.
Idaho Department of Fish and Game. 2005. Idaho Comprehensive Wildlife Conservation Strategy. Boise: Idaho Dept. of Fish and Game.
Illinois Department of Natural Resources. 2005. The Illinois Comprehensive Wildlife Conservation Plan & Strategy. Springfield: Illinois Dept. of Natural Resources.
InfoPEI. 2008. Prince Edward Island Wildlife. Government of Prince Edward Island. www.gov.pe.ca/infopei/index.php3?number=42215.
Ingold, D. J. 1987. Documented double-broodedness in Red-headed Woodpeckers. *J. of Field Ornith.* 58:234–235.
———. 1989. Nesting phenology and competition for nest sites among Red-headed and Red-bellied Woodpeckers and European Starlings. *Auk* 106:209–217.
———. 1990. Simultaneous use of nest trees by breeding Red-headed and Red-bellied Woodpeckers and European Starlings. *Condor* 92:252–253.
———. 1991. Nest-site fidelity in Red-headed and Red-bellied Woodpeckers. *Wilson Bull.* 103:118–122.
Ingold, J. L., and C. M. Weise. 1985. Observations on feather color variation in a presumed Common Flicker intergrade. *J. of Field Ornith.* 56(4):403–405.
Inouye, D. W. 1976. Nonrandom orientation of entrance holes to woodpecker nests in aspen trees. *Condor* 78:101–102.
Inouye, R. S., N. J. Huntly, and D. W. Inouye. 1981. Non-random orientation of Gila Woodpecker nest entrances in saguaro cacti. *Condor* 83:88–89.
Jackson, J. A. 1970. Predation of a black rat snake on Yellow-shafted Flicker nestlings. *Wilson Bull.* 82:329–330.
———. 1971. The adaptive significance of reversed sexual dimorphism in tail length of woodpeckers: an alternative hypothesis. *Bird-Banding* 42:18–20.
———. 1974. Gray rat snakes versus Red-cockaded Woodpeckers: predator-prey adaptations. *Auk* 91:342–347.
———. 1976. A comparison of some aspects of the breeding ecology of Red-headed and Red-bellied Woodpeckers in Kansas. *Condor* 78:67–76.
———. 1977a. Red-cockaded Woodpeckers and pine red heart disease. *Auk* 94:160–163.
———. 1977b. Competition for cavities and Red-cockaded Woodpecker Management. Pp. 103–112 in S. A. Temple, ed., *Management Techniques for Preserving Endangered Species.* Madison: Univ. of Wisconsin Press.
———. 1978. Predation by a gray rat snake on Red-cockaded Woodpecker nestlings. *Bird-Banding* 49:187–188.
———. 1979. Age characteristics of Red-cockaded Woodpeckers. *Bird-Banding* 50:23–29.
———. 1990. Intercolony movements of Red-cockaded Woodpeckers in South Carolina. *J. of Field Ornith.* 61:149–155.
———. 1994. Red-cockaded Woodpecker (*Picoides borealis*). No. 85 in *The Birds of North America,* A. Poole and F. Gill, eds. Philadelphia, PA: Acad. of Nat. Sci., and Washington, D.C: AOU.
———. 2002a. The truth is out there. *Birder's World* 16:40–47.
———. 2002b. Ivory-billed Woodpecker (*Campephilus principalis*). *The Birds of North America Online,* A. Poole, ed. Ithaca, NY: Cornell Lab of Ornithology. Retrieved from The Birds of North America Online database: http://bna.birds.cornell.edu/.
———. 2004. *In Search of the Ivory-billed Woodpecker.* Washington, D.C: Smithsonian Institution Press.
———. 2006. Ivory-billed Woodpecker (*Campephilus principalis*): hope, and the interfaces of science, conservation, and politics. *Auk* 123:1–15.
Jackson, J. A., G. Hammerson, and F. Dirrigl. 1998. Species Management. Abstract, Pileated Woodpecker (*Dryocopus pileatus*). Arlington, VA: The Nature Conservancy.
Jackson, J. A., and E. E. Hoover. 1975. A potentially harmful effect of suet on woodpeckers. *Bird-Banding* 46:131–134.
Jackson, J. A., and H. R. Ouellet. 2002a. Downy Woodpecker (*Picoides pubescens*). No. 613 in *The Birds of North America,* A. Poole and F. Gill, eds. Philadelphia, PA: Birds of North America, Inc.
———. 2002b. Hairy Woodpecker (*Picoides villosus*). No. 702 in *The Birds of North America,* A. Poole and F. Gill, eds. Philadelphia, PA: Birds of North America, Inc.
Jackson, L. S., C. A. Thompson, and J. L. Dinsmore. 1996. *The Iowa Breeding Bird Atlas.* Iowa City: Univ. of Iowa Press.
Jacobs, B., and J. D. Wilson. 1996. *Missouri Breeding Bird Atlas, 1986–1992.* Natural History Series no. 6. Jefferson City: Missouri Dept. of Conservation.
Jacques, K. 2005. Birders allowed into ivory-billed territory. Delta Traveler, November 2005. Supplement to *The Brinkley Argus*, Brinkley, AR.
Janisch, K. 2008. Woodpecker trouble for St. John's; anonymous donor flies in. *Stillwater Gazette* Aug. 21, 2008.

Janssen, R. B. 1987. *Birds in Minnesota.* James Ford Bell Museum of Natural History. Minneapolis: Univ. of Minnesota Press.

Jenkins, H. O. 1906. Variation in the Hairy Woodpecker (*Dryobates villosus* and subspecies). *Auk* 23:161–171.

Jenkins, J. M. 1979. Foraging behavior of male and female Nuttall Woodpeckers. *Auk* 96:418–420.

Jensen, J. K. 1923. Notes on the nesting birds of northern Santa Fe County, New Mexico. *Auk* 45:452–469.

Job, H. K. 1901. The Pileated Woodpecker in Connecticut. *Auk* 18:193.

Johnsgard, P. A. 1979. *Birds of The Great Plains: Breeding Species and Their Distribution.* Lincoln: Univ. of Nebraska Press.

———. 1986. *Birds of the Rocky Mountains.* Lincoln: Univ. of Nebraska Press.

Johnson, C. E. 1934. The chronicle of a flicker's courtship. *Auk* 51:477–481.

Johnson, L. S., and L. H. Kermott. 1994. Nesting success of cavity-nesting birds using natural tree cavities. *J. of Field Ornith.* 65(1):36–51.

Johnson, N. K. 1969. Review: Three papers on variation in flickers by Lester L. Short, Jr. *Wilson Bull.* 81(2):225–230.

Johnson, N. K., and C. B. Johnson. 1985. Speciation in sapsuckers (*Sphyrapicus*) II: Sympatry, hybridization, and mate preference in *S. ruber daggetti* and *S. nuchalis. Auk* 102:1–15.

Johnson, N. K., and M. Zink. 1983. Speciation in sapsuckers (*Sphyrapicus*) I: Genetic differentiation. *Auk* 100:871–884.

Johnson, R. A. 1947. Role of male Yellow-breasted Sapsucker in the care of the young. *Auk* 64:621–623.

Johnson, R. R., L. T. Haight, and J. D. Ligon. 1999. Arizona Woodpecker (*Picoides arizonae*). No. 474 in *The Birds of North America*, A. Poole, ed. Ithaca, NY: Cornell Lab of Ornithology.

Johnson, W. N., and K. McGarigal. 1984. Pileated Woodpecker nest in natural cavity. *J. of Field Ornith.* 55:490.

Jones, B. 2004. History of litigation on the national forests in Texas. *Forest History Today*, Spring/Fall 2004:35–44.

Jones, Z. F., and C. E. Bock. 2003. Relationships between Mexican Jays (*Aphelocoma ultramarine*) and Northern Flickers (*Colaptes auratus*) in an Arizona oak savanna. *Auk* 120:429–432.

Joy, J. B. 2000. Characteristics of nest cavities and nest trees of the Red-breasted Sapsucker in coastal montane forests. *J. of Field Ornith.* 71:525–530.

Judd, W. W. 1956. Red-headed Woodpeckers (*Melanerpes erythrocephalus*) feeding on Carolina locusts. *Auk* 73:285–286.

Jung, C. S. 1927. Additional notes on birds of Vilas County, Wisconsin. *Wilson Bull.* 39(3):173–174.

Kale, H. W., II., and D. S. Maehr. 1990. *Florida's Birds: A Handbook and Reference.* Sarasota, FL: Pineapple Press.

Kansas Bird Records Committee. 2003. Birds of Kansas, Checklist. Winfield: Kansas Ornithological Society.

Kattan, G. 1988. Food habits and social organization of Acorn Woodpeckers in Colombia. *Condor* 90:100–106.

Kattan, G., and C. Murcia. 1985. Hummingbird association with Acorn Woodpecker sap trees in Colombia. *Condor* 87:542–543.

Kaufman, K. 2005. Identification tips: Look-alikes. *Birder's World* Feb. 2005:72–73.

Kays, R. W., and D. E. Wilson. 2002. *Mammals of North America.* Princeton, NJ: Princeton Univ. Press.

Kearney, R. 2003. Partners in Flight Landbird Conservation Plan, Mid-Atlantic Piedmont (Physiographic Area 10). The Plains, VA: ABC.

Kellam, J. S. 2003. Pair bond maintenance in Pileated Woodpeckers at roost sites during autumn. *Wilson Bull.* 115:186–192.

Kemper, J. 1999. *Birding Northern California.* Helena, MT: Falcon Publishing.

———. 2002. *Southern Oregon's Bird Life.* Medford, OR: Outdoor Press.

Kennard, F. H. 1895. January occurrence of the "Sapsucker" in Brookline, Mass. *Auk* 12:301–302.

Kennedy, J. A., P. Dilworth-Christie, and A. J. Erskine. 1999. The Canadian Breeding Bird (Mapping) Census Database. Technical Report no. 342. Ottawa, ON: Canadian Wildlife Service.

Kentucky Department of Fish and Wildlife Resources. 2005. Kentucky's Comprehensive Wildlife Conservation Strategy. Frankfort: Kentucky Dept. of Fish and Wildlife Resources.

Kerpez, T. A., and N. S. Smith. 1990a. Competition between European Starlings and native woodpeckers for nest cavities in saguaros. *Auk* 107:367–375.

———. 1990b. Nest-site selection and nest-cavity characteristics of Gila Woodpeckers and Northern Flickers. *Condor* 92:193–198.

Kessel, B. 1986. Yellow-breasted Sapsucker, *Sphyrapicus varius*, in Alaska. *J. of Field Ornith.* 57:42–47.

Kilham, L. 1953a. Possible commensalisms between Myrtle Warbler and Yellow-bellied Sapsucker. *Wilson Bull.* 65:41.

———. 1953b. Warblers, hummingbird, and sapsucker feeding on sap of yellow birch. *Wilson Bull.* 65:198.

———. 1958a. Pair formation, mutual tapping, and nest hole selection of Red-bellied Woodpeckers. *Auk* 75:318–328.

———. 1958b. Territorial behavior of wintering Red-headed Woodpeckers. *Wilson Bull.* 70:347–358.

———. 1958c. Sealed-in stores of Red-headed Woodpeckers. *Wilson Bull.* 70:107–113.

———. 1958d. Repeated attacks by a Sharp-shinned Hawk on Pileated Woodpecker. *Condor* 60:141.

———. 1959a. Bark-eating of Red-headed Woodpeckers. *Condor* 61:371–372.

———. 1959b. Behavior and methods of communication of Pileated Woodpeckers. *Condor* 61:377–387.

———. 1959c. Early reproductive behavior of flickers. *Wilson Bull.* 71:323–336.

———. 1959d. Head-scratching and wing-stretching of woodpeckers. *Auk* 76:527–528.

———. 1959e. Mutual tapping of the Red-headed Woodpecker. *Auk* 76:235–236.
———. 1959f. Pilot black snake and nesting Pileated Woodpeckers. *Wilson Bull.* 71:191.
———. 1960. Courtship and territorial behavior of Hairy Woodpeckers. *Auk* 77:259–270.
———. 1961a. Reproductive behavior of Red-bellied Woodpeckers. *Wilson Bull.* 73:237–254.
———. 1961b. Aggressiveness of migrant Myrtle Warblers toward woodpeckers and other birds. *Auk* 78:261.
———. 1962a. Breeding behavior of Yellow-breasted Sapsuckers. *Auk* 79:31–43.
———. 1962b. Nest sanitation of Yellow-breasted Sapsucker. *Wilson Bull.* 74:96–97.
———. 1963. Food storing of Red-bellied Woodpeckers. *Wilson Bull.* 75:227–234.
———. 1964. The relations of breeding Yellow-breasted Sapsuckers to wounded birches and other trees. *Auk* 81:520–527.
———. 1966a. Nesting activities of Black-backed Woodpeckers. *Condor* 68:308–310.
———. 1966b. Reproductive behavior of Hairy Woodpeckers. I. Pair formation and courtship. *Wilson Bull.* 78:251–265.
———. 1968. Reproductive behavior of Hairy Woodpeckers. II. Nesting and habitat. *Wilson Bull.* 80:286–305.
———. 1969. Reproductive behavior of Hairy Woodpeckers. III. Agonistic behavior in relation to courtship and territory. *Wilson Bull.* 81:169–183.
———. 1971. Reproductive behavior of Yellow-breasted Sapsuckers. I. Preference for nesting in *Fomes*-infected aspens and nest hole interrelations with flying squirrels, raccoons, and other animals. *Wilson Bull.* 83:159–171.
———. 1973. Colonial-type nesting in Yellow-shafted Flickers as related to staggering of nest times. *Bird-Banding* 44:317–318.
———. 1974a. Play in Hairy, Downy, and other woodpeckers. *Wilson Bull.* 86:35–42.
———. 1974b. Copulatory behavior of Downy Woodpeckers. *Wilson Bull.* 86:23–34.
———. 1974c. Loud vocalizations by Pileated Woodpeckers on approach to roosts or nest holes. *Auk* 91:634–636.
———. 1975. Dirt-bathing by a Pileated Woodpecker. *Bird-Banding* 46:251–252.
———. 1976. Winter foraging and associated behavior of Pileated Woodpeckers in Georgia and Florida. *Auk* 93:15–24.
———. 1977a. Nesting behavior of Yellow-breasted Sapsuckers. *Wilson Bull.* 89:310–324.
———. 1977b. Nest-site differences between Red-headed and Red-bellied Woodpeckers in South Carolina. *Wilson Bull.* 89:164–165.
———. 1977c. Altruism in nesting Yellow-breasted Sapsucker. *Auk* 94:613–614.
———. 1978. Sexual similarity of Red-headed Woodpeckers and possible explanations based on fall territorial behavior. *Wilson Bull.* 90(2):285–287.
———. 1979a. Courtship and the pair-bond of Pileated Woodpeckers. *Auk* 96:587–594.
———. 1979b. Three-week vs. 4-week nestling periods in *Picoides* and other woodpeckers. *Wilson Bull.* 91(2):335–338.
———. 1992. *Woodpeckers of Eastern North America.* New York: Dover Publ. Originally Life History Studies of Woodpeckers of Eastern North America, 1983.
Kingery, H. E. 1998. *Colorado Breeding Bird Atlas.* Colorado Springs: Colorado Bird Atlas Partnership.
Kinsey, J. B. 1955. A recent record of the Pileated Woodpecker in Marin County, California. *Condor* 57:190–191.
Kious, W. J., and R. I. Tilling. 1996. *This Dynamic Earth: The Story of Plate Tectonics.* Denver, CO: U.S. Geological Survey Information Services.
Kirby, V. C. 1980. An adaptive modification in the ribs of woodpeckers and piculets (Picidae). *Auk* 97:521–532.
Kirkpatrick, C., C. J. Conway, and P. B. Jones. 2006. Distribution and relative abundance of forest birds in relation to burn severity in southeastern Arizona. *J. of Wildl. Mgmt.* 70:1005–1012.
Kisiel, D. S. 1972. Foraging behavior of *Dendrocopos villosus* and *D. pubescens* in eastern New York State. *Condor* 74:393–398.
Kleen, V. M., L. Cordle, and R. A. Montgomery. 2004. *The Illinois Breeding Bird Atlas.* Special Publication no. 26. Champaign: Illinois Natural History Survey.
Knutson, M. G., G. Butcher, J. Fitzgerald, and J. Shieldcastle. 2001. Partners in Flight Bird Conservation Plan, Upper Great Lakes Plain (Physiographic Area 16). The Plains, VA: ABC.
Koenig, W. D. 1980a. The incidence of runt eggs in woodpeckers. *Wilson Bull.* 92:169–176.
———. 1980b. Variation and age determination in a population of Acorn Woodpeckers. *J. of Field Ornith.* 51:10–16.
———. 1984. Clutch size of the Gilded Flicker. *Condor* 86:89–90.
———. 1986. Geographic ecology of clutch size variation in North American woodpeckers. *Condor* 88:499–504.
———. 1987. Morphological and dietary correlates of clutch size in North American woodpeckers. *Auk* 104:757–765.
———. 1991. The effects of tannins and lipids on digestion of acorns by Acorn Woodpeckers. *Auk* 108:79–88.
———. 2005. Persistence in adversity: lessons from the Ivory-billed Woodpecker. *Bioscience* 55:646–647.
Koenig, W. D., and L. S. Benedict. 2002. Size, insect parasitism, and energetic value of acorns stored by Acorn Woodpeckers. *Condor* 104:539–547.
Koenig, W. D., P. N. Hooge, M. T. Stanback, and J. Haydock. 2000. Natal dispersal in the cooperatively breeding Acorn Woodpecker. *Condor* 102:492–502.
Koenig, W. D., and R. L. Mumme. 1987. *Population Ecology of the Cooperatively Breeding Acorn Woodpecker.* Monographs in Population Biology no. 24. Princeton, NJ: Princeton Univ. Press.

Koenig, W. D., R. L. Mumme, and F. A. Pitelka. 1983. Females Roles in Cooperatively Breeding Acorn Woodpeckers. Pp. 235–261 in S. K. Wasser, ed., *Social Behavior of Female Vertebrates.* New York: Academic Press.

Koenig, W. D., and F. A. Pitelka. 1979. Relatedness and inbreeding avoidance: counterploys in the communally nesting Acorn Woodpecker. *Science* 206:1103–1105.

Koenig, W. D., M. T. Stanback, P. N. Hooge, and R. L. Mumme. 1991. Distress calls in the Acorn Woodpecker. *Condor* 93:637–643.

Koenig, W. D., E. L. Walters, J. R. Walters, J. S. Kellam, K. G. Michalek, and M. S. Schrader. 2005. Seasonal body weight variation in five species of woodpeckers. *Condor* 107:810–822.

Koenig, W. D., and J. R. Walters. 1999. Sex-ratio selection in species with helpers at the nest: the repayment model revisited. *American Naturalist* 153:124–130.

Koenig, W. D., and P. L. Williams. 1979. Notes on the status of Acorn Woodpeckers in central Mexico. *Condor* 81:317–318.

Kohler, L. S. 1910. Red-headed Woodpeckers vs. Blue Jays. *Wilson Bull.* 71:126.

———. 1911. Additional vernacular names for the Flicker. *Wilson Bull.* 75:134.

Korol, J. J., and R. L. Hutto. 1984. Factors affecting nest site location in Gila Woodpeckers. *Condor* 86:73–78.

Krannitz, P. G. 2006. Recovery strategy for the White-headed Woodpecker (*Picoides albolarvatus*). Species at Risk Recovery Strategy Series. Ottawa, ON: Canadian Wildlife Service and Environment Canada.

Kutac, E. A. 1998. *Birder's Guide to Texas.* 2d ed. Houston, TX: Gulf Publishing.

Kutac, E. A., and S. C. Caran. 1994. *Birds and Other Wildlife of South Central Texas.* Austin: Univ. of Texas Press.

Labisky, R. F., and S. H. Mann. 1961. Observation of avian pox in a Yellow-shafted Flicker. *Auk* 78:642.

LaBranch, M. S. 2005. The Return of "Elvis." *BirdScope* 19(3):9.

Lammertink, M. 2005. Review of *In Search of the Ivory-billed Woodpecker,* by J. A. Jackson. *Condor* 107:726–727.

Landin, M. C. 1978. Screech Owl predation on a Common Flicker nest. *Wilson Bull.* 90:652.

Lane, J. A. 1988. *A Birder's Guide to Southeastern Arizona.* Rev. by H. R. Holt. Denver, CO: L & P Press.

Lanyon, S. M., and R. M. Zink. 1987. Genetic variation in Piciform birds: monophyly and generic and familial relationships. *Auk* 104:724–732.

Latta, M. J., C. J. Beardmore, and T. E. Corman. 1999. Arizona Partners-in-Flight Bird Conservation Plan. Version 1.0. Nongame and Endangered Wildlife Program Technical Report no. 142. Phoenix: Arizona Game and Fish Dept.

Laughlin, S. B., and D. B. Kibbe, eds. 1985. *The Atlas of Breeding Birds of Vermont.* Quechee, VT: Vermont Institute of Natural Science.

Law, J. E. 1916. Odd performance of a Flicker with a malformed bill. *Condor* 18:85.

———. 1929. Another Lewis Woodpecker stores acorns. *Condor* 31:233–238.

Lawrence, L. D. 1966. A Comparative Life-history Study of Four species of Woodpeckers. Ornithological Monograph no. 5.

Lawrence, R. B. 1896. A new Long Island, N.Y. record for the Red-bellied Woodpecker. *Auk* 13:82.

Laybourne, R. C., D. W. Deedrick, and F. M. Hueber. 1994. Feather in amber is earliest New World fossil of Picidae. *Wilson Bull.* 106(1):18–25.

Leach, F. A. 1925. Communism in the California Woodpecker. *Condor* 27:12–19.

Lefebvre, L., N. Juretic, N. Nikolkakis, and S. Timmermans. 2001. Is the link between forebrain size and feeding innovations caused by confounding variables? A study of Australian and North American birds. *Animal Cognition* 4:91–97.

Lefebvre, L., N. Nikolkakis, and D. Boire. 2002. Tools and brains in birds. *Behaviour* 139:939–973.

Lefebvre, L., P. Whittle, E. Lascaris, and A. Finkelstein. 1997. Feeding innovations and forebrain size in birds. *Animal Behavior* 53:549–560.

Leister, C. W. 1919. Aerial evolutions of a Flicker. *Auk* 36:570.

Lennartz, M. R., and R. F. Harlow. 1979. The role of parent and helper Red-cockaded Woodpeckers at the nest. *Wilson Bull.* 91:331–335.

Leonard, D. L., Jr. 2001. Three-toed Woodpecker (*Picoides tridactylus*). No. 588 in *The Birds of North America,* A. Poole and F. Gill, eds. Philadelphia, PA: Birds of North America, Inc.

Leopold, A. 1918. Are Red-headed Woodpeckers moving west? *Condor* 20:122.

———. 1919a. Notes on the Red-headed Woodpecker and Jack Snipe in New Mexico. *Condor* 21:40.

———. 1919b. A breeding record for the Red-headed Woodpecker in New Mexico. *Condor* 21:173–174.

Lester, G. D., S. G. Sorensen, P. L. Faulkner, C. S. Reid, and I. E. Maxit. 2005. Louisiana Comprehensive Wildlife Conservation Strategy. Baton Rouge: Louisiana Dept. of Wildlife and Fisheries.

Levin, A. V. 2007. Animal models for Shaken Baby Syndrome. Farmington, UT: National Center on Shaken Baby Syndrome. www.dontshake.com/.

Lewis, J. C., and E. Rodrick. 2002. White-headed Woodpecker (*Picoides albolarvatus*). Pp. 33-1–33-5 in E. Larsen, J. M. Azerrad, and N. Nordstrom, eds., *Management Recommendations for Washington's Priority Species. Volume IV: Birds.* Olympia: Washington Dept. of Fish and Wildlife.

Lewis, J. C., E. A. Rodrick, and J. M. Azerrad. 2003. Black-backed Woodpecker (*Picoides arcticus*). Pp. 30-1–30-6 in E. Larsen, J. M. Azerrad, and N. Nordstrom, eds., *Management Recommendations for Washington's Priority Species. Volume IV: Birds.* Olympia: Washington Dept. of Fish and Wildlife.

Lewis, S. 1890. The Red-bellied Woodpecker in northwestern New Jersey. *Auk* 7:206.

Li, P., and T. E. Martin. 1991. Nest-site selection and

nesting success of cavity nesting birds in high elevation forest drainages. *Auk* 108:405–418.
Ligon, J. D. 1968. Observations on Strickland's Woodpecker, *Dendrocopos stricklandi*. *Condor* 70:83–84.
———. 1970. Behavior and breeding biology of the Red-cockaded Woodpecker. *Auk* 87:255–278.
———. 1973. Foraging behavior of the White-headed Woodpecker in Idaho. *Auk* 90:862–869.
Ligon, J. D., P. B. Stacey, R. N. Conner, C. E. Bock, and C. S. Adkisson. 1986. Report of the AOU Committee for the Conservation of the Red-cockaded Woodpecker. *Auk* 103:848–855.
Liles, R. 2008. Red-cockaded Woodpecker. [Species account.] Outdoor Alabama Watchable Wildlife. Montgomery: Alabama Dept. of Conservation and Natural Resources.
Lincoln, F. C. 1917. Some notes on the birds of Rock Canyon, Arizona. *Wilson Bull.* 29(2):64–73.
Linder, K. A., and S. H. Anderson. 1998. Nesting habitat of Lewis' Woodpeckers in southeastern Wyoming. *J. of Field Ornith.* 69:109–116.
Linsdale, J. M. 1936a. Habits of Lewis Woodpeckers in winter. *Condor* 38:245–246.
———. 1936b. The Birds of Nevada. *Pacific Coast Avifauna* no. 23.
Little, L. 1920. A peculiar Flicker habit. *Condor* 12:188.
Littlefield, C. D. 1990. *Birds of Malheur National Wildlife Refuge, Oregon*. Corvallis: Oregon State Univ. Press.
Lockwood, M. W. 2001. *Birds of the Texas Hill Country*. Austin: Univ. of Texas Press.
Lockwood, M. W., and B. Freeman. 2004. *The Texas Ornithological Society's Handbook of Texas Birds*. College Station: Texas A&M Univ. Press.
Loftin, R. W. 1991. Ivory-billed Woodpeckers reported on Okefenokee Swamp in 1941–42. *Oriole* 56(4):74–75.
Longcore, T., C. Rich, and S. A. Gauthreaux, Jr. 2005. Scientific basis to establish policy regulating communications towers to protect migratory birds: response to Avatar Environmental, LLC, report regarding migratory bird collisions with communications towers, WT Docket No. 03-187, Federal Communications Comm. Notice of Inquiry. Land Protection Partners. Los Angeles: Land Protection Partners.
Losin, N., C. H. Floyd, T. E. Schweitzer, and S. J. Keller. 2006. Relationship between aspen heartwood rot and the location of cavity excavation by a primary cavity-nester, the Red-naped Sapsucker. *Condor* 108:706–710.
Lowther, P. E. 2000. Nuttall's Woodpecker (*Picoides nuttalli*). No. 555 in *The Birds of North America*, A. Poole and F. Gill, eds. Philadelphia, PA: Birds of North America, Inc.
———. 2001. Ladder-backed Woodpecker (*Picoides scalaris*). No. 565 in *The Birds of North America*, A. Poole and F. Gill, eds. Philadelphia, PA: Birds of North America, Inc.
Lucas, F. A. 1895. The tongues of woodpeckers: relation of the form of the tongue to the character of the food. *USDA Division of Ornithology and Mammalogy Bull.* no. 7:35–41.
Lucas, F. A. 1896. The taxonomic value of the tongue in birds. *Auk* 13:109–115.
MacArthur, R. H., and A. T. MacArthur. 1974. On the use of mist nest for population studies of birds. *Proc. Nat. Acad. Sci.* 71:3230–3233.
Machmer, M., and C. Steeger. 2004. Preliminary Assessment of the Effectiveness of Wildlife Tree Retention on Cutblocks Harvested between 1999–2001 under the Forest Practices Code. Nelson, BC: Pandion Ecological Research Ltd.
MacRoberts, M. H. 1970. Notes on the food habits and food defense of the Acorn Woodpecker. *Condor* 72:196–204.
———. 1975. Food storage and winter territory in Red-headed Woodpeckers in northwestern Louisiana. *Auk* 92:382–385.
MacRoberts, M. H., and B. R. MacRoberts. 1985. Gila Woodpecker stores acorns. *Wilson Bull.* 97:571.
MacTeague, L. 2004. Bat predation by the Acorn Woodpecker. *Western Birds* 35:45–46.
Mailliard, J. 1900. A neglected point concerning the Picidae. *Condor* 2(1):13.
Maine Department of Inland Fisheries and Wildlife. 2007. Maine's Comprehensive Wildlife Conservation Strategy. Augusta: Maine Dept. of Inland Fisheries and Wildlife. http://www.maine.gov/ifw/wildlife/conservation/action_plan.html.
Manitoba Wildlife and Ecosystem Protection Branch. 2008. Managing Animals, Plants, and Habitats: Species at Risk. Winnipeg: Manitoba Dept. of Conservation. http://www.gov.mb.ca/conservation/wildlife/sar/.
Mannan, R.W. 1984. Summer area requirements of Pileated Woodpeckers in western Oregon. *Wildlife Soc. Bull.* 12:265–268.
Manolis, T. 1987. Juvenile plumage of *Picoides* woodpeckers. *North American Bird Bander* 12(3):93.
MapArt. 1999. Alberta. [Road map.] Oshawa, ON: MapArt Publishing Co.
Marqua, D. G. 1963. Red-headed Woodpeckers in southern California. *Condor* 65:332.
Marsden, H. W. 1907. Feeding habits of the Lewis Woodpecker. *Condor* 9:27.
Marshall, D. B. 1992a. Black-backed Woodpecker (*Picoides articus*). in *Sensitive Vertebrates of Oregon*. Portland: Oregon Dept. of Fish and Wildlife.
———. 1992b. Pileated Woodpecker (*Dryocopus pileatus*). in *Sensitive Vertebrates of Oregon*. Portland: Oregon Dept. of Fish and Wildlife.
Marshall, D. B., M. G. Hunter, and A. L. Contreras, eds. 2003. *Birds of Oregon: A General Reference*. Corvallis: Oregon State Univ. Press.
Marshall, J. T. 1957. Birds of pine-oak woodland in southern Arizona and adjacent Mexico. *Pacific Coast Avifauna* no. 32.
Martin, J. W., and J. C. Kroll. 1975. Hoarding of corn by Golden-fronted Woodpeckers. *Wilson Bull.* 87:553.

Martin, K., and S. Ogle. 1998. The use of alpine habitats by fall migrating birds on Vancouver Island. Delta, BC: Department of Forest Sciences, University of British Columbia and Canadian Wildlife Service.
Martindale, S., and D. Lamm. 1984. Sexual dimorphism and parental role switching in Gila Woodpeckers. *Wilson Bull.* 96:116–121.
Maryland Wildlife and Heritage Service. 2005. Wildlife Diversity Conservation Plan. Annapolis: Maryland Dept. of Natural Resources. http://www.dnr.state.md.us/wildlife/Plants_Wildlife/WLDP/divplan_final.asp.
Maslowski, K. H. 1937. Northern Pileated Woodpecker in Hamilton County, Ohio. *Auk* 54:539–540.
Massachusetts Division of Fisheries and Wildlife. 2006, revised. 2005 Massachusetts Comprehensive Wildlife Conservation Strategy. Boston: Massachusetts Executive Office of Environmental Affairs, Department of Fish and Game..
Massey, B. W. 1998. *Guide to Birds of the Anza-Borrego Desert.* Borrego Springs, CA: Anza-Borrego Desert Natural History Assoc.
Maxon, S. J., and G. D. Maxon. 1981. Commensal foraging between Hairy and Pileated Woodpeckers. *J. of Field Ornith.* 52:62–63.
Mayfield, H. 1958. Nesting of the Black-backed Three-toed Woodpecker in Michigan. *Wilson Bull.* 70:195–196.
Mayr, E., and L. L. Short. 1970. *Species Taxa of North American Birds. A Contribution to Comparative Systematics.* Publication of the Nuttall Ornithological Club no. 9.
McAtee, W. L. 1942. Trade value of the beak of the Ivory-billed Woodpecker. *Condor* 44:41.
McAuliffe, J. R., and P. Hendricks. 1988. Determinants of the vertical distributions of woodpecker nest cavities in the saguaro cactus. *Condor* 90:791–801.
McClelland, B. R. 1977. Relationships between Hole-nesting Birds, Forest Snags, and Decay in Western Larch–Douglas-fir Forests of the northern Rocky Mountains. PhD diss., Univ. of Montana.
McClelland, B. R., and P. T. McClelland. 1999. Pileated Woodpecker nest and roost trees in Montana: links with old-growth and forest "health." *Wildlife Soc. Bull.* 27:846–858.
———. 2000. Red-naped Sapsucker nest trees in northern Rocky Mountain old-growth forest. *Wilson Bull.* 112:44–50.
McCreedy, C. 2006. The Draft Desert Bird Conservation Plan: A Strategy for Protecting and Maintaining Desert Habitats and Associated Birds in the Mojave and Colorado Deserts. Ver. 1.0. California Partners in Flight and PRBO Conservation Science. http://www.prbo.org/calpif/plans.html.
McFarlane, R. W. 1992. *A Stillness in the Pines.* New York: W. W. Norton.
McGregor, R. C. 1900. Discoloration of plumage in certain birds. *Auk* 2:18.
———. 1901. A list of the land birds of Santa Cruz County, California. *Pacific Coast Avifauna* no. 2.
McGuire, N. M. 1932. A Red-bellied Woodpecker robs a sapsucker. *Wilson Bull.* 44:39.
McHugh, J. J. 2005. Alabama Comprehensive Wildlife Conservation Strategy. Montgomery: Alabama Dept. of Conservation and Natural Resources.
McIlhenny, E. A. 1941. The passing of the Ivory-billed Woodpecker. *Auk* 58:582–584.
McIver, D. E. 1998. *Birding Utah.* Helena, MT: Falcon Publishing.
McKeever, S., and L. Adams. 1960. Acorn Woodpecker resident east of the Sierra Nevada in California. *Condor* 62:297.
McKerrow, A. J., S. G. Williams, and J. A. Colazzo. 2006. The North Carolina Gap Analysis Project, Final Report. Raleigh, NC: Biodiversity and Spatial Info. Ctr., North Carolina State Univ.
McKinley, D. 1958. Early record for the Ivory-billed Woodpecker in Kentucky. *Wilson Bull.* 70:380–381.
Mead, G. S. 1900. The Red-headed Woodpecker near Chicago, Ill. *Auk* 17:97.
Meanley, B. 1936. Albino Red-headed Woodpecker. *Auk* 53:100–101.
———. 1943. Red-cockaded Woodpecker breeding in Maryland. *Auk* 60:105.
Mearns, E. A. 1890a. Descriptions of a new species and three new subspecies of birds from Arizona. *Auk* 7(3):243–264.
———. 1890b. Observations on the avifauna of portions of Arizona. *Auk* 7(1):45–55.
Merriam, C. H. 1903. Some little-known basket materials. *Science* 17:826.
Meyer, E. J. M. 1981. The capture efficiency of Flickers preying on larval tiger beetles. *Auk* 98:189–191.
Michael, C. W. 1921. Pileated Woodpecker versus Cooper Hawk. *Condor* 23:68.
———. 1926. Acorn storing methods of the California and Lewis Woodpeckers. *Condor* 28:68–69.
———. 1928. The Pileated Woodpecker feeds on berries. *Condor* 30(2):157.
———. 1930. Prodigious drillings of a Williamson's Sapsucker. *Condor* 32:119–120.
———. 1935. Nesting of the Williamson Sapsucker. *Condor* 37:209–210.
———. 1936a. A late nesting record of the California Woodpecker. *Condor* 38:125.
———. 1936b. California Woodpeckers storing walnuts. *Condor* 38:177–178.
Michael, E. R. 1925. Remarkable work of the Pileated Woodpecker. *Condor* 27:174–175.
Miller, A. H. 1933. The inner abdominal feather region in brooding woodpeckers. *Condor* 35:78–79.
———. 1947. Arizona race of Acorn Woodpecker vagrant in California. *Condor* 49:171.
———. 1955a. Acorn Woodpecker on Santa Catalina Island, California. *Condor* 57:373.
———. 1955b. A hybrid woodpecker and its significance in speciation in the genus *Dendrocopos. Evolution* 9:317–321.
Miller, A. H., and C. E. Bock. 1972. Natural history of the Nuttall Woodpecker at the Hastings Reservation. *Condor* 74:284–294.
Miller, E. H., E. L. Walters, and H. Ouellet. 1999. Plum-

age, size, and sexual dimorphism in the Queen Charlotte Islands Hairy Woodpecker. *Condor* 101:86–95.

Miller, G. S. 1892. *Melanerpes carolinus* in Madison County, New York, in winter. *Auk* 9:201.

———. 1894. *Dryobates scalaris lucasanus* in San Diego County, California. *Auk* 11:178.

Miller, R. F. 1918. Early nesting of the Northern Pileated Woodpecker in Pennsylvania. *Auk* 35:479–480.

———. 1923. The Yellow-bellied Sapsucker in Philadelphia in August. *Auk* 40:694.

Mills, T. R., M. A. Rumble, and L. D. Flake. 2000. Habitat of birds in ponderosa pine and aspen/birch forest in the Black Hills, South Dakota. *J. of Field Ornith.* 71(2):187–206.

Milne, K. A., and S. J. Hejl. 1989. Nest-site characteristics of White-headed Woodpeckers. *J. of Wildl. Mgmt.* 53(1):50–55.

Milne, L., and M. Milne. 1980. *The Audubon Society Field Guide to North American Insects and Spiders.* New York: Alfred A. Knopf.

Minnesota Department of Natural Resources. 2006. Tomorrow's Habitat for the Wild and Rare: An Action Plan for Minnesota Wildlife, Minnesota's Comprehensive Wildlife Conservation Strategy. St. Paul: Division of Ecological Services, Minnesota Dept. of Natural Resources.

Mironov, S. V., J. Dabert, and R. Ehrnsberger. 2005. A new species of the feather mite genus *Pterotrogus* Gaud (Analgoidea: Pteronyssidae) from the Ivory-billed Woodpecker *Campephilus principalis* L. (Aves: Piciformes). *Ann. Entomol. Soc. Am.* 98(1): 3–17.

Mississippi Museum of Natural Science. 2005. Mississippi's Comprehensive Wildlife Conservation Strategy. Jackson: Mississippi Dept. of Wildlife, Fisheries and Parks.

Missouri Bird Records Committee. 2008. Annotated Checklist of Missouri Birds. Columbia, MO: Audubon Society of Missouri. http://mobirds.org/RecordsCommittee/MOChecklist.aspx.

Missouri Natural Heritage Program. 2008. Missouri Species and Communities of Concern Checklist. Jefferson City: Missouri Dept. of Conservation.

Mitchell, H. H. 1915. Lewis's Woodpecker taken in Saskatchewan. *Auk* 32:228–229.

Molhoff, W. J. 2000. *Nebraska Breeding Bird Atlas.* Lincoln: Nebraska Game and Parks Comm. and Nebraska Ornithologists' Union.

Monson, G. 1942. Notes on some birds of southeastern Arizona. *Condor* 44:222–225.

Montana Fish, Wildlife and Parks. 2005. Montana's Comprehensive Fish and Wildlife Conservation Strategy. Helena: Montana Fish, Wildlife and Parks.

Montgomery, T. H., Jr. 1905. Summer resident birds of Brewster County, Texas. *Auk* 22(1):12–15.

Moore, W. S. 1987. Random mating in the Northern Flicker hybrid zone: implications for the evolution of bright and contrasting plumage patterns in birds. *Evolution* 41:539–546.

———. 1995. Northern Flicker (*Colaptes auratus*). No. 166 in *The Birds of North America*, A. Poole and F. Gill, eds. Philadelphia, PA: Acad. of Nat. Sci., and Washington, D.C: AOU.

Moore, W. S., and D. B. Buchanan. 1985. Stability of the Northern Flicker hybrid zone. *Evolution* 39: 135–151.

Moore, W. S., J. H. Graham, and J. T. Price. 1991. Mitochondrial DNA variation in the Northern Flicker (*Colaptes auratus*, Aves). *Molec. Biol. and Evol.* 8:327–344.

Moore, W. S., and W. D. Koenig. 1986. Comparative reproductive success of Yellow-shafted, Red-shafted, and hybrid flickers across a hybrid zone. *Auk* 103: 42–51.

Moore, W. S., K. J. Miglia, M. Lammertink, and R. C. Fleischer. 2005. Phylogeny and Evolutionary History of Large Woodpeckers in the Genera *Campephilus, Dryocopus,* and *Mullerpicus.* Detroit, MI: Wayne State Univ.

Moore, W. S., and J. T. Price. 1993. The nature of selection in the Northern Flicker hybrid zone and its implications for speciation theory. Pp. 196–225 in R. G. Harrison, ed., *Hybrid Zones and the Evolutionary Process.* Oxford, UK: Oxford Univ. Press.

Morlan, J. 2002. November 2002 Mystery Bird. http://fog.ccsf.cc.ca.us/~jmorlan/nov02.htm.

Morris, C. H. 1905. Winter notes on the Yellow-bellied Sapsucker (*Sphyrapicus varius*). *Wilson Bull.* 51:56–57.

Morris, E. L. 1905. Societies and Academies: The Biological Society of Washington. *Science* 21(541): 744–747.

Morrison, M. L., and K. A. With. 1987. Interseasonal and intersexual resource partitioning in Hairy and White-headed Woodpeckers. *Auk* 104:225–233.

Morse, B. 2001. *A Birder's Guide to Coastal Washington.* Olympia, WA: R. W. Morse Co.

Moskovits, D. 1978. Winter territorial and foraging behavior of Red-headed Woodpeckers in Florida. *Wilson Bull.* 90:521–535.

Mueller, H. C. 1971. Sunflower seed carrying by Red-bellied Woodpeckers (*Centurus carolinus*). *Bird-Banding* 42:46–47.

Munro, J. A. 1943. Competition between Mountain Bluebirds and Hairy Woodpeckers. *Condor* 45:74.

Myers, H. W. 1915. A late nesting record for the California Woodpecker. *Condor* 17:183–185.

Nappi, A., P. Drapeau, J.-F. Giroux, and J.-P. L. Savard. 2003. Snag use by foraging black-backed woodpeckers (*Picoides arcticus*) in a recently burned eastern boreal forest. *Auk* 120:505–511.

Nature Conservancy of Arkansas. 2005. Conservation in the Big Woods leads to rediscovery of the ivory-bill in Arkansas. aboard the ark, 2005 Member Report. Little Rock: The Nature Conservancy of Arkansas.

Nauman, E. D. 1932. The Red-headed Woodpecker as a mouser. *Wilson Bull.* 44:44.

Neel, L. A., ed. 1999. Nevada Partners in Flight Bird Conservation Plan. Nevada Partners in Flight. http://www.partnersinflight.org/bcps/plan/pl-nv-10.pdf.

Nehls, H. 1985. Distribution of the Yellow-bellied type Sapsuckers in Oregon. *Oregon Birds* 11(4):155–158.

Neill, A. J., and R. G. Harper. 1990. Red-bellied Woodpecker predation on nestling House Wrens. *Condor* 92:789.

Neill, R. L. 1975. *The Birds of the Buescher Division.* Publication no. 3. Austin: Univ. of Texas System Cancer Center, Extramural Programs Division, Environmental Science Park.

Nesbitt, S. A., D. T. Gilbert, and D. B. Barbour. 1978. Red-cockaded Woodpecker fall movements in a Florida flatwoods community. *Auk* 95:145–151.

Nesom, G. 2002. Quaking Aspen Plant Guide. Davis, CA: USDA, Natural Resources Conservation Service, National Plant Data Center..

New Brunswick Natural Resources. 2015. General Status of Wild Species, Species and Status Databases. Fredericton: New Brunswick Dept. of Natural Resources. http://www1.gnb.ca/0078/WildlifeStatus/search-e.asp.

Newfoundland and Labrador Department of Environmental Conservation. 2008. Wildlife at Risk. St. John's, NL: Newfoundland and Labrador Department of Environment and Conservation. http://www.env.gov.nl.ca/env/wildlife/endangeredspecies/birds.html.

New Hampshire Fish and Game Department. 2005. New Hampshire Wildlife Action Plan. Concord: New Hampshire Fish and Game Dept.

New Jersey Department of Environmental Protection. 2004. New Jersey Wildlife Action Plan for Wildlife of Greatest Conservation Need. Trenton: New Jersey Dept. of Environmental Protection.

Newlon, K. R. 2005. Demography of Lewis's Woodpecker, Breeding Bird Densities, and Riparian Aspen Integrity in a Grazed Landscape. Master's thesis, Montana State Univ.

New Mexico Department of Game and Fish. 2006. Comprehensive Wildlife Conservation Strategy for New Mexico. Santa Fe: New Mexico Dept. of Game and Fish.

Nice, M. M. 1927. Pileated Woodpeckers wintering in Cleveland County, Oklahoma. *Auk* 44:103.

Nicholls, T., and M. Ostry. 2003. Dead trees bring life to forest critters. *Minnesota Better Forests* 7(4):14–15.

Nicholoff, S. H., compiler. 2003. Wyoming Bird Conservation Plan. Ver. 2.0. Lander: Wyoming PIF and Wyoming Game and Fish Dept.

Nichols, J. T. 1904. Black-backed Three-toed Woodpecker and Evening Grosbeak at Wellfleet, Mass. *Auk* 21:81–82.

Nicholson, C. P. 1997. *Atlas of the Breeding Birds of Tennessee.* Knoxville: Univ. of Tennessee Press.

Nielsen-Pincus, N. 2005. Nest Site Selection, Nest Success, and Density of Selected Cavity-nesting Birds in Northeastern Oregon with a Method for Improving the Accuracy of Density Estimates. Master's thesis, Univ. of Idaho.

Niskanen, C. 2005. Old friend found. *Birder's World* 19(4):20–23.

Noble, G. K. 1936. Courtship and sexual selection of the flicker (*Colaptes auratus luteus*). *Auk* 53:269–282.

Noecker, R. J., and M. L. Corn. 1997. The Red-cockaded Woodpecker: federal protection and habitat conservation plans. Washington, D.C.: National Council for Science and the Environment.

Nolan, V., Jr. 1959. Pileated Woodpecker attacks pilot black snake at tree cavity. *Wilson Bull.* 71:381–382.

North Carolina Wildlife Resources Committee. 2005. North Carolina Wildlife Action Plan. Raleigh: North Carolina Wildlife Resources Comm.

Northwest Territories Department of Environment and Natural Resources. 2014. Species at Risk in the Northwest Territories. Yellowknife: Government of the Northwest Territories.

Norton, A. H. W. 1896. Queer actions of Golden-fronted Woodpeckers. *Wilson Bull.* 8:4.

Nova Scotia Wildlife Division. 2007. Nova Scotia Endangered Species Act: Legally Listed Species as of 2007. Kentville: Province of Nova Scotia, Dept. of Natural Resources. http://novascotia.ca/natr/wildlife/biodiversity/species-list.asp.

Nunavut Department of Environment. 2007. Statutory Report on Wildlife to the Nunavut Legislative Assembly (Section 176 of the Wildlife Act). Iqaluit: Nunavut Dept. of Environment.

Nye, H. A. 1918. The sapsucker wintering in central Maine. *Auk* 35:353–354.

Oberholser, H. C. 1911. A revision of the forms of the Hairy Woodpecker (*Dryobates scalaris* [Wagler]). *Proc. U.S. Nat. Mus.* 41:139–159.

———. 1918. *Picoides arcticus* in Florida. *Auk* 35:479.

———. 1930. The migration of North American birds, second series: XLIII. Ivory-billed Woodpecker. *Bird-Lore* 32(4):265.

Oberle, M. W. 1970. Endangered species: Congress curbs international trade in rare animals. *Science* 167:152–154.

Ohio Bird Records Committee. 2007. Annotated Checklist of Ohio Birds. Westerville, OH: Ohio Ornithological Society.

Ohio Department of Natural Resources. 2006. Ohio Comprehensive Wildlife Conservation Strategy. Columbus: Ohio Dept. of Natural Resources.

Oklahoma Department of Wildlife Conservation. 2005. Oklahoma Comprehensive Wildlife Conservation Strategy. Oklahoma City: Oklahoma Dept. of Wildlife Conservation.

Oliver, W. W. 1970. The feeding pattern of sapsuckers on ponderosa pine in northeastern California. *Condor* 72:241.

Olson, J. A. 2002. Special-animal abstract for *Picoides arcticus* (black-backed woodpecker). Lansing, MI: Michigan Natural Features Inventory.

Olson, S. L. 1983. Evidence for a polyphyletic origin of the Piciformes. *Auk* 100:126–133.

Opperman, H. 2003. *A Birder's Guide to Washington.* Colorado Springs, CO: American Birding Assoc.

Orcutt, C. R. 1884. Stones placed in pine-trees by birds. *Science* 3:305.

Oregon Department of Fish and Wildlife. 2006. Oregon Conservation Strategy. Salem: Oregon Dept. of Fish and Wildlife.

Ostry, M. E., K. Daniels, and N. A. Anderson. 1982. Downy Woodpeckers—a missing link in a forest disease life cycle? *Loon* 54:170–175.

Ouachita National Forest. 2005. Renewal and recovery: renewal of the shortleaf pine-bluestem grass ecosystem, recovery of Red-cockaded Woodpeckers. Guide to Pine-bluestem Project. Hot Springs, AR: USDA, Ouachita Nat. Forest.

Palmer, T. S. 1928. Notes on persons whose names appear in the nomenclature of California birds: a contribution to the history of West Coast ornithology. *Condor* 30(5):261–307.

Palmer-Ball, B., Jr. 1996. *The Kentucky Breeding Bird Atlas.* Lexington: Univ. Press of Kentucky.

Palouse Audubon Society. 2003. Palouse Audubon Society Conservation Plan. Moscow, ID, and Pullman, WA: Palouse Audubon Society.

Parks, G. H. 1944. Unusual behavior of Hairy Woodpeckers. *Bird-Banding* 23–24.

Parmeter, J., B. Neville, and D. Emkalns. 2002. *New Mexico Bird Finding Guide.* 3d ed. Albuquerque: New Mexico Ornithithological Soc.

Parr, M. 2005. Bird's eye view. *Bird Conservation* June 2005:2.

Parrish, J. R., F. Howe, and R. Norvell, eds. 2002. Utah Partners in Flight Avian Conservation Strategy. Ver. 2.0. UDWR Publ. No. 02-27. Salt Lake City, UT: Utah PIF Program, Utah Div. of Wildlife Res.

Paulson, D. 2013. *Birds of Washington.* Rev. ed. Tacoma, WA: Slater Museum of Natural History, University of Puget Sound.

Pearson, T. G., ed. 1936. *Birds of America.* Garden City, NY: Garden City Publ. Co.

Peck, E. B. 1890. The Red-bellied Woodpecker. *Wilson Bull.* 2:15.

Peck, M. E. 1921. On the acorn-storing habit of certain woodpeckers. *Condor* 23:131.

Pemberton, J. R. 1923. Lewis Woodpecker in eastern Oklahoma. *Condor* 25:107.

Pennsylvania Game Commission. 2005. Pennsylvania Comprehensive Wildlife Conservation Strategy, Version 1.0. Harrisburg: Penn. Game Comm. and Penn. Fish and Boat Comm.

Petersen, W. R. 2007. The regional reports: New England. *North American Birds* 60(4):507.

Peterson, B., and G. Gauthier. 1985. Nest site use by cavity-nesting bird of Cariboo Parkland, British Columbia. *Wilson Bull.* 97:319–331.

Peterson, J., and B. R. Zimmer. 1998. *Birds of the Trans-Pecos.* Austin: Univ. of Texas Press.

Peterson, J. G. 1946. Red-naped Sapsucker in Santa Clara County, California. *Condor* 48:95.

Peterson, R. T. 1948. *Birds Over America.* New York: Dodd, Mead & Co.

———. 1957. *The Bird Watcher's Anthology.* New York: Harcourt, Brace and Co.

Petit, D. R., K. E. Petit, T. C. Grubb, Jr., and L. J. Reichhardt. 1985. Habitat and snag selection by woodpeckers in a clear-cut: an analysis using artificial snags. *Wilson Bull.* 97:525–533.

Pettingill, O. S., Jr. 1970. *Ornithology in Laboratory and Field.* 4th ed. Minneapolis, MN: Burgess Publ. Co.

Peyton, S. B. 1917. Large set of eggs of the California Woodpecker. *Condor* 19:103.

Phillips, A. R., J. T. Marshall, Jr., and G. Monson. 1964. *The Birds of Arizona.* Tucson: Univ. of Arizona Press.

Phillips, L. C., and B. S. Hall. 2000. A historical view of Red-cockaded Woodpecker habitat on Fort Polk, Louisiana. *J. of Field Ornith.* 71:585–596.

Pilliod, D. S., E. L. Bull, J. L. Hayes, and B. C. Wales. 2006. Wildlife and invertebrate response to fuel reduction treatments in dry coniferous forests of the western United States: a synthesis. Gen. Tech. Rept. RMRS-GTR-173. Ft. Collins, CO: USDA Forest Service, Rocky Mtn. Res. Sta.

Pinkowski, B. C. 1977. Food storage and re-storage in the Red-headed Woodpecker. *Bird-Banding* 48:74–75.

Podulka, S., R. W. Rohrbaugh, Jr., and R. Bonney. 2004. *Handbook of Bird Biology.* Ithaca, NY: Cornell Lab of Ornithology.

Potter, L. B. 1930. Drilling habits of the Flicker. *Condor* 32:125.

Powers, M. 2005. Hope knocks. *BirdScope* 19(3):14.

Pranty, B. 2002. The regional reports: Florida. *North American Birds* 56(1):46.

Prychitko, T. M., and W. S. Moore. 2000. Comparative evolution of the mitochondrial b gene and nuclear beta-fibrinogen intron 7 in woodpeckers. *Molec. Biol. and Evol.* 17:1101–1111.

Pulich, W. M. 1988. *The Birds of North Central Texas.* College Station: Texas A&M Univ. Press.

Pyle, P. 1997. *Identification Guide to North American Birds. Part I: Columbidae to Ploceidae.* Bolinas, CA: Slate Creek Press.

Pyle, P., and S. N. G. Howell. 1995. Flight-feather molt patterns and age in North American Woodpeckers. *J. of Field Ornith.* 66:564–581.

Pyle, P., and G. McCaskie. 1992. Thirteenth report of the California Bird Records Committee. *Western Birds* 23:97–132.

Raitt, R. J. 1959. Rocky Mountain race of the Williamson Sapsucker wintering in California. *Condor* 62:142.

Ramp, W. K. 1965. The auditory range of a Hairy Woodpecker. *Condor* 67:183–185.

Raphael, M. G. 1985. Use of *Arbutus menziesii* by cavity-nesting Birds. Gen. Tech. Rept. PSW-GTR-100:19–26. Portland, OR: USDA Forest Service, Pacific Northwest Research Station.

Raphael, M. G., M. L. Morrison, and M. P. Yoder-Williams. 1987. Breeding bird populations during twenty-five years of postfire succession in the Sierra Nevada. *Condor* 89:614–626.

Rappole, J. H., and G. W. Blacklock. 1985. *Birds of the Texas Coastal Bend, Abundance and Distribution.* College Station: Texas A&M Univ. Press.

Rathbun, S. F. 1911. Northern Flicker (*Colaptes auratus luteus*) in San Juan County, Washington. *Auk* 28:486.

Rathcke, B. J., and R. W. Poole. 1974. Red squirrel attacks a Pileated Woodpecker. *Wilson Bull.* 86:465–466.

Reed, C. E. 1929. Yellow-bellied Sapsucker winters in Brookfield, Mass. *Auk* 46:114.

Reinking, D. L. 2004. *Oklahoma Breeding Bird Atlas*. Norman: Univ. of Oklahoma Press.

Renken, R. B., and E. P. Wiggers. 1989. Forest characteristics related to Pileated Woodpecker territory size in Missouri. *Condor* 91:642–652.

———. 1993. Habitat characteristics related to Pileated Woodpecker densities in Missouri. *Wilson Bull.* 105:77–83.

Repasky, R. R., R. J. Blue, and P. D. Doerr. 1991. Laying Red-cockaded Woodpeckers cache bone fragments. *Condor* 93:458–461.

Rett, E. 1918. Downy Woodpecker in Colorado. *Auk* 35:223–224.

Reynolds, P., and S. Lima. 1994. Direct use of wings by foraging woodpeckers. *Wilson Bull.* 106:408–411.

Rhode Island Division of Fish and Wildlife. 2005. Rhode Island's Comprehensive Wildlife Conservation Strategy. Providence: Rhode Island Dept. of Environmental Management.

Rice, D., and B. Peterjohn. 1991. *Ohio Breeding Bird Atlas*. Columbus: Ohio Dept. of Natural Resources.

Rich, T. D., C. J. Beardmore, H. Berlanga, P. J. Blancher, M. S. W. Bradstreet, G. S. Butcher, et al. 2004. Partners in Flight North American Landbird Conservation Plan. Ver.: March 2005. Ithaca, NY: Cornell Lab of Ornithology. www.partnersinflight.org/cont_plan/.

Richardson, C. H. 1910. *Colaptes auratus luteus* in Los Angeles County, Cal. *Condor* 7:53.

Richer, C. 1996. *San Francisco Peninsula Birdwatching*. San Mateo, CA: Sequoia Audubon Society.

Richmond, J. 1985. *Birding Northern California: Site Guides to 72 of the Best Birding Spots*. Walnut Creek, CA: Mt. Diablo Audubon Society.

Ridgway, R. 1887. *A Manual of North American Birds*. Philadelphia, PA: J. B. Lippencott.

———. 1914. Birds of North and Middle America. Pt. 6. *U.S. National Museum Bull.* 50.

Righter, R., R. Levad, C. Dexter, and K. Porter. 2004. *Birds of Western Colorado: Plateau and Mesa Country*. Grand Junction, CO: Grand Valley Audubon Society.

Rinker, G. C. 1941. Boreal Flicker in Kansas. *Auk* 58:581–582.

Rissler, L. J., D. N. Karowe, F. Cuthbert, and B. Scholtens. 1995. The influence of Yellow-breasted Sapsuckers on local insect community structure. *Wilson Bull.* 107:746–752.

Ritter, S. 2000. Idaho Bird Conservation Plan. Ver.1.0. Hamilton, MT: Idaho Partners in Flight.

Ritter, W. E. 1921. Acorn-storing by the California Woodpecker. *Condor* 23:3–14.

———. 1922. Further observations on the activities of the California Woodpecker. *Condor* 24:109–122.

———. 1938. *The California Woodpecker and I*. Berkeley: Univ. of California Press.

Robbins, C. S., ed., and E. A. T. Blom, proj. coord. 1996. *Atlas of the Breeding Birds of Maryland and the District of Columbia*. Pittsburgh, PA: Univ. of Pittsburgh Press.

Robbins, M. B., and D. A. Easterla. 1992. *Birds of Missouri: Their Distribution and Abundance*. Columbia: Univ. of Missouri Press.

Robbins, R. C. 1900. The Arctic Three-toed Woodpecker in Beverly, Mass. *Auk* 27:173.

Roberson, D. 2002. *Monterey Birds*. Carmel, CA: Monterey Peninsula Audubon Society.

Roberts, R. C. 1979. Habitat and resource relationships in Acorn Woodpeckers. *Condor* 81:1–8.

Robertson, J. M. 1935. Lewis Woodpecker in Death Valley. *Condor* 37:173.

Robinson, B. R., and K. V. Rosenberg. 2003. Partners in Flight Landbird Conservation Plan, Allegheny Plateau (Physiographic Area 24). Ver. 1.1, August 2003. The Plains, VA: ABC.

Robinson, G. 1957. Observations of pair relations of White-headed Woodpeckers in winter. *Condor* 59:339–340.

Robinson, J., and J. Alexander. 2002. The draft coniferous forest bird conservation plan: a strategy for protecting and managing coniferous forest habitats and associated birds in California. Ver. 1.0. Stinson Beach, CA: California Partners in Flight and Point Reyes Bird Observatory.

Robinson, W. 1926. Arctic Three-toed Woodpecker at West Point, N.Y. *Auk* 43:98.

Robinson, W. D. 1996. *Southern Illinois Birds: An Annotated List and Site Guide*. Carbondale: Southern Illinois Univ. Press.

Rodewald, P. G., M. J. Santiago, and A. D. Rodewald. 2005. Habitat use of breeding Red-headed Woodpeckers on golf courses in Ohio. *Wildlife Soc. Bull.* 33:448–453.

Rogers, D. T., Jr., J. A. Jackson, B. J. Schardien, and M. S. Rogers. 1979. Observations at a nest of a partial albino Red-headed Woodpecker. *Auk* 96:206–207.

Rogers, J. 1978. Brief notes: White-headed Woodpecker in Curry County. *Oregon Birds* 4(5):34.

Rogers, M. M., and A. Jaramillo. 2002. Report of the California Bird Records Committee: 1999 records. *Western Birds* 33:1–33.

Rohrbaugh, R. 2005. The Search for the Ivory-bill. *BirdScope* 19(3):6–7.

———., proj. dir. 2007. Final Report: 2006–2007 Ivory-billed Woodpecker Surveys and Equipment Loan Program. Ithaca, NY: Cornell Lab of Ornithology.

Rosen, V. H. 1925. Northern Pileated Woodpecker, Cummington, Mass. *Auk* 42:586.

Rosenberg, K. V. 2000a. Partners in Flight Landbird Conservation Plan, Adirondack Mountains (Physiographic Area 26). The Plains, VA: ABC.

———. 2000b. Partners in Flight Landbird Conservation Plan, St. Lawrence Plain (Physiographic Area 18). Draft ver. 1.0. The Plains, VA: ABC.

———. 2003. Partners in Flight Landbird Conservation Plan, Mid-Atlantic Ridge and Valley (Physiographic Area 12). The Plains, VA: ABC.

———. 2005. From Big Woods partnership to Ivory-bill recovery. *BirdScope* 19(3):17.

Rosenberg, K. V., and R. Dettmers. 2004. Partners in Flight Landbird Conservation Plan, Ohio Hills

(Physiographic Area 22). Ver. 1.1. The Plains, VA: ABC.

Rosenberg, K. V., and T. P. Hodgman. 2000. Partners in Flight Landbird Conservation Plan, Eastern Spruce-Hardwood Forest (Physiographic Area 28). Draft 1.0. The Plains, VA: ABC.

Rosenberg, K. V., R. D. Ohmart, W. C. Hunter, and B. W. Anderson. 1991. *Birds of the Lower Colorado River Valley*. Tucson: Univ. of Arizona Press.

Rosenberg, K. V., K. Radamaker, and M. M. Stevenson. 2007. Arizona Bird Committee report, 2000–2004 records. *Western Birds* 38(2):74–101.

Rosenberg, K. V., and B. Robertson. 2003. Partners in Flight Landbird Conservation Plan, Northern Ridge and Valley (Physiographic Area 17). Ver. 1.1. The Plains, VA: ABC.

Rosenberg, K. V., R. W. Rohrbaugh, and M. Lammertink. 2005. An overview of Ivory-billed Woodpecker (*Campephilus principalis*) sightings in eastern Arkansas in 2004–2005. *North American Birds* 59(2):199–206.

Rosenberg, K. V., and J. L. Witzeman. 1998. Arizona Bird Committee report, 1974–1996: Part I (nonpasserines). *Western Birds* 29:199–224.

Rosene, W. M. 1936. The Red-shafted Flicker in Boone County, Iowa. *Wilson Bull.* 48:219–220.

Roth, R. R. 1978. Attacks on Red-headed Woodpeckers by flycatchers. *Wilson Bull.* 90:450–451.

Rothenbach, C. A., and C. Opio. 2005. Sapsuckers usurp a nuthatch nest. *Wilson Bull.* 117:101–103.

Rothschild, B. 2005. Fortuitous photograph of the most arthritis-afflicted of the woodpeckers. *Arch. of Int. Med.* 165:2037.

Rottenborn, S. C., and J. Morlan. 2000. Report of the California Bird Records Committee: 1997 records. *Western Birds* 31(1):1–37.

Royal Alberta Museum. n.d. Project Sapsucker Instruction Booklet. Edmonton: Royal Alberta Museum.

Royall, W. C., Jr., and O. E. Bray. 1980. A study of radio-equipped flickers. *North American Bird Bander* 5:47–50.

Rudolph, D. C., R. N. Conner, and R. R. Schaefer. 1991. Yellow-breasted Sapsuckers feeding at Red-cockaded Woodpecker resin wells. *Wilson Bull.* 103:122–123.

Rumble, M. A., and J. E. Gobeille. 1995. Wildlife associations in Rocky Mountain juniper in the northern Great Plains, South Dakota. Pp. 80–90 in D. W. Shaw, E. F. Aldon, and C. LoSapio, eds., Desired future conditions for pinon-juniper ecosystems: August 8–12, Flagstaff, Arizona. USDA Forest Service Gen. Tech. Rept. RM-258. Ft. Collins, CO: USDA Forest Service, Rocky Mtn. Res. Sta.

Russell, W. C. 1947. Mountain Chickadees feeding young Williamson's Sapsuckers. *Condor* 49:83.

Rustay, C., and S. Norris, compilers. 2007. New Mexico Bird Conservation Plan. Ver. 2.1. Albuquerque: New Mexico Partners in Flight.

Ryser, F. A. 1963. Prothonotary Warbler and Yellow-shafted Flicker in Nevada. *Condor* 65:334.

———. 1985. *Birds of the Great Basin: A Natural History*. Reno: Univ. of Nevada Press.

Saab, V., R. Brannon, J. Dudley, L. Donohoo, D. Vanderzanden, V. Johnson, and H. Lachowski. 2002. Selection of fire-created snags at two spatial scales by cavity-nesting birds. Gen. Tech. Rept. PSW-GTR-181. Albany, CA: USDA Forest Service, Pacific Southwest Research Station.

Saab, V. A., J. Dudley, and W. L. Thompson. 2004. Factors influencing occupancy of nest cavities in recently burned forests. *Condor* 106:20–36.

Saab, V. A., H. D. W. Powell, N. B. Kotliar, and K. R. Newlon. 2005. Variation in fire regimes in the Rocky Mountains: implications for avian communities and fire management. *Studies in Avian Biology* no. 30:76–96.

Saab, V. A., R. E. Russell, and J. G. Dudley. 2007. Nest densities of cavity-nesting birds in relation to post-fire salvage logging and time since wildfire. *Condor* 109:97–108.

Saab, V. A., and K. T. Vierling. 2001. Reproductive success of Lewis's Woodpecker in burned pine and cottonwood riparian forest. *Condor* 103:491–501.

Saenz, D., R. N. Conner, C. E. Shackelford, and D. C. Rudolph. 1998. Pileated Woodpecker damage to Red-cockaded Woodpecker cavity trees in eastern Texas. *Wilson Bull.* 110:362–367.

Salafsky, S. R., R. T. Reynolds, and B. R. Noon. 2005. Patterns of temporal variation in Goshawk reproduction and prey resources. *J. of Raptor Res.* 39:237–246.

San Diego Natural History Museum. 2008. Fossil Mysteries: Geologic Timeline. San Diego: San Diego Nat. History Museum. https://www.sdnhm.org/archive/exhibits/mystery/fg_timeline.html.

Santa Clara Valley Audubon Society. 2002. *Birding at the Bottom of the Bay: South San Francisco Bay*. 3d ed. Cupertino, CA: Santa Clara Valley Audubon Society.

Saskatchewan Conservation and Data Centre. 2008. Species at Risk in Saskatchewan. Regina: Saskatchewan Ministry of Environment.

Sauer, J. R., J. E. Hines, and J. Fallon. 2008. The North American Breeding Bird Survey, Results and Analysis 1966–2007. Ver. 5.15.2008. Laurel, MD: USGS Patuxent Wildlife Research Center.

Saul, L. J. 1983. Red-bellied Woodpecker responses to accipiters. *Wilson Bull.* 95:490–491.

Saunders, A. A. 1921. A distributional list of the birds of Montana. *Pacific Coast Avifauna* no. 14.

Savignac, C., A. Desrochers, and J. Huot. 2000. Habitat use by Pileated Woodpeckers at two spatial scales in eastern Canada. *Can. J. of Zool.* 78:219–221.

Sawyer, E. J. 1916. Arctic Three-toed Woodpecker in Jefferson Co., N.Y. *Auk* 34:88.

Schemnitz, S. D. 1964. Nesting association of Pileated Woodpecker and Yellow-shafted Flicker in a utility pole. *Wilson Bull.* 76:95.

Schneider, R., M. Humpert, K. Stoner, and G. Steinauer. 2005. The Nebraska Natural Legacy Project: A Comprehensive Wildlife Conservation Strategy. Lincoln: Nebraska Game and Parks Comm.

Schoch, D. T. 2005. Forest management for Ivory-billed Woodpeckers (*Campephilus principalis*): a case in

managing an uncertainty. *North American Birds* 59(2):216–221.
Schorger, A. W. 1940. The Arctic Three-toed Woodpecker as a breeding bird in Wisconsin. *Auk* 52:309.
———. 1949. An early record and description of the Ivory-billed Woodpecker in Kentucky. *Wilson Bull.* 61(4):235.
Schram, B. 1998. *A Birder's Guide to Southern California.* Colorado Springs, CO: American Birding Assoc.
Schroeder, R. L. 1983. Habitat suitability index models: Downy Woodpecker. Fort Collins, CO: USFWS.
Schwab, F. E., N. P. P. Simon, and A. R. E. Sinclair. 2006. Bird-vegetation relationships in southeastern British Columbia. *J. of Wildl. Mgmt.* 70:189–197.
Schwab, I. R. 2002. Cure for a headache. *Royal British J. of Ophthalmology* 86:843.
Scotese, C. R. 2008. PALEOMAP Project. Arlington, TX: PALEOMAP Project. www.scotese.com/earth.htm.
Scott, D. M., C. D. Ankney, and C. H. Jarosch. 1976. Sapsucker hybridization in British Columbia: changes in 25 years. *Condor* 78:253–257.
Scott, V. E., K. E. Evans, D. R. Patton, and C. P. Stone. 1977. Cavity-nesting birds of North American forests. Agricultural Handbook no. 511. Washington, DC: USDA.
Scoville, S., Jr. 1923. Pileated Woodpecker in Connecticut and New Jersey. *Auk* 40:533–534.
Seaman, G. A. 1954. Yellow-bellied Sapsucker on Anegada, British West Indies. *Wilson Bull.* 66:61.
Sedgwick, J. A., and F. L. Knopf. 1991. The loss of avian cavities by injury compartmentalization. *Condor* 93:781–783.
Selander, R. K. 1965. Sexual dimorphism in relation to foraging behavior in the Hairy Woodpecker. *Wilson Bull.* 77:416.
Selander, R. K., and D. R. Giller. 1959. Interspecific relations of woodpeckers in Texas. *Wilson Bull.* 71:106–124.
———. 1963. Species limits in the woodpecker genus *Centurus. Bull. of AMNH* 124:213–274.
Semenchunk, G. P. 1992. *The Atlas of Breeding Birds of Alberta.* Edmonton: Federation of Alberta Naturalists.
Semo, L. S., and T. Leukering. 2004. Corrigenda and additions to "Amendments to the state review list." *Colorado Birds* 38:23–28.
Semple, J. B. 1930. Red-headed Woodpeckers in migratory flight. *Auk* 47:84–85.
Sennett, G. B. 1878. Notes on the ornithology of the lower Rio Grande of Texas from observations made during the season of 1877. *Bull. U.S. Geol. and Geog. Survey* 4(1):1–39.
Servin, J., S. L. Lindsey, and B. A. Loiselle. 2001. Pileated Woodpecker scavenges on a carcass in Missouri. *Wilson Bull.* 113:249–250.
Seyffert, K. D. 2001. *Birds of the Texas Panhandle: Their Status, Distribution, and History.* College Station: Texas A&M Univ. Press.
Shackelford, C. E. 2000. Woodpecker Damage: A Simple Solution to a Common Problem. Texas Parks and Wildlife PWD BK W7000-616. Austin: Texas Parks and Wildlife.
Shackelford, C. E., R. E. Brown, and R. N. Conner. 2000. Red-bellied Woodpecker (*Melanerpes carolinus*). No. 500 in *The Birds of North America,* A. Poole and F. Gill, eds. Philadelphia, PA: Birds of North America, Inc.
Shackelford, C. E., and R. N. Conner. 1997. Woodpecker abundance and habitat use in three forest types in eastern Texas. *Wilson Bull.* 109:614–629.
Shainin, V. 1939. American and Arctic Three-toed Woodpeckers in the Adirondacks. *Auk* 56:84.
Shaw, W. T. 1924. Occurrence of the Alpine Three-toed Woodpecker in Washington. *Condor* 27:36.
Shelley, L. O. 1933. Some notes on the Hairy Woodpecker. *Bird-Banding* 4:204–205.
———. 1934. Some notes on the Yellow-bellied Sapsucker in southwestern New Hampshire. *Auk* 51:523–524.
———. 1935. Flickers attacked by starlings. *Auk* 52:93.
———. 1938. The Eastern Hairy Woodpecker (*Dryobates v. villosus*) as a migrant. *Bird-Banding* 9:48–50.
Sherman, A. 1907. Another provident *Melanerpes erythrocephalus. Wilson Bull.* 59:72.
———. 1910. At the sign of the Northern Flicker. *Wilson Bull.* 22:135–171.
Sherrill, D. M., and V. M. Case. 1980. Winter home ranges of four clans of Red-cockaded Woodpeckers in the Carolina sandhills. *Wilson Bull.* 92:369–375.
Sherwood, W. E. 1927. Feeding habits of Lewis Woodpecker. *Condor* 29:171.
Shire, G. G., K. Brown, and G. Winegrad. 2000. Communication Towers: A Deadly Hazard to Birds. The Plains, VA: ABC.
Short, L. L. 1965a. A melanistic Pileated Woodpecker specimen from Georgia. *Wilson Bull.* 77:404–405.
———. 1965b. Hybridization in the flickers (*Colaptes*) of North America. *Bull. of AMNH* 129:307–428.
———. 1965c. Variation in West Indian Flickers (Aves, *Colaptes*). *Bull. Florida St. Mus. Bio. Sci.* 10:1–42.
———. 1965d. Specimens of Nuttall Woodpecker from Oregon. *Condor* 67:269–270.
———. 1967. Variation in Central American flickers. *Wilson Bull.* 79:5–21.
———. 1969a. An apparently melanic Hairy Woodpecker from New Mexico. *Bird-Banding* 40: 145–146.
———. 1969b. Taxonomic aspects of avian hybridization. *Auk* 86:84–105.
———. 1970. Reversed sexual dimorphism in tail length and foraging differences in woodpeckers. *Bird-Banding* 41:85–92.
———. 1971. Systematics and behavior of some North American woodpeckers, genus *Picoides* (Aves). *Bull. of AMNH* 145:1–118.
———. 1972. Systematics and behavior of South American flickers (Aves, *Colaptes*). *Bull. of AMNH* 149:1–109.
———. 1974. Habits and Interactions of North American Three-toed Woodpeckers (*Picoides arcticus* and *Picoides tridactylus*). American Museum Novitates 2,547:1–42.
———. 1982. *Woodpeckers of the World.* Greenville, DE:

Delaware Museum of Natural History, Monograph series no. 4.
Short, L. L., and J. J. Morony, Jr. 1970. A second hybrid Williamson's x Red-naped Sapsucker and an evolutionary history of sapsuckers. *Condor* 72:310–315.
Shuford, W. D. 1985. Acorn Woodpecker mutilates nestling Red-breasted Sapsuckers. *Wilson Bull.* 97:234–236.
———. 1986. Have ornithologists or breeding Red-breasted Sapsuckers extended their range in coastal California? *Western Birds* 17:97–105.
———. 1993. *The Marin County Breeding Bird Atlas: A Distributional and Natural History of Coastal California Birds.* California Avifauna Series 1. Bolinas, CA: Bushtit Books.
Shunk, S. A. 2004a. Woodpecker wonderland: there's a whole lotta flakin' going on in Oregon's eastern Cascades. *Winging It* 16(3):1–4.
———. 2004b. Woodpecker conservation in central Oregon. *Winging It* 16(3):5.
———. 2005. *Sphyrapicus* anxiety: identifying hybrid sapsuckers. *Birding* 37(3):289–298.
Sibley, C. G. 1957. The abbreviated inner primaries of nestling woodpeckers. *Auk* 74:102–103.
Sibley, C. G., and B. L. Monroe, Jr. 1990. *Distribution and Taxonomy of Birds of the World.* New Haven, CT: Yale Univ. Press.
Sibley, D. A. 2002. *The Sibley Guide to Birds.* New York: Alfred A. Knopf.
Siegel, R. B., and D. F. DeSante. 1999. Draft Avian Conservation Plan for the Sierra Nevada Bioregion: Conservation Priorities and Strategies for Safeguarding Sierra Nevada Bird Populations. Ver. 1.0. Institute for Bird Populations report to California PIF. Point Reyes Station, CA: Institute for Bird Populations.
Simon, S. 2005. Saving the Big Woods. *BirdScope* 19(3):18.
Simpson, S. F., and J. Cracraft. 1981. The phylogenetic relationships of the Piciformes (Class Aves). *Auk* 98:481–494.
Sinclair, P. H. 2003. *Birds of the Yukon Territory.* Vancouver: Univ. of British Columbia Press.
Skorupa, J. P., and R. W. McFarlane. 1976. Seasonal variation in foraging territory of Red-cockaded Woodpeckers. *Wilson Bull.* 88:662–665.
Skutch, A. F. 1933. Male woodpeckers incubating at night. *Auk* 50:437.
———. 1985. *Life of the Woodpecker.* Santa Monica, CA: Ibis Publishing Co.
Slater, E. M. 1945. *El Paso Birds.* El Paso, TX: Carl Herzog.
Small, A. 1974. *The Birds of California.* New York: Winchester Press.
Smith, A. P. 1912. Status of the Picidae in the lower Rio Grande Valley. *Auk* 29(2):241.
Smith, A. R. 1996. *Atlas of Saskatchewan Birds.* Regina: Saskatchewan Natural History Society.
Smith, J. H. 1954. Ruby-throated Hummingbird feeding at Yellow-bellied Sapsucker holes. *Auk* 71:316.
Smith, J. I. 1987. Evidence of hybridization between Red-bellied and Golden-fronted Woodpeckers. *Condor* 89:377–386.
Smith, K. G. 1986. Winter population dynamics of three species of mast-eating birds in the eastern United States. *Wilson Bull.* 98:407–418.
Smith, K. G., J. H. Withgott, and P. G. Rodewald. 2000. Red-headed Woodpecker (*Melanerpes erythrocephalus*). No. 518 in *The Birds of North America,* A. Poole and F. Gill, eds. Philadelphia, PA: Birds of North America, Inc.
Smith, M. R., P. W. Mattocks, Jr., and K. M. Cassidy. 1997. Breeding birds of Washington State. In K. M. Cassidy, C. E. Grue, M. R. Smith, and K. M. Dvornich, eds., Washington State Gap Analysis: Final Report. Vol. 4. *Seattle Aud. Soc. Publ. in Zool.* no. 1.
Smith, S. C. 1940. The Alabama Academy of Science. *Science* 92:240–241.
Smith, W. G. 1890. Nesting of the Williamson's Sapsucker, *Sphyrapicus thyroideus. Wilson Bull.* 2:15–16.
Snyder, N., and C. Houston. 2000. Army helping endangered woodpecker recovery. *Environmental Update* [USAEC] Fall 2000.
Soelner, G. W. H. 1904. The Pileated Woodpecker in the District of Columbia. *Auk* 21:79–80.
Soule, C. G. 1899. Birds and caterpillars. *Bird-Lore* 1:166.
Sousa, P. J. 1983. Habitat suitability models: Lewis' Woodpecker. Washington, D.C.: USFWS, Div. of Biol. Sci.
South Carolina Department of Natural Resources. 2005. South Carolina Comprehensive Wildlife Conservation Strategy. Columbia: South Carolina Dept. of Natural Resources.
———. 2014. Rare, Threatened, and Endangered Species and Communities Known to Occur in South Carolina. Columbia: South Carolina Dept. of Natural Resources.
South Dakota Department of Game, Fish, and Parks. 2006. South Dakota Comprehensive Wildlife Conservation Plan. Wildlife Division Report 2006–2008. Ft. Pierre: South Dakota Dept. of Game, Fish, and Parks.
Southern, W. E. 1960. Copulatory behavior of the Red-headed Woodpecker. *Auk* 77:218–219.
Southgate, J., Y. Hoyt, and S. F. Hoyt. 1951. Age records of Pileated Woodpeckers. *Auk* 22:125.
Spahr, T. 2005. Searches for Ivory-billed Woodpecker (*Campephilus principalis*) in the Apalachicola River basin of Florida in 2003. *North American Birds* 59(2):210–215.
Speich, S., and W. J. Radke. 1975. Opportunistic feeding of the Gila Woodpecker. *Wilson Bull.* 87:275–276.
Spingarn, E. D. W. 1924. Northern Pileated Woodpecker in Dutchess County, N.Y. *Auk* 41:480.
Spring, L. W. 1965. Climbing and pecking adaptations in some North American woodpeckers. *Condor* 67:457–488.
Springston, R. 2004. Bird sits on a precarious perch. *Richmond Times-Dispatch,* June 14, 2004.
Stallcup, P. L. 1969. Hairy Woodpeckers feeding on pine seeds. *Auk* 86:134–135.

Stark, R. D., D. J. Dodenhoff, and E. V. Johnson. 1998. A quantitative analysis of woodpecker drumming. *Condor* 100:350–356.
Stauffer, D. F., and L. B. Best. 1982. Nest-site selection by cavity nesting birds of riparian habitats in Iowa. *Wilson Bull.* 94:329–337.
Stefferud, A., ed. 1966. *Birds in Our Lives.* Washington, D.C: USFWS.
Stevenson, H. M., and B. H. Anderson. 1994. *The Birdlife of Florida.* Gainesville: Univ. Press of Florida.
Stewart, P. A. 1931. A Red-headed Woodpecker (*Melanerpes erthrocephalus*) destroys its own eggs. *Auk* 48:122.
Stewart, R. E. 1975. *Breeding Birds of North Dakota.* Fargo, ND: Tri-College Center for Environmental Studies.
Stickel, D. W. 1962. Predation on Red-bellied Woodpecker nestlings by a black rat snake. *Auk* 79:118–119.
———. 1963. Interspecific relations among Red-bellied and Hairy Woodpeckers and a flying squirrel. *Wilson Bull.* 75:203–204.
———. 1964. Roosting habits of Red-bellied Woodpeckers. *Wilson Bull.* 76:382–383.
———. 1965. Wing-stretching of Red-bellied Woodpeckers. *Auk* 82:503.
Stolzenburg, W. 2002. Swan Song of the Ivory-bill. *Nature Conservancy* 52(3):38–47.
Stone, R. 1993. Can logging save old-growth forests? *Science* 262:19.
———. 1995a. Court upholds need to protect habitat. *Science* 269:23.
———. 1995b. Incentives offer hope for habitat. *Science* 269:121–123.
Stone, W. 1907. Some changes in the current generic names of North American birds. *Auk* 24:189–199.
———. 1915. Type locality of Lewis's Woodpecker and Clarke's Nutcracker. *Auk* 32:371–372.
———. 1937. *Bird Studies at Old Cape May: An Ornithology of Coastal New Jersey.* Vol. 1. Reprint, Dover: New York, 1965.
Stoner, D. 1925. The toll of the automobile. *Science* 61:56–57.
Stoner, E. A. 1922. A study of roosting holes of the Red-shafted Flicker. *Condor* 14:54–57.
Styrsky, J. D., and J. N. Styrsky. 2003. Golden-fronted Woodpecker provisions nestlings with small mammal prey. *Wilson Bull.* 115:97–98.
Sutton, G. M. 1967. *Oklahoma Birds: Their Ecology and Distribution, with Comments on the Avifauna of the Southern Great Plains.* Norman: Univ. of Oklahoma Press.
Sutton, G. M., and E. P. Edwards. 1941. Does the southern Hairy Woodpecker occur in Oklahoma? *Wilson Bull.* 53:127–128.
Swarth, H. S. 1904. Birds of the Huachuca Mountains, Arizona. *Pacific Coast Avifauna* no. 4.
———. 1911. Publication review: A revision of the forms of the Hairy Woodpecker (*Dryobates villosus*), Oberholser, 1911. *Condor* 13:169–170.
———. 1914. A distributional list of the birds of Arizona. *Pacific Coast Avifauna* no. 10.
———. 1917. Geographical variation in *Sphyrapicus thyroideus. Condor* 19:62–65.
Swarthout, E., and R. Rohrbaugh. 2005. Now what? *Birder's World* 19(4):26–29.
Sweet, P. W. 2001. Illinois' First Red-cockaded Woodpecker. *Meadowlark* 10(2):42–45.
Swierczewski, E. V., and R. J. Raikow. 1981. Hind limb morphology, phylogeny, and classification of the Piciformes. *Auk* 98:466–480.
Taggart, M. W. 1912. On the factors that determine the location of the borings of the Yellow-bellied Sapsucker on the paper birch. *Science* 35:461.
Tanner, J. T. 1942a. The Ivory-billed Woodpecker. Research Report no. 1. New York: National Audubon Society.
———. 1942b. Present status of the Ivory-billed Woodpecker. *Wilson Bull.* 54:57–58.
Taverner, P. A. 1906. A tagged Flicker. *Wilson Bull.* 54:21–22.
———. 1934a. *Birds of Canada.* Ottawa, ON: National Museum of Canada Bull. no. 72, Biological Series no. 19.
———. 1934b. Flicker hybrids. *Condor* 36:34–35.
Taylor, H. J. 1920. Habits of a Red-breasted Sapsucker. *Condor* 22:158.
Taylor, P., ed. 2003. *The Birds of Manitoba.* Winnipeg: Manitoba Avian Research Committee and Manitoba Naturalists Society.
Teachenor, D. 1929. Red-headed Woodpecker in New Mexico. *Auk* 46:114.
Tekin, B. 2006. Great Dismal Swamp — April 2. Posting to va-bird listserve, Apr. 3, 2006. www.freelists.org/archives/va-bird/04-2006/msg00039.html.
Tennessee Wildlife Resources Agency. 2005. Tennessee's Comprehensive Wildlife Conservation Strategy. Nashville: Tennessee Wildlife Resources Agency.
Test, F. 1939. The form and pigmentation of a supernumerary secondary of a Flicker. *Condor* 41:30–32.
———. 1940. Effects of natural abrasion and oxidation on the coloration of Flickers. *Condor* 42:76–80.
———. 1942. The nature of the red, yellow and orange pigments in woodpeckers of the genus *Colaptes. Univ. Calif. Publ. Zool.* 46:371–390.
———. 1945. Molt in flight feathers of flickers. *Condor* 47:63–72.
———. 1969. Relation of wing and tail color of the woodpeckers *Colaptes auratus* and *C. cafer* to their food. *Condor* 71:206–211.
Thayer, J. E. 1911. A Northern Pileated Woodpecker in Massachusetts. *Auk* 28:266.
Thompson, E. E. 1891. The birds of Manitoba. 2d ed. *Proc. U.S. Nat. Mus.* 13:457–643.
Thompson, M. 1889. A Red-headed family. *Oologist* 6(2):23–29.
Thompson, M. C., and C. Ely. 1989. *Birds in Kansas.* Vol. 1. Lawrence: Univ. of Kansas Museum of Natural History.
Thompson, R. L. 1976. Change in status of Red-cockaded Woodpecker colonies. *Wilson Bull.* 88:491–492.
Thone, T. 1936. Life given to museum groups of birds. *Science* 84:6–8.

Thornburgh, M. E. 1937. The Pileated Woodpecker in Clayton County, Iowa. *Wilson Bull.* 49:302–303.
Tobalske, B. W. 1992. Evaluating habitat suitability using relative abundance and fledging success of Red-naped Sapsuckers. *Condor* 94:550–553.
———. 1996. Scaling of muscle composition, wing morphology, and intermittent flight behavior in woodpeckers. *Auk* 113:151–177.
———. 1997. Lewis's Woodpecker (*Melanerpes lewis*). No. 284 in *The Birds of North America*, A. Poole and F. Gill, eds. Philadelphia, PA: Acad. of Nat. Sci., and Washington, D.C.: AOU.
Tobish, T. 2000. The regional reports: Alaska. *North American Birds* 55(1):90.
———. 2001. The regional reports: Alaska. *North American Birds* 55(3):340.
———. 2002. The regional reports: Alaska. *North American Birds* 56(2):211.
———. 2003. The regional reports: Alaska. *North American Birds* 58(1):127.
———. 2005. The regional reports: Alaska. *North American Birds* 59(3):481.
Torrey, B. 1901. *Melanerpes erythrocephalus* breeding near Boston. *Auk* 18:394.
Trochet, J., J. Morlan, and D. Roberson. 1988. First record of the Three-toed Woodpecker in California. *Western Birds* 19:109–115.
Trombino, C. 2000. Helping behavior within sapsuckers (*Sphyrapicus* spp.). *Wilson Bull.* 112:273–275.
Tucker, M. B. 1926. Arctic Three-toed Woodpecker in West Chester Co., N.Y. *Auk* 43:235.
Tyler, J. G. 1913. Some birds of the Fresno district, California. *Pacific Coast Avifauna* no. 9.
U. S. Army Environmental Center. 2006. Fort Bragg achieves endangered species milestone. *Environmental Update* [USAEC] Winter 2006.
———. 2008. Fort Benning shares longleaf pine restoration costs. *Environmental Update* [USAEC] Spring 2008.
U. S. Fish and Wildlife Service. 1999. Red-cockaded Woodpecker (*Picoides borealis*). [Species account.] Pp.4-473–4-502 in South Florida Multi-species Recovery Plan. Atlanta, GA: USFWS.
———. 2001a. Kachemak Bay Birds Annotated Species List. Homer, AK: Kachemak Bay Shorebird Fest. Steering Comm.
———. 2001b. Piedmont National Wildlife Refuge Bird List. Round Oak, GA: Piedmont National Wildlife Refuge.
———. 2003. Recovery Plan for the Red-cockaded Woodpecker (*Picoides borealis*): Second Revision. Atlanta, GA: USFWS, Southeast Region.
———. 2005. Extinct no more: Ivory-billed Woodpecker rediscovered. *People, Land, & Water* 11(5):24–25.
———. 2006. Checklist of the birds of Joshua Tree National Park. Twenty-nine Palms, CA: Joshua Tree Natural History Assoc. and National Park Service.
———. 2007a. Draft Recovery Plan for the Ivory-billed Woodpecker (*Campephilus principalis*). Atlanta, GA:USFWS, Southeast Region.
———. 2007b. Tetlin National Wildlife Refuge Bird Checklist. Tok, AK: Tetlin National Wildlife Refuge.
U. S. Forest Service. 2003. Management Indicator Species Population and Habitat Trends Gainesville, GA: USFS, Chattahoochee-Oconee National Forests.
U. S. Forest Service. 2011. Forest Service Manual, Supplement No. 2600-2011-1. Rocky Mountain Region (Region 2), Denver, CO.
U. S. National Park Service. 2007. Wrangell-St. Elias Bird Checklist. Copper Center, AK: Wrangell-St. Elias National Park and Preserve.
Unitt, P. 1986. Another hybrid Downy x Nuttall's Woodpecker from San Diego County. *Western Birds* 17:43–44.
Utah Bird Records Committee. 2008. Comprehensive list of rare bird sightings. Ogden: Utah Ornithological Society. www.utahbirds.org/RecCom/RareBirdsIndex.html.
van Rossem, A. J. 1933. The Gila Woodpecker in the Imperial Valley of California. *Condor* 35:74.
———. 1936a. Birds of the Charleston Mountains, Nevada. *Pacific Coast Avifauna* no. 24.
———. 1936b. Remarks stimulated by Brodkorb's "Two new subspecies of the Red-shafted Flicker." *Condor* 38:40.
———. 1942. Four new woodpeckers from the western United States and Mexico. *Condor* 44:22–26.
———. 1944. The Santa Cruz Island Flicker. *Condor* 46:245.
Van Tyne, J. 1926. An unusual flight of Arctic Three-toed Woodpeckers. *Auk* 43:469–474.
Van Tyne, J., and G. M. Sutton. 1937. *The Birds of Brewster County, Texas*. Misc. Publ. no. 37., Univ. of Michigan Museum of Zoology.
Varner, J. M., III, J. S. Kush, and R. S. Meldahl. 2006. Characteristics of sap trees used by overwintering *Sphyrapicus varius* (Yellow-bellied Sapsuckers) in an old-growth pine forest. *Southeastern Naturalist* 5:127–134.
Vasquez, M. 2005. Red-naped Sapsucker (*Sphyrapicus nuchalis*) species assessment. Draft. Prepared for the Grand Mesa, Uncompahgre, and Gunnison National Forests, Gunnison, CO.
Velland, M., and V. Connolly. 1999. COSEWIC assessment and status report on the Lewis's woodpecker (*Melanepes lewis*) in Canada. Ottawa, ON: Comm. on the Status of Endangered Wildlife in Canada.
Vermont Fish and Wildlife Department. 2005. Vermont Wildlife Action Plan. Waterbury: Vermont Fish and Wildlife Dept.
Verner, J. 1965. Northern limit of the Acorn Woodpecker. *Condor* 67:265.
Vickers, E. W. 1914. The roll of the Pileated Woodpecker. *Wilson Bull.* 26:15–17.
Vierling, K. T. 1997. Habitat selection of Lewis's Woodpeckers in southeastern Colorado. *Wilson Bull.* 109:121–130.
Vierling, K. T., L. B. Lentile, and N. Neilsen-Pincus. 2008. Preburn characteristics and woodpecker use of burned coniferous forests. *J. of Wildl. Mgmt.* 72(2):422–427.

Viet, R., and W. Petersen. 1993. *Birds of Massachusetts*. Lincoln, MA: Massachusetts Audubon Society.
Villard, P., and C. W. Beninger. 1993. Foraging behavior of male Black-backed and Hairy Woodpeckers in a forest burn. *J. of Field Ornith.* 64:71–76.
Villard, P., and J. Cuisin. 2004. How do woodpeckers extract grubs with their tongues? A study of the Guadaloupe Woodpecker (*Melanerpes herminieri*) in the French West Indies. *Auk* 121(2):509–514.
Virginia Department of Game and Inland Fisheries. 2005. Virginia's Comprehensive Wildlife Conservation Strategy. Richmond: Virginia Dept. of Game and Inland Fisheries.
Visher, S. S. 1910. Notes on the birds of Pima County, Arizona. *Auk* 27:279–288.
von Bloecker, J. C. 1927. A strange meeting with a Flicker. *Condor* 29:200.
———. 1935. Flickers and jays feeding on scarab beetles in flight. *Condor* 37:288–289.
———. 1936. Red-shafted Flickers feeding on aphids. *Condor* 38:90.
Voous, K. H. 1947. On the history of the distribution of the genus *Dendrocopos. Limosa* 20:1–142.
Wahl, T. R., and D. R. Paulson. 1991. *A Guide to Bird Finding in Washington*. Bellingham, WA: T. R. Wahl.
Walker, K. M. 1952. Northward extension of range of the Acorn Woodpecker in Oregon. *Condor* 54:315.
Wallace, C. R. 1926. The Pileated Woodpecker in Tuscarawas County, Ohio. *Wilson Bull.* 38:232–233.
Walter, S. T., and C. C. McGuire. 2005. Snags, cavity-nesting birds, and silvicultural treatments in western Oregon. *J. of Wildl. Mgmt.* 69(4):1578–1591.
Walters, E. L. 1990. Habitat and Space Use of the Red-naped Sapsucker (*Sphyrapicus nuchalis*) in the Hat Creek Valley, South-central British Columbia. Master's thesis, Univ. of Victoria.
Walters, E. L., E. H. Miller, and P. E. Lowther. 2002a. Red-breasted Sapsucker (*Sphyrapicus ruber*) and Red-naped Sapsucker (*Sphyrapicus nuchalis*). No. 663 in *The Birds of North America,* A. Poole and F. Gill, eds. Philadelphia, PA: Birds of North America, Inc.
———. 2002b. Yellow-bellied Sapsucker (*Sphyrapicus varius*). No. 662 in *The Birds of North America,* A. Poole and F. Gill, eds. Philadelphia, PA: Birds of North America, Inc.
Warren, S. V. 1916. The Northern Pileated Woodpecker and Pine Grosbeak in northwestern Ohio. *Wilson Bull.* 28:92.
Wasson, T., L. Yasui, K. Brunson, S. Amend, and V. Ebert. October 2005. A Future for Kansas Wildlife, Kansas' Comprehensive Wildlife Conservation Strategy. Topeka: Dynamic Solutions, Inc., in cooperation with Kansas Dept. of Wildlife and Parks.
Waters, C. 2005. Birds and birders at Corn Creek Station, Nevada. *Winging It* 17:3–5.
Watt, D. J. 1980. Red-bellied Woodpecker predation on nestling American Redstarts. *Wilson Bull.* 92:249.
Watts, B. D. 1999. Partners in Flight Landbird Conservation Plan, Mid-Atlantic Coastal Plain (Physiographic Area 44). Ver. 1.0. The Plains, VA: ABC.
———. 2005. VA Red-cockaded male translocated. Posting to va-bird listserve, Apr. 19, 2005. www.freelists.org/archives/va-bird/04-2005/msg00246.html.
———. 2006. Five Red-cockaded Woodpeckers translocated into VA population. Posting to va-bird listserve, Oct. 3, 2005. www.freelists.org/archives/va-bird/10-2005/msg00026.html.
Wauer, R. H. 1965. Intraspecific relationship in Red-shafted Flickers. *Wilson Bull.* 77:404.
———. 1973. *Birds of Big Bend National Park and Vicinity*. Austin: Univ. of Texas Press.
———. 2001. *Naturally ... South Texas: Nature Notes from the Coastal Bend*. Austin: Univ. of Texas Press.
Wauer, R. H., and M. A. Elwonger. 1998. *Birding Texas*. Helena, MT: Falcon Publishing.
Wayne, A. T. 1905. A rare plumage of the Ivory-billed Woodpecker (*Campephilus principalis*). *Auk* 22:414.
Weathers, W. W. 1983. *Birds of Southern California's Deep Canyon*. Berkeley: Univ. of California Press.
Webb, D. W., and W. S. Moore. 2005. A phylogenetic analysis of woodpeckers and their allies using 12S, Cyt b, and COL nucleotide sequences (Class Aves; Order Piciformes). *Mol. Phylogen. and Evol.* 36:2.
Weikel, J. M., and J. P. Hayes. 1999. The foraging ecology of cavity-nesting birds in young forests of the northern coast range of Oregon. *Condor* 101:58–66.
Weisser, W. 1973. A mixed pair of sapsuckers in the Sierra Nevada. *Western Birds* 4:107–108.
Welch, J. M. 1899. Notes on Lewis' Woodpecker. *Condor* 1:29.
———. 1900. Lewis' Woodpecker as a flycatcher. *Condor* 2(4):89.
West, D. W., and J. M. Speirs. 1959. The 1956–1957 invasion of the Three-toed Woodpeckers. *Wilson Bull.* 71:348–363.
West, G. C. 2002. *A Birder's Guide to Alaska*. Colorado Springs, CO: American Birding Assoc.
Wetmore, A. 1931. Record of an unknown woodpecker from the lower Pliocene: *Condor* 33:255–256.
———. 1940. Notes on the woodpeckers from West Virginia. *Auk* 57:113–114.
———. 1943. Evidence of the former occurrence of the Ivory-billed Woodpecker in Ohio. *Wilson Bull.* 55:55.
———. 1948. The Golden-fronted Woodpeckers of Texas and northern Mexico. *Wilson Bull.* 60:185–186.
Wetmore, A., and H. Friedmann. 1949. The name for the Wryneck recorded from Alaska. *Condor* 51:103.
White, M. 1995. *A Birder's Guide to Arkansas*. Colorado Springs, CO: American Birding Assoc.
Whitney, T. H. 1917. Red-bellied Woodpecker at Atlantic, Iowa. *Wilson Bull.* 29:106.
Wiebe, K. L. 2000a. Assortative mating by color in a population of hybrid Northern Flickers. *Auk* 117:525–529.
———. 2000b. Northern Flicker incubates Hooded Merganser egg. *BC Birds* 10:13–15.
———. 2001. Microclimate of tree cavity nests: is it important for reproductive success in Northern Flickers? *Auk* 118:412–421.

———. 2002. First reported case of classical polyandry in a North American woodpecker, the Northern Flicker. *Wilson Bull.* 114:401–403.
———. 2004. Innate and learned components of defence by Flickers against a novel nest competitor, the European Starling. *Ethology* 110:779–791.
———. 2008. Division of labour during incubation in a woodpecker *Colaptes auratus* with reversed sex roles and facultative polyandry. *Ibis* 150:115–224.
Wiebe, K. L., and G. R. Bortolotti. 2001. Variation in colour within a population of Northern Flickers: a new perspective on an old hybrid zone. *Can. J. of Zool.* 79:1046–1052.
———. 2002. Variation in carotenoid-based color in Northern Flickers in a hybrid zone. *Wilson Bull.* 114:393–400.
Wiebe, K. L., W. D. Koenig, and K. Martin. 2006. Evolution of clutch size in cavity-excavating birds: the nest site limitation hypothesis revisited. *American Naturalist* 167(3):343–353.
Wiedenfeld, D. A., and M. M. Swan. 2000. *Louisiana Breeding Bird Atlas.* Louisiana Sea Grant College Prog. Baton Rouge: Louisiana State Univ.
Wiggins, D. A. 2004. American Three-toed Woodpecker (*Picoides dorsalis*): A Technical Conservation Assessment. Species Conservation Project. USDA Forest Service, Rocky Mtn. Region.
Wildlife Action Plan Team. 2006. Nevada Wildlife Action Plan. Reno: Nevada Dept. of Wildlife.
Wilds, C. 1992. *Finding Birds in the National Capital Area.* 2d. Washington, D.C.: Smithsonian Institution Press.
Wilkins, H. D, and G. Ritchison. 1999. Drumming and tapping by Red-bellied Woodpeckers: description and possible causation. *J. of Field Ornith.* 70:578–586.
Willett, G. 1912. Birds of the Pacific slope of southern California. *Pacific Coast Avifauna* no. 7.
———. 1933. A revised list of the birds of southwestern California. *Pacific Coast Avifauna* no. 21.
Williams, J. B. 1900.The Flicker wintering in Montreal. *Auk* 17:174–175.
Williams, J. B. 1980. Foraging by Yellow-breasted Sapsuckers in central Illinois during spring migration. *Wilson Bull.* 92:519–523.
Williams, J. J. 1905. Notes on the Lewis Woodpecker. *Condor* 7:56.
Williams, L. 2002. A helping hand for birds. *Mississippi Outdoors* July 2002:4–9.
Wilson, M. 2006. 2006. Red-cockaded Woodpecker breeding results. Posting to va-bird listserve, July 26, 2006. www.freelists.org/archives/va-bird/07-2006/msg00113.html.
Wilson, M. 2008. CCB continues work with the endangered Red-cockaded Woodpecker. Center for Conservation Biology. www.ccbbirds.org/2008/06/15/.
Wilson, M., and A. M. Rea. 1976. Late Pleistocene Williamson's Sapsucker from Wyoming. *Wilson Bull.* 89:622.
Winkler, H., D. A. Christie, and D. Nurney. 1995. *Woodpeckers, a Guide to the Woodpeckers of the World.* New York: Houghton Mifflin.
Winkler, H., and L. L. Short, Jr. 1978. A comparative analysis of acoustical signals in pied woodpeckers (Aves: *Picoides*). *Bull. of AMNH* 160:1–110.
Wisconsin Department of Natural Resources. 2005. Wisconsin's Strategy for Wildlife Species of Greatest Conservation Need. Madison: Wisconsin Dept of Natural Resources.
Withgot, J. 2002. Observations. *The Gull* 87(7):12–13.
Wood, J. C. 1905a. Some nesting sites of the Hairy Woodpecker (*Dryobates villosus*). *Wilson Bull.* 17:66.
———. 1905b. Nesting of the Yellow-bellied Sapsucker. *Wilson Bull.* 51:58–59.
Wood, M. 1959. Black-backed Three-toed Woodpecker in central Pennsylvania. *Auk* 76:361.
Wood, N. A. 1913. First Michigan specimen of the Three-toed Woodpecker. *Auk* 30:272–273.
———. 1921. Rare records for Ann Arbor and the State of Michigan. *Auk* 38:282–283.
Woodbury, A. M. 1938. Red-naped Sapsucker and Rufous Hummingbird. *Condor* 40:125.
Wright, H. W. 1905. The Arctic Three-toed Woodpecker in Melrose, Mass. *Auk* 22:80.
———. 1919. Arctic Three-toed Woodpecker (*Picoides arcticus*) at Belmont, Mass. *Auk* 36:110–111.
Wygnanski-Jaffe, T., C. J. Murphy, C. Smith, M. Kubai, P. Christopherson, C. R. Ethier, et al. 2007. Protective ocular mechanisms in woodpeckers. *Eye* 21:83–89.
Wyman, L. E. 1923. An albino Nuttall Woodpecker. *Condor* 25:139.
Wyoming Game and Fish Commission. 2005. A Comprehensive Wildlife Strategy for Wyoming. Cheyenne: Wyoming Game and Fish Comm.
Yocom, C. F. 1960. Records of Lewis Woodpecker for Humboldt County, California. *Condor* 62:410.
Yom-Tov, Y., and A. Ar. 1993. Incubation and fledging duration of woodpeckers. *Condor* 95:282–287.
Yunick, R. P. 1985. A review of recent irruptions of the Black-backed Woodpecker and Three-toed Woodpecker in eastern North America. *J. of Field Ornith.* 56:138–152
Zack, S., lead author. 2002. The Oak Woodland Bird Conservation Plan: A Strategy for Protecting and Managing Oak Woodland Habitats and Associated Birds in California. Ver. 2.0. Stinson Beach, CA: California PIF and Point Reyes Bird Observatory.
Zeranski, J. D., and T. R. Baptist. 1990. *Connecticut Birds.* Hanover, NH: Univ. Press of New England.
Zink, R. M., S. Rohwer, S. Drovetski, R. C. Blackwell-Rago, and S. L. Farrell. 2002. Holarctic phylogeography and species limits of Three-toed Woodpeckers. *Condor* 104:167–170.
Zohrer, J. J. 2007. Securing a Future for Fish and Wildlife: A Conservation Legacy for Iowans. Rev. 2006. Des Moines: Iowa Dept. of Natural Resources.
Zwartjes, P. W., and S. E. Nordell. 1998. Patterns of cavity-entrance orientation by Gilded Flickers (*Colaptes chrysoides*) in cardon cactus. *Auk* 115:119–126.

# INDEX

Page numbers in **bold** refer to text graphics.


adaptations. *See* anatomy/adaptations overview; *specific adaptations*
albinism defined, 14
anatomy/adaptations overview, 2–9
  *see also specific adaptations*; feathers
anisodactyl foot, 7–9
anting, 28
artificial nest boxes, 33, **33**, 261

barbs on tongues, 2–3, **3**
behavior overview, 14–28, **15–28**
bill specialization, 6, **6**
book use guidelines, 38–41
Borelli, Giovanni Alfonso, **3**
breeding behavior, 14–19, **15–19**
Burt, William, 5

*Campephilus*
  *guatemalensis*, **9**
  *principalis*, 244–51, **244–47**, **249–51**
  *principalis bairdii*, 248
  *principalis principalis*, 248
cavities
  competition for, 34, 36, **37**
  description/preparation, 16, **16–17**
  entering/exiting, 17
  need for, 35–37, **36–38**, 261
  other animals and, 28–29, **29**
  sanitation, 19–20
cerebrospinal fluid, 5
chestnut blight, 30
climbing description, 27
*Colaptes*
  *auratus*, **2**, **3**, **5–7**, **14**, **18**, **28**, 214–25, **214–21**, **223–24**
  *auratus auratus*, 219, **221**
  *auratus "borealis,"* 219, **221**
  *auratus cafer*, 219, **221**
  *auratus "canescens,"* 219, **221**
  *auratus "chihuahuae,"* 219, **221**
  *auratus chrysocaulosus*, 220, **221**
  *auratus collaris*, 219, **221**
  *auratus gundlachi*, 220, **221**
  *auratus luteus*, 219, **221**
  *auratus "martirensis,"* 219, **221**
  *auratus mexicanoides*, 220, **221**
  *auratus mexicanus*, 219–20, **221**
  *auratus nanus*, 219, **221**
  *auratus "pinicolus,"* 220, **221**
  *auratus rufipileus*, 220, **221**
  *auratus "sedentarius,"* 219, **221**
  *auratus* x *Colaptes chrysoides*, 222, 226, 230
  *chrysoides*, **29**, **221**, 226–33, **226–32**
  *chrysoides brunnescens*, **221**, 230
  *chrysoides chrysoides*, **221**, 230
  *chrysoides mearnsi*, **221**, 230
  *chrysoides tenebrosus*, **221**, 230
  *chrysoides* x *Colaptes auratus*, 222, 226, 230
communication overview, 22–25, **24**
conservation. *See* ecology/conservation
Coues, Elliott, 12
courtship/pair bonding behavior, 14–15, **15**, 24, **25**
cranium/adaptations
  forebrain size, 20
  hammering trees and, 5–6
  as model for human applications, 4–6
  overview, 4–7, **5–6**

decay facilitation, 30
drumming
  description/overview, 22–24, **24**, 261
  human conflict and, 261
*Dryocopus*
  *pileatus*, **5–6**, **21**, **32**, **34**, 234–43, **234–36**, **238–42**, **247**
  *pileatus abieticola*, 237–38
  *pileatus floridanus*, 237
  *pileatus picinus*, 237
  *pileatus pileatus*, 237, 238

ear adaptations, 6–7
ecology/conservation overview
  artificial nest boxes, 33, **33**, 261
  cavity competition, 34, 36, **37**
  fire and woodpeckers, 31–34, **32–34**
  habitat conservation, 34–38, **35–38**, 261
  habitat conservation changes needed, 38
  habitat loss, 34–35, 261

overview, 28–38, **29–38**
snags/cavities need and, 35–37, **36–38**, 261
*See also* keystone organisms
ectropodactyl toe configuration, 8
emerald ash borer, 29–30, **31**
Endangered Species Act (U.S./1973), 37
European Starling introduction/affects, 36
eye adaptations, 6

feathers
bristled feathers, 12, **13**
colors/environmental discoloration, 13–14, **14**
down and, 9
hatching through fledging, 9–11
molts (adults), 11–13
nasal tufts, **12**, 13
overview, 9–14, **10–14**
preformative molt, 9, 10, 11
tail prop, 7, 12, **12**, 27
feeding
adult behavior overview, 19–22, **21–22**
artificial feeding stations, 22, **23**
fledging and, 19
flycatching, 22, **22**, 27
insect control, 29–30, **30**
insects after fires, 31–34, **32–34**, 34
for nestlings, 17, 18**20**
sap for other species, 30, **31**
sap wells, 20–22, **21**, 30, **31**
subsurface excavation, 20, **21**
surface feeding, 22
feet and legs adaptations, 7–9, **7–9**
fire and woodpeckers
human fire suppression effects, 34
insect food and, 31–34, **32–34**, 34
overview, 31–34, **32–34**
fledging description, 19
Flicker
Gilded, **29**, 226–33, **226–32**
Gilded x Northern Flicker, "Red-shafted," 222, 226, 230
"Guatemalan," 220
Mearns's Gilded, 230
Northern, **2**, **3**, **5–7**, **14**, **18**, **28**, 214–25, **214–21**, **223–24**
Northern "Red-shafted," 214, **214**, **217**, 218, 219–20, **219**, 222
Northern "Red-shafted" intergrades with "Yellow-shafted," **14**, **220**, **221**, 222, **223**
Northern "Red-shafted" x Gilded Flicker, 222, 226, 230
Northern "Yellow-shafted," **28**, 214, **214–16**, 218, **218**, 219, 222, 223, **224**
flight description, 27

guidelines to book use, 38–41

habitat conservation, 34–38, **35–38**, 261
head scratching description, 27
helmet design, 4
hickory borer, 29
Hopkins, A.D., 29
human conflicts overview, 261
hummingbird feeders, 22, **23**
hybridization vs. introgression/intergrade, 40–41
hyoid apparatus, 3–4, **4–5**

incubation/brooding, 16–17
insect control, 29–30, **30**
introgression/intergrade vs. hybridization, 40–41

jaw structure, 6

keystone organisms
cavity excavations and, 28–29, **29**
insect control, 29–30, **30**
woodpeckers as, 28–30, **29–31**, 34

legs and feet adaptations, 7–9, **7–9**
leucism defined, 14
locomotion overview, 27, **28**
Lucas, Frederic, **3**

maps key, **39**
*Melanerpes*
*aurifrons*, **4**, **5**, **10**, **23**, 78–85, **78–80**, **82**, **84–85**, **138**
*aurifrons canescens*, 81
*aurifrons dubius*, 81
*aurifrons "frontalis,"* 81
*aurifrons grateloupensis*, 81
*aurifrons hughlandi*, 81
*aurifrons incanescens*, 81
*aurifrons insulanus*, 81
*aurifrons leei*, 81
*aurifrons pauper*, 81
*aurifrons polygrammus*, 81
*aurifrons santacruzi*, 81
*aurifrons turneffensis*, 81
*aurifrons veraecrucis*, 81
*aurifrons* x *Melanerpes carolinus*, 78, 81–82, 89
*aurifrons* x *Melanerpes hoffmannii*, 82
*aurifrons* x *Melanerpes uropygialis*, 75, 81, 82
*carolinus*, **5–6**, **23**, 86–93, **86–90**, **92**
*carolinus* x *Melanerpes aurifrons*, 78, 81–82, 89
*erythrocephalus*, **3**, 52–59, **52–58**
*formicivorus*, **16**, **25**, 60–69, **60–68**
*formicivorus "aculeatus,"* 63
*formicivorus albeolus*, 64
*formicivorus angustifrons*, 64
*formicivorus bairdi*, **62**, 63–64, **63**, **65**, **67**
*formicivorus flavigula*, 64
*formicivorus formicivorus*, **61**, **62**, 63, **63**
*formicivorus lineatus*, 64

*Melanerpes (cont.)*
*formicivorus "martirensis,"* 63–64
*formicivorus striatipectus*, 64, **64**
*hoffmannii* x *Melanerpes aurifrons*, 82
*lewis*, **5–6**, **13**, **22**, **33**, 44–51, **44–47**, **49–50**
*uropygialis*, **23**, 70–77, **70–74**, **76**
*uropygialis albescens*, 75
*uropygialis brewsteri*, 75
*uropygialis cardonensis*, 75
*uropygialis "fuscescens,"* 75
*uropygialis "sulfuriventer,"* 75
*uropygialis "tiburonensis,"* 75
*uropygialis uropygialis*, 75
*uropygialis* x *Melanerpes aurifrons*, 75, 81, 82
*Melanerpes aurifrons aurifrons*, 81
Migratory Bird Treaty Act (U.S./1918), 37
Migratory Birds Convention Act (Canada/1994), 37
molts
adults, 11–13
preformative molt, 9, 10, 11
muscles
hyoid apparatus/tongue, 3–4, **3**, 5
musculoskeletal features, 7

nesting calendar/schedule, 15–17
nestling development, 17–18, **18–19**
nests. *See* cavities
nictitating membrane, 6

pamprodactyl toe configuration, 8, **9**
phloem wells, 21, **21**, **31**
*Picoides*
*albolarvatus*, **4**, **12**, **33**, 188–97, **188–95**, **197**
*albolarvatus albolarvatus*, 192, 195
*albolarvatus gravirostris*, 192, 195–96
*arcticus*, **3**, **6**, **27**, **35**, 206–13, **206–10**, **212**
*arizonae*, 168–75, **168–74**
*arizonae arizonae*, 172–73
*arizonae fraterculus*, 173
*arizonae "websteri,"* 173
*arizonae* x *Picoides stricklandi*, 173
*borealis*, **33**, 176–87, **176–87**
*borealis borealis*, 180
*borealis "hylonomus,"* 180
*dorsalis*, **31**, 198–205, **198–200**, **202–05**
*dorsalis bacatus*, 201, **202**
*dorsalis dorsalis*, **199**, 201
*dorsalis fasciatus*, **200**, 201, **203**, **204**
*nuttallii*, **15**, 140–47, **140–46**
*nuttallii* x *Picoides pubescens*, 140, 143, 144, 154
*nuttallii* x *Picoides scalaris*, 133, 137, 140, 143–44
*pubescens*, **8**, **17**, **29**, 148–57, **148–53**, **155–56**
*pubescens "fumidus,"* 154
*pubescens gairdnerii*, **148**, 153–54, **153**
*pubescens glacialis*, 153
*pubescens "homorus,"* 153
*pubescens leucurus*, 153, **155**
*pubescens medianus*, **151**, 153
*pubescens "meridionalis,"* 153
*pubescens "microleucus,"* 153
*pubescens "minimus,"* 153
*pubescens nelsoni*, **152**, 153
*pubescens "oreoecus,"* 153
*pubescens "parvirostris,"* 153
*pubescens pubescens*, **150**, 153, **156**
*pubescens turati*, 154
*pubescens "verus,"* 153
*pubescens* x *Picoides scalaris*, **136**, 137, 154
*pubescens* x *Picoides nuttallii*, 140, 143, 144, 154
*scalaris*, **3**, **14**, **26**, 132–39, **132–38**
*scalaris "agnus,"* 137
*scalaris "azelus,"* 137
*scalaris "bairdi,"* 137
*scalaris cactophilus*, 133, 136–37
*scalaris centrophilus*, 137
*scalaris eremicus*, 137
*scalaris "giraudi,"* 137
*scalaris graysoni*, 137
*scalaris "lambi,"* 137
*scalaris leucoptilurus*, 137
*scalaris lucasanus*, 137
*scalaris "mohavensis,"* 137
*scalaris parvus*, 137
*scalaris "percus,"* 137
*scalaris "ridgwayi,"* 137
*scalaris scalaris*, 137
*scalaris sinaloensis*, 137
*scalaris "soulei,"* 137
*scalaris "symplectus,"* 136–37
*scalaris* x *Picoides nuttallii*, 133, 137, 140, 143–44
*scalaris* x *Picoides pubescens*, **136**, 137, 154
*scalaris* x *Picoides villosus*, 137, 165
*scalaris "yumanensis,"* 137
*stricklandi*, **173**
*stricklandi* x *Picoides arizonae*, 173
*villosus*, **3**, **5**, **11**, **30**, 158–67, **158–66**
*villosus audubonii*, 162, 163
*villosus enissomenus*, 163
*villosus extimus*, **164**, 165
*villosus fumeus*, 165
*villosus harrisi*, 162, **162**, 163
*villosus harrisi* x *Picoides villosus orius*, **161**
*villosus hylobatus*, 163
*villosus hyloscopus*, 162–63
*villosus icastus*, 163
*villosus intermedius*, 163
*villosus jardinii*, 165
*villosus leucothorectis*, 163
*villosus maynardi*, 165
*villosus monticola*, **161**, 163
*villosus orius*, **160**, 163

*villosus orius* x *Picoides villosus harrisi*, **161**
*villosus picoideus*, 162
*villosus piger*, 165, **165**
*villosus sanctorum*, 165
*villosus septentrionalis*, 162, 163
*villosus sitkensis*, 162
*villosus terraenovae*, 163
*villosus villosus*, **158**, 160–61, 162, 163, **163**
*villosus* x *Picoides scalaris*, 137, 165
predators/defense, 16, 26–27, **27**
preening, 28
preformative molt, 9, 10, 11

"Red-shafted." *See* Flicker

salivary glands, submaxillary, 3
sanitation of cavities, 19–20
sap wells
other animals using, 30, **31**
woodpeckers and, 20–22, **21**, 30
Sapsucker
Red-breasted, **12**, **21**, **24**, 122–31, **122–31**
Red-breasted x Red-naped Sapsucker, **19**, **107**, 123, 126, 127–28, **129–31**
Red-breasted x Yellow-bellied Sapsucker, 107, **107**
Red-naped, **3**, 112–21, **112–16**, **118–19**
Red-naped x Red-breasted Sapsucker, **19**, **107**, 123, 126, 127–28, **129–31**
Red-naped x Williamson's Sapsucker, 99
Red-naped x Yellow-bellied Sapsucker, 107, **107–09**, 112, 117
Williamson's, 94–101, **94–98**, **100–101**
Williamson's x Red-naped Sapsucker, 99
Yellow-bellied, **5–6**, **21**, **35**, 102–11, **102–04**, **106–10**
Yellow-bellied x Red-breasted Sapsucker, 107, **107**
Yellow-bellied x Red-naped Sapsucker, 107, **107–09**, 112, 117
Shaken Baby Syndrome, 4–5
skeleton
CT scan, **4**
musculoskeletal features, 7
social interactions overview, 25–27, **25–27**
*Sphyrapicus*
*nuchalis*, **3**, 112–21, **112–16**, **118–19**
*nuchalis* x *Sphyrapicus ruber*, **19**, **107**, 123, 126, 127–28, **129–31**
*nuchalis* x *Sphyrapicus thyroideus*, 99
*nuchalis* x *Sphyrapicus varius*, 107, **107–09**, 112, 117
*ruber*, **12**, **21**, **24**, **31**, 122–31, **122–31**
*ruber daggetti*, **122**, **125**, 126–27, **126–28**
*ruber ruber*, **124**, 126–27, **126–28**
*ruber* x *Sphyrapicus nuchalis*, **19**, **107**, 123, 126, 127–28, **129–31**
*ruber* x *Sphyrapicus varius*, 107, **107**
*thyroideus*, 94–101, **94–98**, **100–101**
*thyroideus nataliae*, 99, 101
*thyroideus thyroideus*, 99, 101
*thyroideus* x *Sphyrapicus nuchalis*, 99
*varius*, **5–6**, **21**, **31**, **35**, 102–11, **102–04**, **106–10**
*varius* x *Sphyrapicus nuchalis*, 107, **107–09**, 112, 117
*varius* x *Sphyrapicus ruber*, 107, **107**
suet feeding, 22, **23**

tail adaptations/uses, 7, 12, **12**, 27
tongues/adaptations, 2–3, **2–3**, **5**
tossing wood chips (excavating), 16, **17**

vocalizations overview, 24–25

water drinking, 22, **23**
website for reporting behavior, 28, 41
Wildlife Conservation and Restoration Program (U.S.), 38
window collisions, **14**
wing stretching description, 27–28
Woodpecker
Acorn, **16**, **25**, 60–69, **60–68**
American Three-toed, **31**, 198–205, **198–200**, **202–05**
Arizona, 168–75, **168–74**
Arizona x Strickland's Woodpecker, 173
"Batchelder's" Downy, 153
Black-backed, **3**, **6**, **27**, **35**, 206–13, **206–10**, **212**
Downy, **8**, **17**, **29**, 148–57, **148–53**, **155–56**
Downy x Ladder-backed Woodpecker, **136**, 137, 154
Downy x Nuttall's Woodpecker, 140, 143, 144, 154
Gila, **23**, 70–77, **70–74**, **76**
Gila x Golden-fronted Woodpecker, 75, 81, 82
Golden-fronted, **4**, **5**, **10**, **23**, 78–85, **78–80**, **82**, **84–85**, **138**
Golden-fronted x Gila Woodpecker, 75, 81, 82
Golden-fronted x Hoffmann's Woodpecker, 82
Golden-fronted x Red-bellied Woodpecker, 78, 81–82, 89
Hairy, **3**, **5**, **11**, **30**, 158–67, **158–66**
Hairy x Ladder-backed Woodpecker, 137, 165
Hoffmann's x Golden-fronted Woodpecker, 82
Imperial, **5**, 248
Ivory-billed, 244–51, **244–47**, **249–51**
Ladder-backed, **3**, **14**, **26**, 132–39, **132–38**
Ladder-backed x Downy Woodpecker, **136**, 137, 154
Ladder-backed x Hairy Woodpecker, 137, 165
Ladder-backed x Nuttall's Woodpecker, 133, 137, 140, 143–44
Lewis's, **5–6**, **13**, **22**, **33**, 44–51, **44–47**, **49–50**
"Mearns's," 63

Woodpecker (*cont.*)
Nuttall's, **15**, 140–47, **140–46**
× Downy Woodpecker, 140, 143, 144, 154
× Ladder-backed Woodpecker, 133, 137, 140, 143–44
Pale-billed, **9**
Pileated, **5–6**, **21**, **32**, **34**, 234–43, **234–36**, **238–42**, **247**
Red-bellied, **5–6**, **23**, 86–93, **86–90**, **92**
Red-bellied × Golden-fronted Woodpecker, 78, 81–82, 89
Red-cockaded, **33**, 176–87, **176–87**
Red-headed, **3**, 52–59, **52–58**
Strickland's, **173**
Strickland's × Arizona Woodpecker, 173
White-headed, **4**, **12**, **33**, 188–97, **188–95**, **197**

xylem wells, 20–21, **21**

"Yellow-shafted." *See* Flicker

zygodactyl toe configuration, **7**, 8

Purchase Peterson Field Guide titles wherever books are sold.
For more information on Peterson Field Guides, visit **www.petersonfieldguides.com.**

## PETERSON FIELD GUIDES®

Roger Tory Peterson's innovative format uses accurate, detailed drawings to pinpoint key field marks for quick recognition of species and easy comparison of confusing look-alikes.

**BIRDS**

*Birds of North America*

*Birds of Eastern and Central North America*

*Western Birds*

*Eastern Birds*

*Feeder Birds of Eastern North America*

*Hawks of North America*

*Hummingbirds of North America*

*Warblers*

*Eastern Birds' Nests*

**PLANTS AND ECOLOGY**

*Eastern and Central Edible Wild Plants*

*Eastern and Central Medicinal Plants and Herbs*

*Western Medicinal Plants and Herbs*

*Eastern Forests*

*Eastern Trees*

*Western Trees*

*Eastern Trees and Shrubs*

*Ferns of Northeastern and Central North America*

*Mushrooms*

*North American Prairie*

*Venomous Animals and Poisonous Plants*

*Wildflowers of Northeastern and North-Central North America*

**MAMMALS**

*Animal Tracks*

*Mammals*

*Finding Mammals*

**INSECTS**

*Insects*

*Eastern Butterflies*

*Moths of Northeastern North America*

**REPTILES AND AMPHIBIANS**

*Eastern Reptiles and Amphibians*

*Western Reptiles and Amphibians*

**FISHES**

*Freshwater Fishes*

**SPACE**

*Stars and Planets*

**GEOLOGY**

*Rocks and Minerals*

## PETERSON FIRST GUIDES®

The first books the beginning naturalist needs, whether young or old. Simplified versions of the full-size guides, they make it easy to get started in the field, and feature the most commonly seen natural life.

*Astronomy*

*Birds*

*Butterflies and Moths*

*Caterpillars*

*Clouds and Weather*

*Fishes*

*Insects*

*Mammals*

*Reptiles and Amphibians*

*Rocks and Minerals*

*Seashores*

*Shells*

*Trees*

*Urban Wildlife*

*Wildflowers*

## PETERSON FIELD GUIDES FOR YOUNG NATURALISTS

This series is designed with young readers ages eight to twelve in mind, featuring the original artwork of the celebrated naturalist Roger Tory Peterson.

*Backyard Birds*

*Birds of Prey*

*Songbirds*

*Butterflies*

*Caterpillars*

## PETERSON FIELD GUIDES® COLORING BOOKS®

Fun for kids ages eight to twelve, these color-your-own field guides include color stickers and are suitable for use with pencils or paint.

*Birds*

*Butterflies*

*Dinosaurs*

*Reptiles and Amphibians*

*Wildflowers*

*Seashores*

*Shells*

*Mammals*

## PETERSON REFERENCE GUIDES®

Reference Guides provide in-depth information on groups of birds and topics beyond identification.

*Behavior of North American Mammals*

*Birding by Impression*

*Molt in North American Birds*

*Owls of North America and the Caribbean*

*Seawatching: Eastern Waterbirds in Flight*

## PETERSON AUDIO GUIDES

*Birding by Ear: Eastern/Central*

*Bird Songs: Eastern/Central*

## PETERSON FIELD GUIDE / *BIRD WATCHER'S DIGEST* BACKYARD BIRD GUIDES

*Identifying and Feeding Birds*

*Hummingbirds and Butterflies*

*Bird Homes and Habitats*

*The Young Birder's Guide to Birds of North America*

*The New Birder's Guide to Birds of North America*

## DIGITAL

App available for Apple and Android.

*Peterson Birds of North America*

## E-books

*Birds of Arizona*

*Birds of California*

*Birds of Florida*

*Birds of Massachusetts*

*Birds of Minnesota*

*Birds of New Jersey*

*Birds of New York*

*Birds of Ohio*

*Birds of Pennsylvania*

*Birds of Texas*

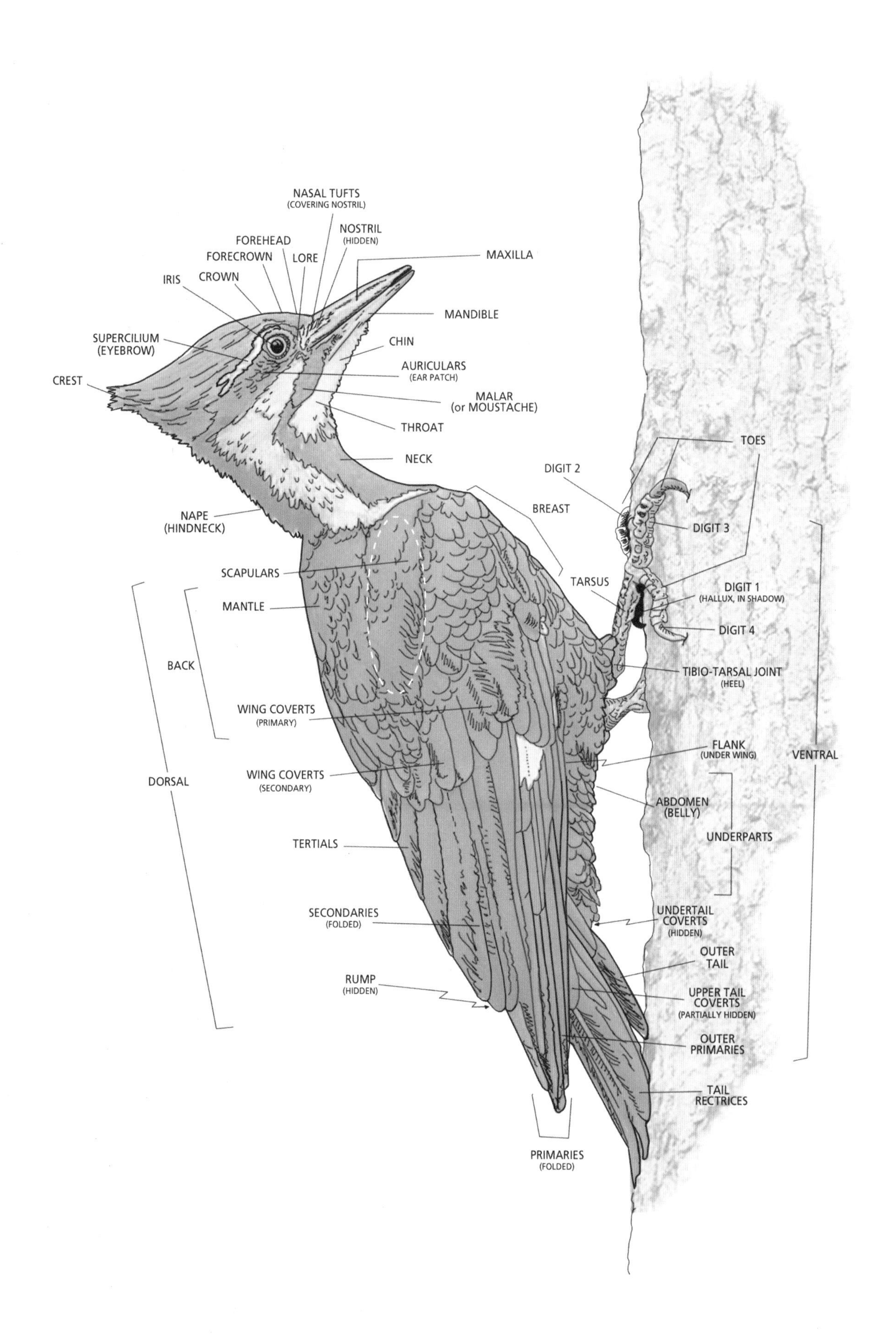
NASAL TUFTS
(COVERING NOSTRIL)
NOSTRIL
(HIDDEN)
FOREHEAD
FORECROWN
LORE
MAXILLA
IRIS
CROWN
MANDIBLE
SUPERCILIUM
(EYEBROW)
CHIN
AURICULARS
(EAR PATCH)
CREST
MALAR
(or MOUSTACHE)
THROAT
NECK
TOES
DIGIT 2
BREAST
NAPE
(HINDNECK)
DIGIT 3
SCAPULARS
TARSUS
DIGIT 1
(HALLUX, IN SHADOW)
MANTLE
DIGIT 4
BACK
TIBIO-TARSAL JOINT
(HEEL)
WING COVERTS
(PRIMARY)
FLANK
(UNDER WING)
VENTRAL
DORSAL
WING COVERTS
(SECONDARY)
ABDOMEN
(BELLY)
UNDERPARTS
TERTIALS
SECONDARIES
(FOLDED)
UNDERTAIL
COVERTS
(HIDDEN)
OUTER
TAIL
RUMP
(HIDDEN)
UPPER TAIL
COVERTS
(PARTIALLY HIDDEN)
OUTER
PRIMARIES
TAIL
RECTRICES
PRIMARIES
(FOLDED)